Graduate Texts in Mathematics
48

Editorial Board

F. W. Gehring

P. R. Halmos
Managing Editor

C. C. Moore

R. K. Sachs
H. Wu

General Relativity for Mathematicians

Springer-Verlag
New York   Heidelberg   Berlin

Dr. Rainer K. Sachs
University of California at
  Berkeley
Department of Physics
Berkeley, California 94720

Dr. Hung-Hsi Wu
University of California at
  Berkeley
Department of Mathematics
Berkeley, California 94720

Editorial Board

P. R. Halmos
Managing Editor
University of California
Mathematics Department
Santa Barbara,
California 93106

F. W. Gehring
University of Michigan
Department of Mathematics
Ann Arbor,
Michigan 48104

C. C. Moore
University of California at Berkeley
Department of Mathematics
Berkeley, California 94720

AMS classifications: 53C50, 5302, 83C99, 83F05, 8302 (Primary)
　　　　　　　　　 53C20, 53B30, 83C05, 8502, 85A40 (Secondary)

Library of Congress Cataloging in Publication Data

Sachs, Rainer Kurt, 1932–
　General relativity for mathematicians.
　(Graduate texts in mathematics; 48)
　Bibliography: p.
　Includes indexes.
　1. Relativity (Physics)　I. Wu, Hung-Hsi, 1940–
joint author.　II. Title.　III. Series.
QC173.55.S34　　530.1′1′02451　　76-47697

All rights reserved.

No part of this book may be translated or reproduced
in any form without written permission from Springer-Verlag

© 1977 by Springer-Verlag, New York Inc.

Printed in the United States

9 8 7 6 5 4 3 2 1

ISBN 0–387–90218–X　Springer-Verlag　New York
ISBN 3–540–90218–X　Springer-Verlag　Berlin　Heidelberg

Preface

This is a book about physics, written for mathematicians. The readers we have in mind can be roughly described as those who:

1. are mathematics graduate students with some knowledge of global differential geometry
2. have had the equivalent of freshman physics, and find popular accounts of astrophysics and cosmology interesting
3. appreciate mathematical clarity, but are willing to accept physical motivations for the mathematics in place of mathematical ones
4. are willing to spend time and effort mastering certain technical details, such as those in Section 1.1.

Each book disappoints some readers. This one will disappoint:

1. physicists who want to use this book as a first course on differential geometry
2. mathematicians who think Lorentzian manifolds are wholly similar to Riemannian ones, or that, given a sufficiently good mathematical background, the essentials of a subject like cosmology can be learned without some hard work on boring details
3. those who believe vague philosophical arguments have more than historical and heuristic significance, that general relativity should somehow be "proved," or that axiomatization of this subject is useful
4. those who want an encyclopedic treatment (the books by Hawking–Ellis [1], Penrose [1], Weinberg [1], and Misner–Thorne–Wheeler [1] go further into the subject than we do; see also the survey article, Sachs-Wu [1]).
5. mathematicians who want to learn quantum physics or unified field theory (unfortunately, quantum physics texts all seem either to be for physicists, or merely concerned with formal mathematics).

Preface

While using this book in classes, we found that our canonical reader can learn nonquantum physics rather quickly. Indeed, equipped with geometric intuition and a facility with abstract arguments, he is in a position to deal directly with the general, currently accepted models used in relativity without being handicapped by the prejudices that inevitably come with years of Newtonian training in the standard physics curriculum. However, this shortcut does involve a price: one cannot really see the diversity of special cases behind the deceptively simple foundation without spending more time than a mathematics student normally can or should.

We have felt for a long time that a serious effort should be made by physicists to communicate with mathematicians somewhat along the line of this book. We started with the aim of keeping the physics honest, keeping the mathematics honest, and keeping the logical distinction between the two straight. But we were ill-prepared for the attendant trauma of such an undertaking. In particular, the third point proved to be a veritable nightmare. We managed to emerge from our many moments of doubt to complete this book with the original plan intact, not the least because we were sustained from time to time by the encouragement of some of our friends and colleagues, particularly S. S. Chern and B. O'Neill. Nevertheless, we are pessimistic about further attempts at explaining genuine physics to mathematicians using only prerequisites familiar to them.

Many people believe that current physics and mathematics are, on balance, contributing usefully to the survival of mankind in a state of dignity. We disagree. But should humans survive, gazing at stars on a clear night will remain one of the things that make existence nontrivial. We hope that at some point this book will remind you of the first time you looked up.

Through the several drafts of this book as classroom notes, we were fortunate to have the excellent secretarial assistance of Joy Kono, Nora Lee, and Marnie McElhiney. A philosophical remark from Professor S. S. Chern was responsible for an overhaul of our overall presentation. Many minor and quite a few major improvements were due to suggestions by J. Arms, J. Beem, K. Sklower and T. Langer. But for the warm hospitality of the DAMTP of Cambridge University and the unswerving support of Kuniko Weltin under rather trying circumstances, the final stage of the book-writing would have been interminable and insufferable. Finally, support from the National Science Foundation greatly facilitated the preparation of the manuscript. To all of them, we wish to express our deep appreciation.

Contents

Guidelines for the reader   xi

Chapter 0
Preliminaries   1

0.0   Review and notation   1
0.1   Physics background   7
0.2   Preview of relativity   12

Chapter 1
Spacetimes   17

1.0   Review and notation   17
1.1   Causal character   20
1.2   Time orientability   24
1.3   Spacetimes   27
1.4   Examples of spacetimes   29

Chapter 2
Observers   36

2.0   Mathematical preliminaries   36
2.1   Observers and instantaneous observers   41
2.2   Gyroscope axes   50
2.3   Reference frames   52

Chapter 3
Electromagnetism and matter — 60

PART ONE: BASIC CONCEPTS
3.0 Review and notation — 61
3.1 Particles — 66
3.2 Particle flows — 69
3.3 Stress-energy tensors — 71
3.4 Electromagnetism — 74
3.5 Matter and relativistic models — 76

PART TWO: INTERACTIONS
3.6 Some mathematical methods — 78
3.7 Maxwell's equations — 83
3.8 Particle dynamics — 88
3.9 Matter equations: an example — 95
3.10 Energy-momentum 'conservation' — 96
3.11 Two initial value theorems — 98
3.12 Appropriate matter equations — 102

PART THREE: OTHER MATTER MODELS
3.13 Examples — 103
3.14 Normal stress-energy tensors — 104
3.15 Perfect fluids — 107

Chapter 4
The Einstein field equation — 111

4.0 Review and notation — 111
4.1 The Einstein field equation — 111
4.2 Ricci flat spacetimes — 113
4.3 Gravitational attraction and the phenomenon of collapse — 117

Chapter 5
Photons — 124

5.0 Mathematical preliminaries — 124
5.1 Photons — 129
5.2 Light signals — 133
5.3 Synchronizable reference frames — 136
5.4 Frequency ratio — 137
5.5 Photon distribution functions — 140
5.6 Integration on lightcones — 148
5.7 A photon gas — 152

Chapter 6
Cosmology — 159

6.0 Review, notation and mathematical preliminaries — 160
6.1 Data — 168
6.2 Cosmological models — 176

6.3	The Einstein–de Sitter model	184
6.4	Simple cosmological models	196
6.5	The early universe	203
6.6	Other models	208
6.7	Appendix: Luminosity distance in the Einstein–de Sitter model	210

Chapter 7
Further applications　　　216

7.0	Review and notation	216
7.1	Preview	217
7.2	Stationary spacetimes	218
7.3	The geometry of Schwarzschild spacetimes	220
7.4	The solar system	228
7.5	Black holes	237
7.6	Gravitational plane waves	242

Chapter 8
Optional exercises: relativity　　　250

8.1	Lorentzian algebra	250
8.2	Differential topology and geometry	254
8.3	Chronology and causality	257
8.4	Isometries and characterizations of gravitational fields	259
8.5	The Einstein field equation	262
8.6	Gases	264

Chapter 9
Optional exercises: Newtonian analogues　　　266

9.0	Review and notation	266
9.1	Maxwell's equations	267
9.2	Particles	269
9.3	Gravity	271

Glossary of symbols　　　273
Bibliography　　　274
Index of basic notations　　　277
Index　　　281

Guidelines for the reader

1. All indented fine-print portions of the book are optional; they may be skipped without loss of mathematical continuity. Some of these fine-print paragraphs are proofs that we consider noninstructive. But the majority of them contain comments that presuppose a knowledge of physics (and on a few occasions, of mathematics) beyond the level of our formal prerequisites. Nevertheless, we urge the reader to at least glance through those dealing with physics; they may be read with the assurance that each has been revised many times to minimize distortion of the physics.
2. The remainder of the text should be treated as straight mathematics, though one should keep in mind the following peculiarity: There will be no all-encompassing mathematical abstractions; instead, the emphasis throughout is on simple definitions and propositions that have a multitude of physical implications.

    Physics attempts to describe certain aspects of nature mathematically. Now, nature is not a mathematical object, much less a theorem. There is no overriding mathematical structure that covers all of physics. Since the subject matter of this book is physics, the reader will find here not a coherent and profound mathematical study of general Lorentzian manifolds culminating in a *Hauptsatz*, but rather a disjointed collection of propositions about a special class of four-dimensional Lorentzian manifolds. Mathematics plays a subordinate role; it is a tool rather than the ultimate object of interest. For example, in Chapter 3 the emphasis is not on the coordinate-free version of Stokes' theorem, which is taken for granted. Instead, this theorem is used to define and analyze many physical concepts: the conservation of electric charge; the creation and annihilation of matter; the hypothesis that magnetic monopoles don't exist; relativistic versions of Gauss' law for electric flux, Faraday's law of magnetic induction, and Maxwell's displacement current hypothesis; the special relativistic laws for conservation of energy, momentum, and angular momentum; and so on. These concepts in turn apply to a very rich variety of known phenomena. Our aim in discussing the theorem is merely to indicate how it can manage to say so much about the world so concisely.

In brief: economy will be central, mathematical generality will be irrelevant.

The reader wishing to pursue the deeper mathematical theorems of relativity should consult Hawking–Ellis [1].

3. The expository style of the book is strictly mathematical: all concepts are explicitly defined and all assertions precisely proved. Now, in a serious physics text basic physical quantities are almost never explicitly defined. The reason is that the primary definitions are actually obtained by showing photographs, by pointing out of the window, or by manipulating laboratory equipment. The more mathematically explicit a definition, the less accurate it tends to be in this primary sense. The reader is therefore forewarned that on this one point we have intentionally distorted an essential feature of physics in order to accommodate the mathematician's intolerance of theorems about mathematically undefined terms.
4. The exercises at the end of each section are, at least in principle, an integral part of the text. We have been very conscientious in making sure that each is workable within a reasonable amount of time.
5. Chapters 0 through 5 are meant to be read consecutively. The remaining chapters are independent.

Preliminaries 0

This chapter is intended mainly to clear the boards for action. A reader with a solid background might try just skimming the chapter. Section 0.0 reviews some of the differential geometry we shall need. Section 0.1 gives some physics background. Section 0.2 gives an intuitive discussion of the transition from Newtonian physics to relativity. A reader who has never studied relativity should work all the exercises for Section 0.2.

0.0 Review and notation

This section sets the notation. Definitions not explicitly stated and theorems not explicitly proved are all discussed, for example, in the text by Bishop and Goldberg [1], referred to as Bishop–Goldberg throughout. We follow the Bishop–Goldberg notation as closely as feasible.

0.0.1 Sets, maps, and topology

Suppose A and B are sets, and $i: A \to B$ is a map, the image of $a \in A$ is written either ia or $i(a)$. For example, suppose C is a set and $k: B \to C$ is a map; then $(k \circ i)(a) = (k \circ i)a = k(ia) = kia$ with $k(ia)$ preferred. Suppose D is a subset of A. We write $D \subset A$ and write $A - D = \{a \in A \mid a \notin D\}$; $i|_D$ is the restriction of i to D. Suppose $E \subset B$; we write $i^{-1}E \subset A$ for the complete inverse image. Suppose A above is a topological space; then D^- and ∂D will denote the closure and boundary, respectively, of D.

$\mathbb{Z}$ denotes the integers and $\mathbb{R}$ the reals. If $\mathscr{E} \subset \mathbb{R}$ is connected and open, we sometimes write $\mathscr{E} = (a, b)$, with $a = -\infty$ and/or $b = \infty$ allowed. $f: \mathscr{E} \to \mathbb{R}$ is called *positive affine* iff $fu = cu + d$, where $c > 0$ and $d \in \mathbb{R}$.

0 Preliminaries

0.0.2 Tensor algebra

If V is a vector space (always understood to be over $\mathbb{R}$), V^* denotes its dual. The image of $v \in V$ under $\omega \in V^*$ is denoted by $\omega(v)$ or ωv. By a *subspace* of V we mean a vector subspace. If $V_1, \ldots, V_N$ are finite dimensional vector spaces, then $V_1 \oplus \cdots \oplus V_N$ will denote their direct sum and $V_1^* \otimes \cdots \otimes V_N^*$ the vector space of multilinear maps $V_1 \times \cdots \times V_N \to \mathbb{R}$. The space of (r, s) *tensors* over V_1 is $T_s^r(V_1) = V_1 \otimes \cdots \otimes V_1 \otimes V_1^* \otimes \cdots \otimes V_1^*$, where there are r unstarred and s starred factors. (r, s) is the *type* of each tensor in $T_s^r(V_1)$. Suppose $S \in T_s^r(V_1)$ and $T \in T_q^p(V_1)$; then $T \otimes S \in T_{s+q}^{r+p}(V_1)$ denotes the tensor product.

> We are following the convention of Bishop–Goldberg in placing the contravariant variables in front of all the covariant variables for each tensor in $T_s^r(V_1)$. This is aimed at facilitating any discussion concerning tensors when no mention of indices is allowed.

0.0.3 Inner products

Let V be a finite dimensional vector space. A nondegenerate symmetric bilinear form g on V is called an *inner product* on V (Bishop–Goldberg 2.21). Let $S = \{W \,|\, W \text{ is a subspace of } V \text{ and } g|_W \text{ is negative definite}\}$. The *index* I of g is the integer $I = \max_{W \in S}$ (dimension W). Define the *norm* of $v \in V$ as $|v| = [|g(v, v)|]^{1/2}$; define $v \in V$ as a *unit* vector iff $|v| = 1$; define $v, w \in V$ as *orthogonal* iff $g(v, w) = 0$.

Let $N = \dim V$, $B = (e_1, \ldots, e_N)$ be an ordered basis of V, and $(\varepsilon^1, \ldots, \varepsilon^N)$ be the dual basis of V^*. B is called ("ordered," "semi-") *orthonormal* iff $g = \sum_{A=1}^{N-I} \varepsilon^A \otimes \varepsilon^A - \sum_{A=N-I+1}^{N} \varepsilon^A \otimes \varepsilon^A$, where the appropriate sum is zero if $I = 0$ or $I = N$. Equivalently, B is orthonormal iff: $g(e_A, e_A) = 1$ for $1 \le A \le N - I$, $g(e_A, e_A) = -1$ for $N - I + 1 \le A \le N$, and $g(e_A, e_B) = 0$ for $A \ne B$. A basis of pairwise orthogonal unit vectors can always be made an orthonormal basis by appropriate reordering. If $e \in V$ is a unit vector, there exists an orthonormal basis that contains e.

We shall call the pair (V, g) a *Lorentzian vector space* and g a *Lorentzian inner product* iff $\dim V \ge 2$ and $I = 1$.

> This is the case of main interest. The reader should not assume it is essentially similar to the positive definite case. The differences are central in physics, as the rest of this book shows. For example, suppose g is an inner product on V. The subset $\{v \in V \,|\, g(v, v) < 0\}$ has two connected components iff (V, g) is Lorentzian. Locally, these components correspond to the physical past and physical future. When the algebraic structure of a Lorentzian (V, g) is unwrapped from tangent spaces into a manifold, a rich structure results (Penrose [1], Hawking and Ellis [1]). See Optional exercises 8.3.

0.0.4 C^∞ Manifolds and maps

Unless specifically denied, all manifolds, all objects on them, and all maps from one manifold into another will be C^∞; however, we sometimes redundantly write "a C^∞ manifold," and so on, for emphasis. A manifold M introduced by a definition need not be connected, but will always be finite-dimensional, real, Hausdorff, and paracompact. Throughout the remainder of this book, M is a manifold. M_x denotes the tangent space at $x \in M$. The *tangent bundle TM* is $\{(x, X) \mid x \in M \text{ and } X \in M_x\}$ with its standard C^∞ manifold structure (Bishop–Goldberg 3A); the projection $\Pi \colon TM \to M$ has the rule $\Pi(x, X) = x$. As in Bishop–Goldberg, M_x will be identified with the *fibre* $\Pi^{-1}x$ over x.

Let N be a manifold and $\phi \colon N \to M$ be a map. Then the map $\phi_* \colon TN \to TM$ between tangent bundles denotes the differential and ϕ^* denotes the pullback. Thus $(\phi \circ \psi)^* = \psi^* \circ \phi^*$. ϕ is an *immersion* iff $\forall n \in N$, ϕ_* restricted to N_n is one-one. An immersion ϕ is an *imbedding* iff ϕN, with the topology induced by that of M, is homeomorphic to N under ϕ. Then ϕ is one-one and ϕN is called an *imbedded submanifold*. Any open subset of M is an imbedded submanifold. A *diffeomorphism* is an onto imbedding.

0.0.5 Tensor fields

Let $T_s^r M$ be the bundle of (r, s) tensors over M, and $P \colon T_s^r M \to M$ be the standard projection (Bishop–Goldberg 3A). An (r, s)-*tensor field* $\mathbf{B}$ on $\mathscr{U} \subset M$ is a map $\mathbf{B} \colon \mathscr{U} \to T_s^r M$ such that $P \circ \mathbf{B} = $ identity on $\mathscr{U}$. Thus for each $x \in \mathscr{U}$, $\mathbf{B}x \in T_s^r(M_x)$. If $\mathscr{U}$ is a submanifold of M, then $\mathbf{B}$ being C^∞ makes sense, and this property will then be automatically assumed by our convention.

We follow the standard definitions of the usual tensor formalism (Bishop–Goldberg 2, 3, 4). For example, suppose $f \colon M \to \mathbb{R}$ is a function on M; V and W are vector fields on M; and φ and ψ are 1-forms on M. Then: (a) L_V denotes the Lie derivative with respect to V; (b) $Vf = df(V) = L_V f$ is a function on M; (c) $[V, W]$ denotes the Lie bracket so that $[V, W] = L_V W$; (d) $\varphi \wedge \psi = \frac{1}{2}(\varphi \otimes \psi - \psi \otimes \varphi)$; and (e) $2d\varphi(V, W) = V\varphi(W) - W\varphi(V) - \varphi([V, W])$. A q-form τ on M is called *closed* iff $d\tau = 0$, *exact* iff there is a $(q-1)$-form μ on M such that $\tau = d\mu$. An exact q-form is closed.

We use the usual swindle for domains of definition. For example, let g be a $(0, 2)$ tensor field on M, and V be a vector field on M; suppose $W \subset M_x$ for some $x \in M$. Then $g(V, W)$ means $gx(Vx, W) \in \mathbb{R}$ and $g(\cdot, W)$ means $gx(\cdot, W) \in M_x^*$. As another example, if U is a vector field defined on an open submanifold $\mathscr{N}$ of M, then $g(U, V)$ means $g|_{\mathscr{N}}(U, V|_{\mathscr{N}})$, which is a function on $\mathscr{N}$.

An n-dimensional manifold M is called *orientable* iff there is a nowhere zero n-form ω on M; any such ω is called a *volume element* and determines an *orientation* (Bishop–Goldberg 3C and p. 185). If M is an oriented manifold and $\mathscr{U} \subset M$ is open, we always assign the consistent orientation to $\mathscr{U}$.

If, furthermore, $\partial \mathcal{U}$ is a submanifold of M, then $\partial \mathcal{U}$ inherits an *induced orientation* from M in the following manner: if $x \in \partial \mathcal{U}$ and $\{x^1, \ldots, x^n\}$ are coordinate functions in an open set $\mathcal{A}$ containing x such that $\mathcal{U} \cap \mathcal{A} = \{x^1 < 0\}$ and $dx^1 \wedge \cdots \wedge dx^n$ is consistent with the orientation of M, then $dx^2 \wedge \cdots \wedge dx^n$ restricted to $\partial \mathcal{U}$ is consistent with the induced orientation on $\partial \mathcal{U}$.

0.0.6 Curves

Let $\mathscr{E} \subset \mathbb{R}$ be an interval, which may be infinite, and $\gamma \colon \mathscr{E} \to M$ a map. γ will always be understood to be C^∞ in the following sense: there exists an open set $\hat{\mathscr{E}} \subset \mathbb{R}$ containing $\mathscr{E}$ and a C^∞ map $\hat\gamma \colon \hat{\mathscr{E}} \to M$ such that $\hat\gamma|_{\mathscr{E}} = \gamma$. Such a C^∞ map $\gamma \colon \mathscr{E} \to M$ is called a *curve* in M. We denote the inclusion function $\mathscr{E} \to \mathbb{R}$ by s, t, or u and the distinguished vector field on $\mathscr{E}$ by d/ds, and so on. For example, $du(d/du) = 1$. For each $u \in \mathscr{E}$, $\gamma_* u$ denotes the tangent vector at γu; thus $\gamma_* u = [\gamma_*(d/du)](u) \in M_{\gamma u}$.

A curve $\gamma \colon \mathscr{E} \to M$ is called *inextendible* iff any other curve $\zeta \colon \mathscr{F} \to M$ satisfying $\mathscr{E} \subset \mathscr{F}$ and $\zeta|_{\mathscr{E}} = \gamma$ is the curve $\gamma \colon \mathscr{E} \to M$ itself. A curve $\zeta \colon \mathscr{F} \to M$ is called an (*orientation-preserving*) *reparametrization* of $\gamma \colon \mathscr{E} \to M$ if there exists an onto map $\alpha \colon \mathscr{E} \to \mathscr{F}$ with positive derivative such that $\gamma = \zeta \circ \alpha$. If α is positive affine, then ζ is called a *positive affine reparametrization* of γ.

If X is a vector field on M, the maximal integral curve of X through $x \in M$ is the unique curve $\gamma \colon (a, b) \to M$, $-\infty \le a < b \le \infty$, such that (a) $\gamma 0 = x$; (b) $\gamma_* u = X(\gamma u) \; \forall u \in (a, b)$; and (c) γ is inextendible (Bishop–Goldberg 3.4). The flow of X will be denoted by $\{\mu_s\}$. For example, if X is complete, $\mu_s \colon M \to M$ is obtained by moving each $x \in M$ s parameter units along the maximal integral curve through x (Bishop–Goldberg 3.5).

0.0.7 Metrics and isometries

Let g be a symmetric $(0, 2)$ tensor field on M. g is called a *metric tensor with index I on M* iff gx is nondegenerate and index $(gx) = I \; \forall \, x \in M$. Then (M, g) is called a *Riemannian* manifold iff $I = 0$, *semi-Riemannian* otherwise. We will call a semi-Riemannian manifold *Lorentzian* iff $I = 1$ and the dimension of M is at least 2.

Let (M, g) and (N, h) be Riemannian or semi-Riemannian manifolds. A map $\phi \colon M \to N$ is called an *isometry* iff ϕ is one-one, onto, and $\phi^* h = g$. Then ϕ is a diffeomorphism. (M, g) is then called *isometric* to (N, h) under ϕ. A map $\psi \colon M \to N$ is defined as a *local isometry* iff $\psi^* h = g$.

0.0.8 Geodesics

Throughout Section 0.0.8, (M, g) is a Riemannian or semi-Riemannian manifold. The *Levi–Civita connection* D of (M, g) is that ("linear," "affine") connection on M characterized by: (a) *symmetry*, $D_V W - D_W V = [V, W]$ for all vector fields V, W on M; and (b) *compatibility*, $D_V g = 0$ for all such V (Bishop–Goldberg 5.11). A curve $\gamma \colon \mathscr{E} \to M$ is a *geodesic of (M, g)* iff it is a geodesic of D on M (Bishop–Goldberg 5.12). We shall not count a constant curve, which has $\gamma\mathscr{E} = x \in M$, as a geodesic. If γ is a geodesic of (M, g), there is an $a \in \mathbb{R}$ such that $g(\gamma_*u, \gamma_*u) = a \ \forall u \in \mathscr{E}$. γ is called a *maximal* (or *inextendible*) *geodesic* iff it is both a geodesic and an inextendible curve. Note that if a geodesic γ is a reparametrization of another geodesic ξ, then γ is necessarily a positive affine reparametrization of ξ. Let X be a nowhere zero vector field. X is called a *geodesic vector field* iff $D_X X \equiv 0$. Thus X is geodesic iff each of its integral curves is a geodesic.

The *exponential map* $\exp_x$ at $x \in M$ maps a subset $\mathscr{U}_x \subset M_x$ into M as follows. The zero vector $0 \in M_x$ is in $\mathscr{U}_x$ and $\exp_x 0 = x$. A nonzero vector $X \in M_x$ is in $\mathscr{U}_x$ iff there is a geodesic $\gamma: [0, 1] \to M$ such that $\gamma 0 = x$ and $\gamma_* 0 = X$. For $X \in \mathscr{U}_x$, $X \neq 0$, γ is unique and $\exp_x X = \gamma 1$. $\mathscr{U}_x$ is open and $\exp_x$ is C^∞. For each $x \in M$, there is an open neighborhood $\mathscr{V}_x \subset \mathscr{U}_x$ of 0 such that $\exp_x|_{\mathscr{V}_x}$ is a diffeomorphism. (M, g) is *complete* iff $\mathscr{U}_x = M_x \forall x \in M$. (M, g) is complete iff every geodesic $\gamma: \mathscr{E} \to M$ can be extended to a geodesic $\mathbb{R} \to M$ (Bishop–Goldberg 5.13).

0.0.9 Bases and coordinate maps

Assume dimension $M = n \geq 1$. An ordered set $\{X_1, \ldots, X_n\}$ of vector fields on M is called a basis of vector fields on M iff $\{X_A x\}$ is a basis of $M_x \forall x \in M$. A basis $\{\omega^A\}$ of 1-forms on M is defined similarly. Bases $\{X_A\}$ and $\{\omega^A\}$ are called *dual* iff $\omega^B X_A = \delta_A{}^B \ \forall A, B \in \{1, \ldots, n\}$. Any basis uniquely determines a dual basis. If M is oriented, we assign the consistent orientation to each tangent space; unless explicitly denied, each basis used will then have the consistent orientation. A basis $\{X_A\}$ on a Riemannian or semi-Riemannian manifold (M, g), and its dual, are called *orthonormal* iff $\{X_A x\}$ is an orthonormal basis of $M_x \forall x \in M$ (cf. Exercise 0.0.15). On a given M there usually does not exist a basis of vector fields or 1-forms. However, one can always find such a basis in each coordinate neighborhood, and if g is also given, one can even choose this basis to be orthonormal.

We define $\mathbb{R}^N = \mathbb{R} \times \cdots \times \mathbb{R}$, where there are N factors. $u^A: \mathbb{R}^N \to \mathbb{R}$ denotes projection onto the Ath factor. Thus $\{du^A\}$ is a basis of 1-forms on any open submanifold of $\mathbb{R}^N$; the dual basis will be denoted by $\{\partial_A\}$. If $\mathscr{U} \subset M$ and $x: \mathscr{U} \to \mathbb{R}^N$ is a coordinate map, $x^A = u^A|_{x\mathscr{U}} \circ x$ denotes the Ath *coordinate function*. The basis on $\mathscr{U}$ dual to $\{dx^A\}$ will be denoted also by $\{\partial_A\}$.

The *unit* $(N-1)$-*sphere* $(\mathscr{S}^{N-1}, h, \zeta)$ is $\mathscr{S}^{N-1} = \{x \in \mathbb{R}^N | \ |x| = 1\}$, regarded as a C^∞ manifold, together with the standard induced metric h on $\mathscr{S}^{N-1}$ and the standard volume element ζ on $\mathscr{S}^{N-1}$. Thus if $I: \mathscr{S}^{N-1} \to \mathbb{R}^N$ is the inclusion, $h = I^*(\sum_{A=1}^N du^A \otimes du^A)$. Note that $\mathscr{S}^0$ is just the two points $\{-1, 1\} \subset \mathbb{R}$.

EXERCISE 0.0.10

Let V be a finite dimensional vector space. When V is regarded as a C^∞ manifold, it can be canonically identified with any of its tangent spaces. A basis-free method is part (a) following. (a) Regard $\omega \in V^*$ as a function $\tilde{\omega}: V \to \mathbb{R}$. Show that for each $v \in V$ there is precisely one isomorphism $\phi_v: V_v \to V$ such that $\omega(\phi_v w) = d\tilde{\omega}(w) \forall w \in V_v$ and $\omega \in V^*$. (b) Let g be an inner product on V, and $\tilde{g}: V \to \mathbb{R}$ be the function determined by $\tilde{g}(v) = g(v, v)$. Show $d\tilde{g}(w) = 2g(\phi_v w, v) \forall w \in V_v$ and $v \in V$. (c) Let (V, g) be a Lorentzian vector space and define a $(0, 2)$ tensor field g on V by $g(w, z) = g(\phi_v w, \phi_v z) \forall v \in V$ and $w, z \in V_v$. Show (V, g) is a Lorentzian manifold.

Exercise 0.0.11

Let V be an N-dimensional vector space, g be an inner product on V, and $W \subset V$ be a K-dimensional subspace. We define $W^\perp = \{v \in V \mid g(v, w) = 0 \; \forall w \in W\}$; if w spans W, we shall also write $w^\perp \equiv W^\perp$. Show: (a) $W^\perp$ is an $(N - K)$-dimensional subspace. (b) $W^{\perp\perp} = W$. (c) $V = W \oplus W^\perp$ iff $g|_W$ is nondegenerate.

Exercise 0.0.12

If M is a manifold, show: (a) $\mathscr{U} \subset TM$ open implies $\Pi \mathscr{U} \subset M$ is open; (b) $\mathscr{V} \subset M$ open implies $\Pi^{-1} \mathscr{V} \subset TM$ is open.

Exercise 0.0.13

(a) Show that for Riemannian or semi-Riemannian manifolds, the relation "is isometric to" is an equivalence relation. (b) Let (M, g) and $(\hat{M}, \hat{g})$ be Riemannian or semi-Riemannian manifolds. Show that $\phi: M \to \hat{M}$ is a local isometry iff each $x \in M$ has an open neighborhood $\mathscr{U} \subset M$ such that $(\mathscr{U}, g|_\mathscr{U})$ is isometric to $(\phi \mathscr{U}, \hat{g}|_{\phi \mathscr{U}})$ under $\phi|_\mathscr{U}$. (c) Let $\phi: M \to \hat{M}$ be a local isometry as in (b) and let $\gamma: \mathscr{E} \to M$ be a geodesic of (M, g). Show that $\hat{\gamma} = \phi \circ \gamma$ is a geodesic of $(\hat{M}, \hat{g})$. (d) Show that the set $\mathscr{I}M$ of isometries of a Riemannian or semi-Riemannian manifold (M, g) onto itself forms a group.

Exercise 0.0.14

Let V be a finite dimensional vector space and $\phi: V \to V^*$ a given isomorphism. (a) Show that for $r, s \in \mathbb{Z}$, $r > 0$, $s \geq 0$, ϕ can be extended uniquely to an isomorphism (to be denoted by the same symbol) $\phi: T_s^r(V) \to T_{s+1}^{r-1}(V)$ such that $\phi(v_1 \otimes \cdots \otimes v_r \otimes \omega^1 \otimes \cdots \otimes \omega^s) = v_1 \otimes \cdots \otimes v_{r-1} \otimes \phi(v_r) \otimes \omega^1 \otimes \cdots \otimes \omega^s$, $\forall v_1, \ldots, v_r \in V$ and $\omega^1, \ldots, \omega^s \in V^*$. (b) Show by induction that there is a unique isomorphism $\phi_s^r: T_s^r(V) \to T_{r+s}^0(V)$ for all nonnegative integers r, s such that $\phi_s^r(v_1 \otimes \cdots \otimes v_r \otimes \omega^1 \otimes \cdots \otimes \omega^s) = \phi(v_1) \otimes \cdots \otimes \phi(v_r) \otimes \omega^1 \otimes \cdots \otimes \omega^s$. (c) Suppose p, q and r, s are nonnegative integers such that $p + q = r + s$. For $A \in T_q^p(V)$ and $B \in T_s^r(V)$, define: A is ϕ-equivalent to B (in symbols: $A \sim B$) iff $\phi_q^p(A) = \phi_s^r(B)$. Show that $\sim$ is an equivalence relation.

Exercise 0.0.15

Let V be a finite dimensional vector space and g be an inner product on V. (a) Show that $\phi: V \to V^*$ defined by $(\phi v)w = g(v, w) \; \forall v, w \in V$ is an isomorphism. We shall call this ϕ the *metric isomorphism (induced by g)*. (b) Show that the map

$\hat{g}\colon V^* \times V^* \to \mathbb{R}$ defined by $\hat{g}(\omega, \omega') = g(\phi^{-1}\omega, \phi^{-1}\omega')$, $\forall \omega, \omega' \in V^*$, is an inner product on V^*. (c) Show that index g = index $\hat{g}$. (In particular, $\hat{g}$ is Lorentzian iff g is.) (d) Let $\{e_1, \ldots, e_N\}$ be an orthonormal basis of V with respect to g, and let $\{\varepsilon^1, \ldots, \varepsilon^N\}$ be its dual basis. Show that $\{\varepsilon^1, \ldots, \varepsilon^N\}$ is orthonormal with respect to $\hat{g}$. (This justifies the terminology of "orthonormal basis of 1-forms" introduced in Section 0.0.9.) (e) Show that $\hat{g}$, considered as a (2, 0) tensor, is ϕ-equivalent to g in the sense of Exercise 0.0.14(c). (f) Show that the element of V ϕ-equivalent to an $\omega \in V^*$ is given by $\hat{g}(\omega, \cdot)$.

EXERCISE 0.0.16

Let V be a finite dimensional vector space, $\psi\colon V \to V$ be a given isomorphism, and $\psi^*\colon V^* \to V^*$ be the adjoint isomorphism. Show that for all nonnegative integers r, s there is a unique extension of ψ to an isomorphism $\psi_s^r\colon T_s^r(V) \to T_s^r(V)$ such that $(\psi_s^r A)(\omega^1, \ldots, \omega^r, v_1, \ldots, v_s) = A(\psi^*\omega^1, \ldots, \psi^*\omega^r, \psi v_1, \ldots, \psi v_s) \forall A \in T_s^r(V)$, $\omega^1, \ldots, \omega^r \in V^*$, and $v_1, \ldots, v_s \in V$.

0.1 Physics background

0.1.1 General relativity

No well-defined current physical theory claims to model all nature; each intentionally neglects some effects. Roughly, general relativity is a model of nature, especially of gravity, that neglects quantum effects. Its central assumption is that space, time, and gravity are all aspects of a single entity, called spacetime, which is modelled by a 4-dimensional Lorentzian manifold. It analyzes spacetime, electromagnetism, matter, and their mutual influences. It is used mainly in the study of large-scale phenomena: dense stars, the universe, and so on.

Now in microphysics, gravity counts as a very minor effect. For example the electric repulsion between two electrons is believed to be more than 10^{40} times as large as their mutual gravitational attraction. But gravity is long range and cumulative. In the realm of stars and galaxies it can dominate. For example, the discovery of pulsars has now made it virtually certain that there are some stars that manage to resist total collapse caused by their own gravity only by a last-ditch effort, at a radius of perhaps 10 miles. For such stars, and for the universe as a whole, general relativity is the best available theory. It is also believed that there are stars for which gravity has triumphed completely, collapsing the star to a black hole. If so, general relativity will become very exciting during the next decade.

Since we are giving a mathematical exposition of general relativity, the basic postulates of this branch of physics are of necessity disguised as definitions. The key definitions are given in Sections 1.3.1, 3.3.1, 3.4.2, 3.5, 3.7.1, and 4.1.1. These definitions, not theorems, are central. Such definitions carry the connotation "nature is really somewhat like that," so they require more motivation than purely mathematical definitions. But we shall soft-pedal motivations. Genuine motivations cannot be given piecemeal; they refer to nature as a whole and a physical theory as a whole. Moreover, completely convincing motivations can never be given. Physical theories are guessed, not deduced; if only deductions were required, every competent hack could be an Einstein or a Feynman.

0.1.2 Physical theories

Newtonian physics can handle weak gravitational effects; it cannot adequately handle strong gravitational effects, or high-speed effects which occur when relative speeds comparable to the speed of light are involved. Special relativity can handle high-speed effects but not gravitational ones. General relativity incorporates Newtonian physics and special relativity into a theory that can handle both high-speed effects and gravitational effects of any strength. We give a table that shows the main theories of current physics and serves to interrelate some physics terms here treated as undefinable. The parenthetical phrases merely refer to certain ambiguities in the current physics literature and will often be omitted.

Mathematical theory		Effects included:	Gravity	High-speed	Quantum
A. Basic	(Nonquantum)	General relativity	Yes	Yes	No
	(Special-relativistic)	Quantum theory	No	Yes	Yes
B. Derived	(Nonquantum)	Special relativity	No	Yes	No
		Newtonian physics	Weak	No	No
	(Nongravitational)	Quantum mechanics	No	No	Yes

Each A implies two Bs by appropriate limiting processes. We shall use *relativity* to mean (nonquantum) general relativity or (nonquantum) special relativity, or both. Very roughly, "quantum" refers to the "fuzzy, jumpy" behavior of small objects; we attempt no further definition of "quantum."

As indicated, the two basic theories are quantum theory and general relativity. No one really knows how to combine these, though many attempts have been made. We indicate roughly the domain of validity that these two theories are believed to have in the following table; the table also introduces some more physics terms here treated as undefinable. In the table, "NUC" is an abbreviation for "the strong (nuclear) interaction or electromagnetism or the weak interaction or some combination" (cf. Weinberg [2]). For an explanation of the scale in terms of light-seconds, see Section 0.1.4. In B to D precise dividing lines—for example, between "macrophysics" and "microphysics"—are intentionally omitted.

0.1 Physics background

Scale (light-seconds)	10^{-24} ↓	10^{-3} ↓	10^{18} ↓
A. System	Nucleus	Moon	Observable universe
B. Dominant Force	NUC	Electromagnetism	Gravitational
C. Domain	Microphysics		Macrophysics
D. Theory	Quantum theory		General relativity

0.1.3 History and current status of general relativity

Special relativity was introduced around 1905 by Einstein, Lorentz, Poincaré, Minkowski, and others. Some 10 years later, Einstein introduced general relativity, generalizing from flat to nonflat 4-dimensional Lorentzian manifolds to include gravity in the models. Special relativity and special relativistic quantum theory have been checked literally billions of times. But for many years only small and poorly measured effects within the solar system indicated that general relativity gave better answers than combining its special relativistic and Newtonian limits *ad hoc*.

Today, more accurate measurements within the solar system, the tentative success of general relativistic models for white dwarf stars and pulsars, the possible discovery of the black holes and of the gravitational radiation predicted by general relativity, and the tentative success of general relativistic cosmology have given general relativity a somewhat firmer empirical foundation (Weinberg [1]). No doubt it will eventually have to be scrapped for a more general theory that somehow unifies quantum theory and general relativity. However, its main ideas will almost certainly be instrumental in the formulation of this more accurate theory. This book analyzes (nonquantum) general relativity.

0.1.4 Units

Let c = speed of light $\simeq 3 \times 10^{10}$ cm second^{-1} and let G = Newtonian gravitational constant $\simeq 6.67 \times 10^{-8}$ cm^3 second^{-2} g^{-1}. We shall use units such that $c = 1 = 8\pi G$.

In our system of units, it is possible to quote numerical results in [seconds]^N, and this avoids dimensional juggling. In case the reader is familiar with different systems of units, we give three examples of how ours works. First, speeds are dimensionless. For example, the speed s of the earth with respect to the center of our galaxy is roughly 10^{-3}. This means $s \simeq 10^{-3} = 10^{-3} c \simeq 3 \times 10^7$ cm second^{-1}. The advantage of writing $s \simeq 10^{-3}$ is that one sees explicitly that s is small (compared to c); $s \simeq 10^{-3}$ correctly suggests that cosmological observations made by a hypothetical observer here, at rest with respect to the center of our galaxy, would not differ very significantly from those we actually make. Next, distances can be expressed in seconds; for example, the radius $R_\oplus$ of the earth is roughly 2×10^{-2} seconds $= (2 \times 10^{-2}$ seconds$) c = 2 \times 10^{-2}$ light-seconds $\simeq 6 \times 10^8$ cm. Writing distances in seconds incorporates general relativity's unification of space and time into the numerical estimates. Finally, the mass $M_\oplus$ of the earth is about 4×10^{-10} seconds $= (4 \times 10^{-10}$ seconds$) \times (c^3/8\pi G) \simeq 6 \times 10^{27}$ g. Writing $M_\oplus \simeq 4 \times 10^{-10}$ seconds makes the estimate $M_\oplus \ll R_\oplus$ meaningful. $M_\oplus \ll 8\pi R_\oplus$ is needed to show that the earth can in most discussions be analyzed by Newtonian physics (see Section 0.1.10).

In our units, 1 second $\simeq 3 \times 10^8$ m $\simeq 1.5 \times 10^{34}$ kg.

0.1.5 Newtonian physics

No relativistic model can be deduced from any Newtonian model. No fundamental physics can be deduced from Newtonian physics. The logical and mathematical structure of Newtonian physics is surprisingly complicated, probably more complicated than that of relativity. Allowing Newtonian concepts into a discussion of relativity obscures the mathematics. On the other hand, Newtonian physics is quite indispensable for heuristic and empirical discussions. Hence we shall include some Newtonian physics but keep it carefully isolated from the mathematics.

Alonso and Finn [1] is a straightforward freshman text on Newtonian physics. Feynman *et al.* [1] is a brilliant presentation, ostensibly for freshmen. Such modern texts are careful to avoid assigning a distinguished origin to Euclidean 3-space. But since we only use Newtonian physics heuristically and isolate it from the mathematics, we shall adopt a drastic simplification. We take (Newtonian) space to mean $(\mathbb{R}^3, \sum_{\mu=1}^{3} du^\mu \otimes du^\mu)$, with all the structure of $\mathbb{R}^3$ implied, including a distinguished origin. When we do use Newtonian physics, we use the notation and terminology of the above texts without further apology. For example, we write $(\mathbb{R}^3, \sum_{\mu=1}^{3} du^\mu \otimes du^\mu) \equiv (\mathbb{R}^3, d\vec{x} \cdot d\vec{x})$. Similarly "$\vec{v}$" can mean $\vec{v} \in \mathbb{R}^3$, or $\vec{v} \in \mathbb{R}^3_{\vec{x}}$ for $\vec{x} \in \mathbb{R}^3$, or a vector field $\vec{v} \colon \mathbb{R}^3 \to T\mathbb{R}^3$, or a 1-form $\vec{v} \colon \mathbb{R}^3 \to T_1^0 \mathbb{R}^3$, or a vector field tangent to a curve $\vec{r} \colon \mathscr{E} \to \mathbb{R}^3$, and so on, depending on context.

0.1.6 Newtonian point particles

The Newtonian time axis is $T \equiv \mathbb{R}$. $\vec{\nabla}$, with components $(\partial/\partial x^1, \partial/\partial x^2, \partial/\partial x^3)$, is the gradient operator. A point particle $(\vec{r}, m)$ is a curve $\vec{r} \colon \mathscr{E} \to \mathbb{R}^3$ and an inertial-mass $m \in (0, \infty)$. For $t \in \mathscr{E} \subset T$, $\vec{r}(t) \in \mathbb{R}^3$ is the position of the particle at time t. m is measured by collision experiments that do not involve gravity (Alonso and Finn [1], Chapter 7). $\vec{r}$ is the path, $\dot{\vec{r}} \equiv \vec{v}$ the velocity $|\vec{v}|$ the speed, $\ddot{\vec{r}}$ the acceleration, $m\vec{v}$ the momentum, and $\frac{1}{2}m|\vec{v}|^2$ the kinetic energy. Let $\vec{y} \colon T \to \mathbb{R}^3$ be a curve. Then replacing $\vec{r}$ above by $\vec{r} - \vec{y}$ gives path, velocity, and so on, relative to $\vec{y}$; for example $\ddot{\vec{r}} - \ddot{\vec{y}}$ is the particle's acceleration relative to $\vec{y}$. $m = mc^2$ is sometimes called the rest energy, although

Newtonian physics does not really use this concept. Let $\vec{F}$ be the (Newtonian) force on $(\vec{r}, m)$ (Alonso and Finn [1], Chapter 7). Then $\vec{F} = m\ddot{\vec{r}}$.

0.1.7 Time-independent Newtonian gravitational forces

In Newtonian physics, the gravity of a time-independent source is described by a function $\phi: \mathbb{R}^3 \to \mathbb{R}$, the gravitational potential. $-\vec{\nabla}\phi$ is the gravitational field. ϕ and $\hat{\phi}: \mathbb{R}^3 \to \mathbb{R}$ describe the same gravitational effect iff $\vec{\nabla}\phi = \vec{\nabla}\hat{\phi}$. Every point particle $(\vec{r}, m)$ can be assigned a passive-mass $\tilde{m} \in (0, \infty)$ such that the (Newtonian) gravitational force on $(\vec{r}, m)$ is $\vec{F} = -\tilde{m}\vec{\nabla}\phi$. $\tilde{m}$ is measured by comparing, in a given time-independent gravitational field, the weight of $(\vec{r}, m)$ with the weight of a standard particle. Thereafter, $(\vec{r}, m)$ can be used to determine $\vec{\nabla}\phi$ in other situations (Alonso and Finn [1], Chapter 13). Experiments indicate that $\tilde{m}/m = 1 \pm 10^{-11}$ (Dicke [1]). We henceforth assume $\tilde{m} = m$. Then $m\ddot{\vec{r}} = \vec{F} = -\tilde{m}\vec{\nabla}\phi$ gives $\ddot{\vec{r}} = -\vec{\nabla}\phi$. Thus gravitational acceleration depends only on ϕ; $\tilde{m}$ is irrelevant. In general relativity one never introduces any quantity analogous to $\tilde{m}$ in the first place, although one does use quantities analogous to inertial-mass m and to the active-mass $\bar{m}$ discussed below.

0.1.8 Typical Newtonian gravitational potentials

In our units ϕ is dimensionless and $|\phi|$ is typically much less than 1. For example, one usually takes $\phi = (1000 \text{ cm second}^{-2})h$ near the surface of the earth, where h is height in cm; suppose $h = 9$ cm. Then in our units $\phi = 9000 \text{ cm}^2 \text{ second}^{-2}/c^2 \simeq 10^{-17}$. Let $\phi, \hat{\phi}$ be gravitational potentials. Then $\vec{\nabla}\phi = \vec{\nabla}\hat{\phi}$ iff $\phi = \hat{\phi} + \text{constant}$. Unless explicitly denied, we henceforth assume the constant has been so chosen that $\phi \to 0$ at spatial infinity. This is consistent if the sources of the gravitational field are confined in a compact region (Alonso–Finn [1]).

0.1.9 Newtonian active-mass

Consider an isolated, spherically symmetric, static body centered at the origin of $\mathbb{R}^3$. One can assign an active-mass $\bar{m} \in (0, \infty)$ to the body such that the gravitational potential outside the body is $\phi = -G\bar{m}/|\vec{x}|$. $\bar{m}$ is measured by measuring $\vec{\nabla}\phi$. For example, a point particle half-way between sun and earth suffers a Newtonian gravitational force from the sun about 3.3×10^5 as great as that from the earth; so that one assigns the sun an active-mass 3.3×10^5 times that of the earth. Usually such a gravitating body can be regarded as a point particle with inertial-mass m. Experiments indicate that $\bar{m}/m = 1 \pm 10^{-4}$. If one has n particles per unit volume, each of active-mass $\bar{m}$, Poisson's equation for ϕ is $\nabla^2 \phi = 4\pi G n \bar{m} = \frac{1}{2} n \bar{m}$.

0.1.10 Limitations of Newtonian theory

As long ago as 1799, Laplace suggested there might be bodies so heavy and dense that light could not escape from their surface; a translation of this first "black hole" paper is given in Hawking and Ellis [1]. Though the Newtonian arguments Laplace used are no longer regarded as appropriate, we outline briefly some Newtonian theory, which will facilitate interpretations of the general relativistic models to be presented later.

Consider a spherically symmetric body in $\mathbb{R}^3$, centered at the origin and with Euclidean radius $a \in (0, \infty)$. Let $\rho: \mathbb{R}^3 \to [0, \infty)$ be a C^∞ function, interpreted as Newtonian active-mass per unit Euclidean volume. Because of spherical symmetry, we regard ρ as a function $\rho: \mathbb{R} \to [0, \infty)$ with $\rho(\vec{x}) = \rho(|\vec{x}|)$. Thus the total Newtonian active-mass $\bar{m}$ is $\bar{m} = 4\pi \int_0^a \rho(r) r^2 dr$. Define $f: \mathbb{R}^3 \to [0, \bar{m}/8\pi]$ by $f(\vec{x}) = \frac{1}{2} \int_0^{|\vec{x}|} \rho(r) r^2 dr$. Thus $8\pi f(\vec{x})$ is the active mass within a sphere of radius $|\vec{x}|$. Poisson's equation $\nabla^2 \phi = \frac{1}{2}\rho$, together with the boundary conditions and smoothness requirements mentioned above for the Newtonian gravitational potential ϕ, is equivalent to $\phi(\vec{x}) = \int_\infty^{|\vec{x}|} r^{-2} f(r) dr$ (Alonso and Finn [1]). Thus for $|\vec{x}| \geq a$, $\phi(\vec{x}) = -\bar{m}/8\pi|\vec{x}|$. In particular, at the surface, $\phi = -\bar{m}/8\pi a$. Laplace pointed out that for large $\bar{m}$ and small a, $\phi < -\frac{1}{2}$. He argued that then light could not escape from the surface. Note that for the earth, $\bar{m} \ll 8\pi a$ (see the end of Section 0.1.4) so that $|\phi| \ll 1$ and Laplace's comment is not relevant.

In general, unless a Newtonian model predicts $|\phi| \ll 1$ everywhere and $|\vec{v}| \ll 1$ for all speeds of interest, the model should not be taken seriously. For example, consider a star so dense that Newtonian physics predicts $\phi \leq -(1/2)$ at the surface. Then Laplace's comment is relevant. But nowadays one regards the Newtonian model as self-defeating. In fact, in such a case, one should not attempt to use any Newtonian concept, especially not T, $(\mathbb{R}^3, d\vec{x} \cdot d\vec{x})$, or ϕ. For example, in a general relativistic black hole model (Sections 1.4 and 7.5), the only quantity that could reasonably be regarded as Newtonian time is also the only quantity that could reasonably be regarded as Newtonian radius. Both Newtonian concepts are then so misleading they are worse than useless.

0.2 Preview of relativity

Spacetimes form the "universe of discourse" for general relativity. A spacetime is a 4-dimensional Lorentzian manifold (M, g) satisfying certain technical requirements to be specified in Chapter 1; it usually carries additional structures that model electromagnetism, matter, and so on. By using M, we can describe the complete history of a physical process, viewed as a whole. g carries the essential information about space, time, and gravity. In going from Newtonian physics to relativity, physicists had to forget various Newtonian concepts; g somehow remembers the right things and forgets the wrong ones. Concepts of causality, distance, time, velocity, speed, acceleration, rotation, rigidity, simultaneity, orthogonality, gravity, and so on, are derived from g to the extent they are retained at all. g therefore must play many roles; its unifying power is remarkable.

More specifically, general relativity models an ordinary point particle as a curve $\gamma: \mathscr{E} \to M$ and a rest-mass $m \in (0, \infty)$ (cf. Section 3.1.1). When analyzing a particle, g is used in each of the following ways: (a) together with one choice between "+" and "−", g supplies M with a sense of "future" and "past" (Sections 1.2, 5.0.1, and 8.3); (b) g supplies γ with a kind of arc-length; this arc-length models time on a clock moving with the particle and replaces, to the extent anything does, Newtonian time (Chapter 2); (c) g replaces the Euclidean metric of ordinary 3-space (Section 2.1); (d) g replaces the Newtonian gravitational potential (Chapters 1 to 4); (e) g and its Levi–Civita connection supply a local sense of "no rotation" for the axis of a gyroscope the particle carries (Section 2.2); (f) let $\gamma_* u$ be the curve tangent at $u \in \mathscr{E}$; the condition $g(\gamma_* u, \gamma_* u) < 0$ replaces the Newtonian condition that the particle speed be less than the speed of light; in fact, γ is parametrized so that $g(\gamma_* u, \gamma_* u) = -m^2 \; \forall u \in \mathscr{E}$; $\gamma_* u$ then replaces and unifies the ordinary energy and momentum of the particle (Section 3.1). And so on.

In this section we give a heuristic discussion of a spacetime of particular importance in physics—Minkowski space. To simplify matters, we shall consider a model with only one space dimension. For this purpose, imagine a small body moving in a straight line in the absence of gravity. In Newtonian physics, the body is assigned an inertial-mass $m \in (0, \infty)$. Its motion is described by a function $x: \mathbb{R} \to \mathbb{R}$, with $x(t)$ the position at Newtonian time t. By our conventions, x is C^∞. $v = dx/dt$ is the velocity, $|v|$ is the speed. In our units (Section 0.1.4), $|v(t)| < 1$ iff the speed at t is less than that of light; suppose $|v(t)| < 1$ for all t.

To get a relativistic model now requires three steps: (a) $x: \mathbb{R} \to \mathbb{R}$ is replaced by a curve into $\mathbb{R}^2$ (much as one replaces a function by its graph in freshman calculus); (b) the essential structure on $\mathbb{R}^2$ is assembled into a Lorentzian metric and a "future" on $\mathbb{R}^2$; (c) extraneous structure is thrown away. We first perform steps (a) and (b).

$g = du^1 \otimes du^1 - du^2 \otimes du^2$ is a Lorentzian metric on $\mathbb{R}^2$. For $q \in \mathbb{R}^2$, we call $W \in \mathbb{R}^2_q$ *future pointing* iff $g(W, W) \le 0$ and $g(\partial_2, W) < 0$. For example, at any q, $\partial_2 + \frac{1}{2}\partial_1$ is future pointing while $\partial_2 + 2\partial_1$ and $-\partial_2 + \frac{1}{2}\partial_1$ are not. When supplied with this sense of future pointing, $(\mathbb{R}^2, g)$ is called *2-dimensional Minkowski space*. Let $\gamma: \mathscr{E} \to \mathbb{R}^2$ be a curve. For brevity, we write $\gamma^i = u^i \circ \gamma$, $i = 1, 2$, so that $\gamma = (\gamma^1, \gamma^2)$. To avoid irrelevant ambiguities, we demand that γ be inextendible, that $0 \in \mathscr{E}$, and that $\gamma^2 0 = 0$. Let x and m be the Newtonian quantities above.

Proposition 0.2.1. *For each pair (x, m) there is a unique $\gamma: \mathscr{E} \to \mathbb{R}^2$ with closed image $\gamma\mathscr{E}$ such that $\forall u \in \mathscr{E}$: (a) $\gamma_* u$ is future pointing; (b) $\gamma^1 = x \circ \gamma^2$; (c) $g(\gamma_* u, \gamma_* u) = -m^2$.*

PROOF. Suppose (x, m) is given. The following are asserted or demanded for all $t \in \mathbb{R}$ and/or $u \in \mathscr{E}$. Define $s: \mathbb{R} \to \mathbb{R}$ by $s(t) = (1/m) \int_0^t [1 - v(y)^2]^{1/2} dy$. Then $ds/dt = (1 - v^2)^{1/2}/m$. Since $|v(t)| < 1$, $ds/dt > 0$. Thus s is a diffeomorphism from $\mathbb{R}$ onto $s(\mathbb{R})$. Since $s(0) = 0$, we have $s^{-1}(0) = 0$. Now define $\mathscr{E} = s(\mathbb{R})$, and define $\gamma: \mathscr{E} \to \mathbb{R}^2$ by $\gamma u = ((x \circ s^{-1})u, s^{-1}u)$. Then $\gamma^2 0 = s^{-1} 0 = 0$ as required. Moreover, $(\gamma \circ s)t = (x(t), t)$, so $\gamma\mathscr{E}$ is the graph of a function defined on the whole x^2-axis. Hence $\gamma\mathscr{E}$ is closed and γ is inextendible. By definition of γ, $\gamma^1 = x \circ \gamma^2$, so (b) holds. For (a) and (c), we compute $\gamma_* u$: with $u = s(t) \in \mathscr{E}$, $\gamma_* s(t) = (mv(t)/[1 - v(t)^2]^{1/2}, m/[1 - v(t)^2]^{1/2}) \in \mathbb{R}^2_{(\gamma \circ s)t}$. Thus $(d\gamma^2/du)u = m/[1 - v(t)^2]^{1/2} > 0$, which is equivalent to $g(\partial_2, \gamma_* u) < 0$. Furthermore, $g(\gamma_* u, \gamma_* u) = -[(d\gamma^2/du)u]^2 + [(d\gamma^1/du)u]^2 = -m^2/[1 - v(t)^2] + m^2 v(t)^2/[1 - v(t)^2] = -m^2 < 0$. Thus (a) and (c) both hold.

Finally, we consider uniqueness. Let $\hat{\gamma}: \hat{\mathscr{E}} \to \mathbb{R}^2$ obey (a) to (c) with γ replaced by $\hat{\gamma}$. Writing $\hat{\gamma} = (\hat{\gamma}^1, \hat{\gamma}^2)$, we see from (a) that $d\hat{\gamma}^2/d\hat{u} > 0$, where $\hat{u}$ denotes the coordinate function on $\hat{\mathscr{E}}$. Since for $\gamma = (\gamma^1, \gamma^2)$ as above, $d\gamma^2/du > 0$, γ^2 is a diffeomorphism from $\mathscr{E}$ onto $\mathbb{R}$. Let $\alpha = (\gamma^2)^{-1} \circ \hat{\gamma}^2$, then $\hat{\gamma}^2 = \gamma^2 \circ \alpha$; note that $d\alpha/d\hat{u} > 0$. From (b) we have $\hat{\gamma}^1 = x \circ \hat{\gamma}^2$ and from (c), $-(d\hat{\gamma}^2/d\hat{u})^2 + (d\hat{\gamma}^1/d\hat{u})^2 = -m^2$. Thus we obtain $-(d\gamma^2/du)^2(d\alpha/d\hat{u})^2 + (dx/dt)^2(d\gamma^2/du)^2(d\alpha/d\hat{u})^2 = -m^2$ by the chain rule, or equivalently,

$$(d\alpha/d\hat{u})^2\{-(d\gamma^2/du)^2 + (dx/dt)^2(d\gamma^2/du)^2\} = -m^2.$$

However, we also know that

$$-(d\gamma^2/du)^2 + (d\gamma^1/du)^2 = -m^2 \Rightarrow \{-(d\gamma^2/du)^2 + (dx/dt)^2(d\gamma^2/du)^2\} = -m^2.$$

Consequently, $(d\alpha/d\hat{u})^2 = 1$, and since $d\alpha/d\hat{u} > 0$, $d\alpha/d\hat{u} = 1$. From $\hat{\gamma}^2 = \gamma^2 \circ \alpha$, we see that $\hat{\gamma}^2$ and γ^2 differ by a translation. Since $\hat{\gamma}^2$ is also assumed inextendible and $\hat{\gamma}^2 0 = \gamma^2 0 = 0$, $\hat{\gamma}^2 = \gamma^2$ and $\hat{\mathscr{E}} = \mathscr{E}$. Since $\hat{\gamma} = (x \circ \hat{\gamma}^2, \hat{\gamma}^2)$ and $\gamma = (x \circ \gamma^2, \gamma^2)$, we conclude $\gamma = \hat{\gamma}$. □

The proof of the following is now left as an exercise. It is the converse of Proposition 0.2.1 and shows x can indeed be replaced by a curve.

Proposition 0.2.2. *Let $\gamma: \mathscr{E} \to \mathbb{R}^2$ be an inextendible curve with closed image $\gamma\mathscr{E}$ such that Proposition 0.2.1a and 0.2.1c hold with $m \in (0, \infty)$. Then there is a unique $x: \mathbb{R} \to \mathbb{R}$ such that 0.2.1b also holds; moreover, then $|(dx/dt)(t)| < 1$ $\forall t \in \mathbb{R}$.*

The following figure capsulizes the preceding discussion via a concrete example.

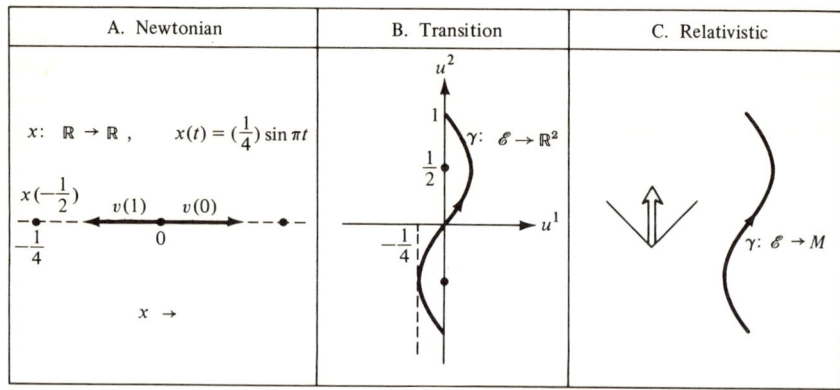

Figure 0.2.3. Three descriptions of motion in a straight line; note the simplicity of C.

0.2 Preview of relativity

These propositions and Figures 0.2.3A and 0.2.3B show roughly the way relativity uses 4 dimensions and some of the roles the Lorentzian metric g plays. But if u^1 and u^2, respectively, could be identified with Newtonian position and time, we would have nothing new. To get to relativity, we must forget.

Let $\Uparrow = \{(q, W) \in T\mathbb{R}^2 | W \text{ is future pointing}\}$. Now regard $\mathbb{R}^2$ as a manifold M; for example $\mathbb{R}^2$ has a distinguished origin and M does not. $(M, g, \Uparrow)$ is the real structure of interest. Now one can start afresh by *defining* a particle (γ, m) as a rest-mass $m \in [0, \infty)$ and a curve $\gamma: \mathscr{E} \to M$ such that $g(\gamma_* u, \gamma_* u) = -m^2$ and $(\gamma u, \gamma_* u) \in \Uparrow \forall u \in \mathscr{E}$. Then $S = ((M, g, \Uparrow), (\gamma, m))$ is a genuine, though very simple, relativistic model (compare Figure 0.2.3c). In principle, a model like S must be interpreted and related to measurements by referring to relativistic physics, not to Newtonian physics as in Proposition 0.2.1. For example u^2 cannot be canonically recovered from $(M, g, \Uparrow)$ so we have lost the Newtonian sense of absolute time (Exercise 0.2.7). One can consider a clock moving with the particle to get some sense of time beyond the very qualitative information in $\Uparrow$ (Exercise 0.2.8). But such a time is dependent on the particle, quite unlike Newtonian time. By ruthlessly exploiting the Lorentzian structure, we shall gradually develop the concepts necessary to interpret relativistic models.

In addition to defining, via $g(\gamma_* u, \gamma_* u) < 0$, what is meant by a speed less than the speed of light, g can also replace the Newtonian gravitational potential. Roughly, the idea is the following. Consider a small body isolated from all external influences except gravity. Galileo was surprised to find that all such bodies accelerate at the same rate in a given gravitational field (cf. Section 0.1.7). Their motion thus depends only on their initial position and velocity, not on their composition or other properties. But the inextendible geodesics of a Lorentzian manifold have a similar property. They are uniquely determined by an initial point and an initial tangent vector. Noticing this similarity, Einstein suggested modelling such bodies as appropriate geodesics. Then the Lorentzian metric replaces the Newtonian gravitational potential. Proposition 0.2.1 actually shows the key idea, albeit in the trivial case of no gravitational field: a wholly isolated small body has no Newtonian acceleration, so $d^2x/dt^2 = 0$, and the reader may check that the latter condition is both necessary and sufficient for the γ of Proposition 0.2.1 to be a geodesic.

Exercise 0.2.4

This exercise shows how $\gamma_* u$ in Proposition 0.2.1 unifies Newtonian kinetic energy and Newtonian momentum. To this end, suppose $|v(t)| \ll 1$. Show that then:

(a) $du^1(\gamma_* s(t)) = mv(t) + O(v^3)$;
(b) $du^2(\gamma_* s(t)) = mc^2 + \frac{1}{2}mv(t)^2 + O(v^4)$.

For the interpretations of the Newtonian quantities on the right, compare Section 0.1.6.

0 Preliminaries

EXERCISE 0.2.5

In Newtonian physics, particles travelling at the speed of light are hard to discuss. In relativity, one simply uses g. To see roughly how, suppose in Proposition 0.2.1 that $|v(t)| = 1$ instead of $|v(t)| < 1\ \forall t \in \mathbb{R}$; let $a \in (0, \infty)$ be given. Show that Proposition 0.2.1 remains valid if 0.2.1c is replaced by $g(\gamma_* u, \gamma_* u) = 0$ and 0.2.1a is replaced by $du^2(\gamma_* u) = a$. Thus $g(\gamma_* u, \gamma_* u) = 0$ becomes the key condition. (Here a corresponds to energy; see Section 3.1).

EXERCISE 0.2.6

(a) In 2-dimensional Minkowski space, let $\mathscr{T} = \{(q, W) \in T\mathbb{R}^2 \mid g(W, W) < 0\}$. Show $\mathscr{T}$ has two connected components. (b) Show that $g(\gamma_* u, \gamma_* u) \leq 0$ has the geometric interpretation: $\gamma_* u$ makes an angle $\geq 45°$ with the u^1 axis.

EXERCISE 0.2.7

Let $(\mathbb{R}^2, g)$ be 2-dimensional Minkowski space. For $\theta \in \mathbb{R}$, define $\hat{u}^1 \colon \mathbb{R}^2 \to \mathbb{R}$ and $\hat{u}^2 \colon \mathbb{R}^2 \to \mathbb{R}$ by $\hat{u}^1 = (\cosh \theta) u^1 + (\sinh \theta) u^2$, $\hat{u}^2 = (\sinh \theta) u^1 + (\cosh \theta) u^2$. (a) Show that $g = d\hat{u}^1 \otimes d\hat{u}^1 - d\hat{u}^2 \otimes d\hat{u}^2$ and that $W\hat{u}^2 > 0$ for all future-pointing vectors W. (b) In Proposition 0.2.1 show $v(t) = du^1(\gamma_* s(t))/du^2(\gamma_* s(t))$. (c) Define $\hat{v}(t) = d\hat{u}^1(\gamma_* s(t))/d\hat{u}^2(\gamma_* s(t))$. Show that for each $t \in \mathbb{R}$, there is exactly one θ such that $\hat{v}(t) = 0$. (d) Show that in (c) the same θ will work for all t iff γ is a geodesic. (e) In Exercise 0.2.5 show that $|\hat{v}(t)| = 1\ \forall t, \theta \in \mathbb{R}$.

> In popularizations, the above results are sometimes referred to as follows: (a) "Space and time are relative." (c) "(Newtonian) velocity is relative." (d) "(Nongravitational) accelerations are absolute." (e) "The speed of light is absolute."

EXERCISE 0.2.8

In Proposition 0.2.1, let $u = ms$. (a) Show that if γ is reparametrized by u then its tangent becomes a unit vector so that u is a kind of arc-length. (b) Show $|u(t)| \leq |t|\ \forall t \in \mathbb{R}$, where equality holds $\forall t \in \mathbb{R}$ iff $v(t) = 0\ \forall t \in \mathbb{R}$. (c) Give an example of an x such that $\mathscr{E} \neq \mathbb{R}$.

> u models time measured on a clock moving with the particle. (b) is sometimes called the time dilation effect—"if the particle is moving, it ages less rapidly."
>
> Forgetting extraneous structure is genuinely painful. Phrases such as those below Exercises 0.2.7 and 0.2.8, or such as "twin-paradox," "Lorentz contraction," and so on, have essentially the same meaning as "ouch!"

Spacetimes 1

We focus attention on Lorentzian manifolds in this chapter. After a brief mathematical review, the fundamental notions of causality and time orientability are introduced. Spacetimes are then defined and several examples of spacetimes important in physics are discussed. These examples will be everywhere dense in this book and we urge the reader to master them before proceeding further.

1.0 Review and notation

Let (M, g) denote an N-dimensional Riemannian or semi-Riemannian manifold; as usual, M_x denotes the tangent space of M at $x \in M$.

1.0.1 Physical equivalence

We will analyze an equivalence relation for tensor fields; various physical quantities will later be represented by corresponding equivalence classes. First recall from Exercise 0.0.15a that $\forall x \in M$ the inner product gx on M_x induces the metric isomorphism $\phi_x : M_x \to M_x^*$ determined by $(\phi_x v)w = g(v, w)$ $\forall v, w \in M_x$. By Exercise 0.0.14, the metric isomorphism ϕ_x gives rise to an equivalence relation $\sim$ among tensors in $\{T_s^r(M_x) | r + s =$ a fixed integer$\}$ as follows. Suppose $r + s = p + q$, $A \in T_q^p(M_x)$, and $B \in T_s^r(M_x)$. Then $A \sim B$ iff $\phi_{xq}{}^p(A) = \phi_{xs}{}^r(B)$, where $\phi_{xq}{}^p$ and $\phi_{xs}{}^r$ are, respectively, the isomorphisms induced by ϕ_x of $T_q^p(M_x)$ and $T_s^r(M_x)$ onto $T_{r+s}^0(M_x)$.

Now let A and B be two tensor fields defined in $\mathscr{U} \subset M$. By definition, A is *physically equivalent* to B iff $Ax \sim Bx$ $\forall x \in \mathscr{U}$. As an example, a vector field X and a 1-form ω are physically equivalent iff $\omega = g(X, \cdot)$ (Exercise 0.0.15f). For two vector fields X and Y, the tensor fields physically equivalent to $X \otimes Y$ are precisely: $g(X, \cdot) \otimes Y$, $X \otimes g(Y, \cdot)$, and $g(X, \cdot) \otimes g(Y, \cdot)$. In

17

relativity, all tensor fields in a physical equivalence class have the same physical interpretation.

The generic symbol of a tensor field physically equivalent to a given tensor field A will be $\tilde{A}$ or $\hat{A}$, but there will be occasional deviations.

> In classical terminology, two tensor fields are physically equivalent iff one is obtained from the other by "raising and/or lowering indices." Compare Section 3.6 following. The term "physically equivalent" is fully appropriate only when (M, g) is a spacetime (Section 1.3 following). However, we shall use the mathematical concept in other situations as well.

1.0.2 Curvature (Bishop–Goldberg 5.11)

Let D be the Levi-Civita connection of (M, g). The *curvature operator* of (M, g) assigns to each ordered pair of vector fields (X, Y) on M an operator $\mathbf{R}_{XY}$ on vector fields as follows: $\mathbf{R}_{XY}Z = D_X D_Y Z - D_Y D_X Z - D_{[X,Y]}Z$. The *curvature tensor* $\mathbf{R}$ of (M, g) is the $(1, 3)$ tensor field $\mathbf{R}$ such that $\mathbf{R}(\omega, Z, X, Y) = \omega(\mathbf{R}_{XY}Z)$, $\forall$ 1-form ω and for all vector fields X, Y, Z. The *Ricci tensor* of (M, g) is a contraction of $\mathbf{R}$: suppose $X \in M_x$, $Y \in M_x$, $\{X_A\}$ is any basis of M_x and $\{\omega^A\}$ is the dual basis, then $\mathbf{Ric}(X, Y) = \sum_{A=1}^{N} \mathbf{R}(\omega^A, X, X_A, Y)$. **Ric** is a symmetric $(0, 2)$ tensor field on M. The *scalar curvature* ("Ricci scalar") $S: M \to \mathbb{R}$ of (M, g) is the contraction of the $(1, 1)$ tensor field physically equivalent to **Ric**. The *Einstein tensor* $\mathbf{G}$ of (M, g) is $\mathbf{G} = \mathbf{Ric} - \tfrac{1}{2} S g$. $\mathbf{G}$ is a symmetric $(0, 2)$ tensor field since **Ric** and g are. (M, g) is *flat* iff $\mathbf{R} = 0$, *Ricci flat* iff $\mathbf{Ric} = 0$.

Let $\hat{\mathbf{R}}$ be the $(0, 4)$ tensor field physically equivalent to $\mathbf{R}$ on (M, g). Then $\hat{\mathbf{R}}(W, Z, X, Y) = g(\mathbf{R}_{XY}Z, W)$. $\hat{\mathbf{R}}$ obeys: (a) $\hat{\mathbf{R}}(X, Y, Z, W) = -\hat{\mathbf{R}}(X, Y, W, Z) = \hat{\mathbf{R}}(Y, X, W, Z)$; (b) $\hat{\mathbf{R}}(X, Y, Z, W) + \hat{\mathbf{R}}(X, Z, W, Y) + \hat{\mathbf{R}}(X, W, Y, Z) = 0$; (c) $\hat{\mathbf{R}}(X, Y, Z, W) = \hat{\mathbf{R}}(Z, W, X, Y)$ (Bishop–Goldberg 5.11).

> We have followed the conventions of Bishop–Goldberg, 5.10, but have altered certain signs for agreement with the physics literature as well as common usage in geometry. These signs have the following consequence: positive scalar curvature will later correspond to a "pulling together" of the interesting geodesics—namely, the timelike and lightlike ones—and to a positive mass density.
>
> In case (M, g) is a spacetime (Section 1.3), both $\mathbf{R}$ and $\hat{\mathbf{R}}$ are interpreted as "gravitational field gradients" (cf. Chapter 4). A Newtonian analogue of $\mathbf{R}$ and $\hat{\mathbf{R}}$ is $\{-\partial^2 \phi / \partial x^\mu \partial x^\nu \mid \mu, \nu = 1, 2, 3\}$, where ϕ is the Newtonian gravitational potential (Section 0.1.8). A Newtonian analogue of $\mathbf{G}$ (the Einstein tensor) is then $2 \sum_{\mu=1}^{3} \partial^2 \phi / (\partial x^\mu)^2$ (cf. Section 9.3).

1.0.3 Computations

The following computational formulae will suffice for Chapters 1 and 2. Let (M, g) be a 4-dimensional Lorentzian manifold, $\{\omega^i\}$ be a local basis of 1-forms on M, $\{X_i\}$ the dual basis, and D the Levi-Civita connection. The *connection forms* $\{\omega_j{}^i\}$ for $\{\omega^i\}$ are characterized by

$$D_{X_i} X_j = \sum_{k=1}^{4} \omega_j{}^k(X_i) X_k,$$

$\forall i, j = 1, \ldots, 4$ (Bishop–Goldberg 5.7). If X and Y are vector fields on M, $D_X Y$ can be computed as follows:

(a) $$D_X Y = \sum_{i=1}^{4} \left\{ X(\omega^i Y) + \sum_{j=1}^{4} \omega_j{}^i(X) \omega^j(Y) \right\} X_i.$$

The *curvature forms* $\Omega_j{}^i$ for (M, g) are defined by

(b) $$\Omega_j{}^i = 2\left(d\omega_j{}^i + \sum_{k=1}^{4} \omega_k{}^i \wedge \omega_j{}^k \right),$$

$\forall i, j = 1, \ldots, 4$. The curvature tensor R can be computed as follows (Bishop–Goldberg 5.10):

(c) $$R = \sum_{i,j=1}^{4} X_i \otimes \omega^j \otimes \Omega_j{}^i.$$

If $\{\omega^i\}$ is an orthonormal basis, the connection forms are uniquely determined by the following two conditions:

(d) $$d\omega^i = -\sum_{j=1}^{4} \omega_j{}^i \wedge \omega^j \qquad i = 1, \ldots, 4;$$

(e) $$\omega_4{}^4 = 0, \; \omega_\mu{}^4 = \omega_4{}^\mu, \; \omega_\nu{}^\mu = -\omega_\mu{}^\nu \qquad \mu, \nu = 1, 2, 3.$$

Equations (d) and (e) are very convenient in computations.

If relative to a basis of 1-forms $\{\omega^i\}$ R is expressed as

$$R = \sum_{i,j,k,l=1}^{4} R^i_{jkl} X_i \otimes \omega^j \otimes \omega^k \otimes \omega^l,$$

then the Ricci tensor is: $\mathbf{Ric} = \sum_{j,l=1}^{4} \sum_{i=1}^{4} R^i_{jil} \omega^j \otimes \omega^l$. We will write $(\mathrm{Ric})_{jl}$ for $\sum_{i=1}^{4} R^i_{jil}$. To get a formula for the scalar curvature S, let us write the metric tensor g as $g = \sum_{i,j=1}^{4} g_{ij} \omega^i \otimes \omega^j$, and the $(2, 0)$ tensor field $\hat{g}$ physically equivalent to g as $\hat{g} = \sum_{i,j=1}^{4} g^{ij} X_i \otimes X_j$. From Exercise 0.0.15e, we know that $\sum_{j=1}^{4} g^{ij} g_{jk} = \delta_k{}^i$ (Kronecker delta). Then

$$S = \sum_{j,l=1}^{4} g^{jl} (\mathrm{Ric})_{jl}.$$

EXERCISE 1.0.4

Suppose: (M, g) is a Lorentzian manifold with scalar curvature S and Einstein tensor G; similarly for $(\hat{M}, \hat{g})$, $\hat{S}$ and $\hat{G}$; and $\phi: M \to \hat{M}$ is an isometry. Show $S = \hat{S} \circ \phi$, $\phi^* \hat{G} = G$.

EXERCISE 1.0.5

Let (M, g) be a Lorentzian manifold, $f: M \to \mathbb{R}$ be a function, X be the vector field on M physically equivalent to df. Show that if $g(X, X)$ is a constant, $D_X X = 0$.

1   Spacetimes

EXERCISE 1.0.6

Let M_x be a tangent space to a Lorentzian manifold (M, g), $L: M_x \to M_x$ be a linear transformation. Recall that L is *self-adjoint* (respectively, *skew-adjoint*) *with respect to gx* iff $g(LX, Y) = g(X, LY)$ (respectively, $g(LX, Y) = -g(X, LY)$) $\forall X, Y \in M_x$. Show L can be regarded as $L \in T_1^1(M_x)$ and is then self-adjoint (respectively, skew-adjoint) iff it is physically equivalent to a symmetric (respectively, skew-symmetric) tensor $S \in T_2^0(M_x)$.

1.1 Causal character

In this section we investigate an N-dimensional Lorentzian vector space (V, g) (Section 0.0.3). The structure is subtler than in the positive definite case; many of the deeper results in relativity hinge on seemingly rather trivial properties of such spaces.

Definition 1.1.1. Let $W \subset V$ be a subspace. The *causal character* of W is: (a) *spacelike* iff g is positive definite on W; (b) *lightlike* iff g is positive semi-definite but not positive definite on W; (c) *timelike* otherwise. Suppose $v \in V$; the *causal character* of v is that of span v; v is defined as *causal* iff v is not spacelike.

EXAMPLE 1.1.2. The definition implies: (a) The zero vector is spacelike; a non-zero vector $v \in V$ is spacelike iff $g(v, v) > 0$, lightlike iff $g(v, v) = 0$, and timelike iff $g(v, v) < 0$. (b) A subspace $W \subset V$ is: spacelike iff all its vectors are spacelike, lightlike iff it contains a lightlike vector but no timelike vector, timelike iff it contains a timelike vector. (c) None of the above cases are empty.

> Causal character is important for physics: a single relativistic concept usually corresponds to two or more Newtonian concepts; it is usually causal character which sorts out the various Newtonian analogues. For example, regard 2-dimensional Minkowski space as a Lorentzian vector space and let L be a 1-dimensional subspace:
>
> *Newtonian analogue.* If L is spacelike, L is like a straight line in Euclidean space. If L is timelike, L is like the complete history of an undisturbed Newtonian point particle. If L is lightlike, L is like the complete history of an undisturbed light signal. Our subsequent discussion will similarly unify many sets of Newtonian concepts: (energy, momentum); (electric field, magnetic field); (simultaneity, orthogonality in Euclidean 3-space); and so on. Physics students usually find such unifications very satisfying.

Proposition 1.1.3. Let $W \subset V$ be a subspace. (a) W timelike $\Leftrightarrow W^\perp$ spacelike, and W spacelike $\Leftrightarrow W^\perp$ timelike. (b) W lightlike $\Leftrightarrow W \cap W^\perp \neq \{$the zero vector$\} \Leftrightarrow W^\perp$ lightlike.

PROOF. The notation $W^\perp$ is defined in Exercise 0.0.11, and we will make use of this exercise without further comments in the following.

Now, W timelike $\Rightarrow W$ contains a timelike vector $\Rightarrow W$ contains a unit timelike vector $w \Rightarrow$ there exists an orthonormal basis $\{e_1, \ldots, e_{N-1}, w\}$. Since $W^\perp \subset \text{span } \{e_i | i = 1, \ldots, N - 1\}$ and g is positive definite on this span, $W^\perp$ is spacelike. Conversely, suppose $W^\perp$ is spacelike. Then $V = W \oplus W^\perp$. Let $v \in V$ be timelike; then $v = w + \hat{w}$ for some $w \in W$, $\hat{w} \in W^\perp$, and $g(w, w) = g(v, v) - g(\hat{w}, \hat{w}) < 0$. Thus w is timelike and hence W is timelike. The rest of (a) follows from $W^{\perp\perp} = W$.

Next, W lightlike implies W contains a lightlike vector w_0, but no timelike vector. Then $\forall a \in \mathbb{R}$ and $\forall w \in W$, $g(w + aw_0, w + aw_0) = g(w, w) + 2ag(w, w_0) \geq 0$. Since a was arbitrary, we have $g(w, w_0) = 0 \; \forall w \in W$, $\Rightarrow w_0 \in W^\perp \Rightarrow W \cap W^\perp \neq \{0\}$. Conversely if $0 \neq w_0 \in W \cap W^\perp$, then w_0 is lightlike. Since W cannot contain a timelike vector by (a), W is lightlike by Example 1.1.2b and the fact that $w_0 \in W$. The rest of (b) follows from $W^{\perp\perp} = W$. $\square$

Corollary 1.1.4. *$w \in W$ is timelike iff $w^\perp \subset V$ is spacelike.*

Corollary 1.1.5. *Two lightlike vectors are orthogonal iff they are proportional.*

PROOF. Let $v, w \in V$ be lightlike, $e \in V$ be timelike, and suppose $g(v, w) = 0$. Then $g(e, v) \neq 0$ by Corollary 1.1.4, so that for some $a \in \mathbb{R}$ $g(e, w + av) = 0$. Then $w + av$ is spacelike, again by Corollary 1.1.4. But $g(w + av, w + av) = g(w, w) + 2ag(v, w) + g(v, v) = 0$, so $w + av = 0$ and the vectors are proportional. The converse is trivial. $\square$

> The physical interpretation of orthogonality is surprisingly subtle. It depends on the dimension and causal character of the subspaces involved. We shall discuss it systematically when we have available the concept of an observer in Chapter 2. We shall use the following special case of Proposition 1.1.3 and Corollary 1.1.5 when we discuss waves travelling at the speed of light. The special case is quite hard to understand intuitively.

EXAMPLE 1.1.6. Let $W \subset V$ be an $(N - 1)$-dimensional lightlike subspace. Then $W^\perp$ is lightlike and 1-dimensional. Moreover, $W^\perp \subset W$. If $w \in W$ and $w \notin W^\perp$, then w is spacelike.

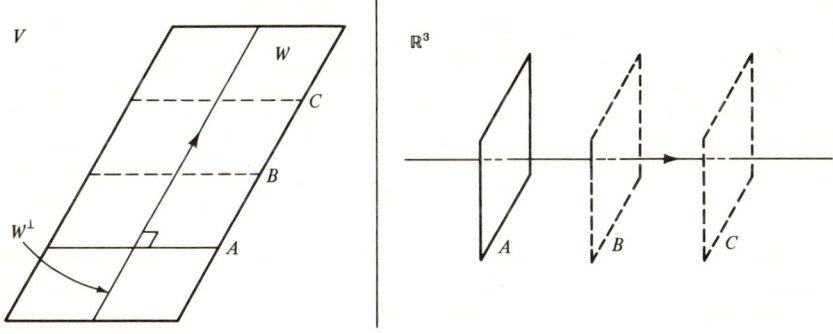

1 Spacetimes

Newtonian analogue. Suppose $N = 4$. Then W is like the complete history of a Euclidean plane that travels at the speed of light in the Euclidean direction perpendicular to itself. $W^\perp$ is like the complete history of a dot painted on the plane. In the preceding diagrams, we take $W^\perp$ oriented for vividness. Also we have suppressed one dimension in drawing W.

Let (M, g) be a Lorentzian manifold and let $\phi: N \to M$ be an immersion of a manifold N. If $\phi_* N_x \subset M_{\phi x}$ has the same causal character $\forall x \in N$, that *causal character* is assigned to the immersion ϕ and to its image ϕN. The corresponding definitions are used for curves into M, vector fields on M, and so on. In particular, a vector field X defined on $\mathcal{U} \subset M$ is *timelike* (respectively, *spacelike, lightlike*) iff $\forall x \in \mathcal{U}$, Xx is timelike (respectively, spacelike, lightlike). Let ω be a 1-form defined on $\mathcal{U} \subset M$ and X the vector field physically equivalent to ω. If X possesses a causal character, this *causal character* is assigned to ω. For instance, if X is a timelike vector field, then $g(X, \cdot)$ is a timelike 1-form.

One can also define the causal character of a 1-form directly by making use of the $(2, 0)$ tensor field $\tilde{g}$ physically equivalent to g. See Exercise 1.1.11.

The set $\mathcal{L}_0 \subset V$ of all lightlike vectors in V is called the *lightcone* in V. We can regard (V, g) as a Lorentzian manifold (V, g) (see Exercise 0.0.10) and investigate whether $\mathcal{L}_0$ has a causal character.

Proposition 1.1.7. *The lightcone $\mathcal{L}_0$ is a lightlike submanifold.*

By taking V to be a 3-dimensional Lorentzian vector space and drawing a picture of $\mathcal{L}_0$, one can easily convince oneself that indeed $\mathcal{L}_0$ is lightlike; for in this case, any tangent plane to $\mathcal{L}_0$ is a 2-dimensional vector subspace containing a generator of the cone $\mathcal{L}_0$ and clearly such a subspace cannot contain any timelike vectors (cf. Example 1.1.8).

Proof. Suppose $v \in \mathcal{L}_0$; then $g(v, v) = 0$ and $v \neq 0$. Let $\mathcal{U}$ be a neighborhood of v that does not contain the origin and define $\tilde{g}: \mathcal{U} \to \mathbb{R}$ by $\tilde{g}w = g(w, w) \forall w \in \mathcal{U}$. Then $\mathcal{L}_0 \cap \mathcal{U}$ is defined by $\tilde{g} = 0$. Moreover, $d\tilde{g}$ is nowhere zero on $\mathcal{U}$ because g is nondegenerate (Exercise 0.0.10). Thus $\mathcal{L}_0$ is a submanifold by the implicit function theorem. To show $\mathcal{L}_0$ is lightlike, suppose that $w \in V_v$ for some $v \in \mathcal{L}_0$, and let $\phi_v: V_v \to V$ be the canonical isomorphism of Exercise 0.0.10. Then w is in the tangent space $(\mathcal{L}_0)_v$ iff $w\tilde{g} = 0$, iff $g(\phi_v w, v) = 0$, and iff $g(w, \phi_v^{-1} v) = 0$. Thus $(\mathcal{L}_0)_v = (\phi_v^{-1} v)^\perp \subset V_v$. But $\phi_v^{-1} v \in V_v$ is lightlike because $g(\phi_v^{-1} v, \phi_v^{-1} v) = g(v, v) = 0$, hence $(\mathcal{L}_0)_v$ is lightlike by Proposition 1.1.3b. Since this holds for all $v \in \mathcal{L}_0$, $\mathcal{L}_0$ is lightlike. $\square$

EXAMPLE 1.1.8. Let $N = 3$ and $\{e_1, e_2, e_3\}$ be an orthonormal basis. Then $v \in \mathcal{L}_0$ iff its components obey $(v^3)^2 = (v^1)^2 + (v^2)^2 > 0$. Thus $\mathcal{L}_0$ is actually represented by a cone with the apex deleted, as shown. The timelike vectors obey $(v^3)^2 > (v^1)^2 + (v^2)^2$; they are represented by the points inside $\mathscr{L}_0$ and are shown as $\mathscr{T}_0^+$ and $\mathscr{T}_0^-$.

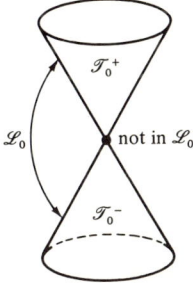

This figure correctly suggests many features of the general case. Indeed (see Exercise 1.1.9), for an arbitrary N-dimensional Lorentzian vector space, the set $\mathscr{T}_0$ of timelike vectors is an open submanifold with two connected components, $\mathscr{T}_0^+$ and $\mathscr{T}_0^-$, each diffeomorphic to $\mathbb{R}^N$, and for $N \geq 3$ the lightcone $\mathscr{L}_0$ also splits into two connected components, $\mathscr{L}_0^+$ and $\mathscr{L}_0^-$, each diffeomorphic to $\mathbb{R} \times \mathscr{S}^{N-2}$.

Newtonian analogue. Suppose $N = 4$. Then one component of $\mathscr{L}_0$ is like the complete history of an "information-gathering" sphere in $\mathbb{R}^3$ which contracts with the speed of light. Again, one dimension is suppressed in the following diagram of $\mathscr{L}_0^-$.

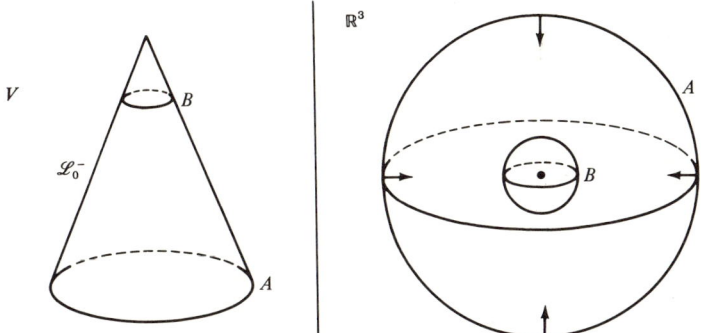

EXERCISE 1.1.9

Let $e \in V$ be timelike. Use an orthonormal basis to prove the following results. (a) The set of timelike vectors, $\mathscr{T}_0$, is an open submanifold with two connected components $\mathscr{T}_0^+ = \{v \in \mathscr{T}_0 | g(e, v) < 0\}$ and $\mathscr{T}_0^- = \{v \in \mathscr{T}_0 | g(e, v) > 0\}$; each component is diffeomorphic to $\mathbb{R}^N$. (b) If $N \geq 3$, the lightcone $\mathscr{L}_0$ has two connected components $\mathscr{L}_0^+ = \{v \in \mathscr{L}_0 | g(e, v) < 0\}$ and $\mathscr{L}_0^- = \{v \in \mathscr{L}_0 | g(e, v) > 0\}$, and each component is diffeomorphic to $\mathbb{R} \times \mathscr{S}^{N-2}$, where $\mathscr{S}^{N-2}$ is the unit $(N-2)$-sphere. (c) If $N \geq 2$, and $v \in \mathscr{L}_0^+ \cup \mathscr{T}_0^+$, $w \in \mathscr{L}_0^+ \cup \mathscr{T}_0^+$, then $g(v, w) \leq 0$. Equality holds iff $v \in \mathscr{L}_0^+$ and w is proportional to v.

1 Spacetimes

EXERCISE 1.1.10

(a) Let $v \in V$, $w \in V$ be causal. Show that the Schwarz inequality now "goes the wrong way": $|g(v, w)| \geq |v||w|$, and equality holds iff v and w are proportional.
(b) Let (M, g) be a Lorentzian manifold, X a timelike vector in M_x, and $\omega \in M_x^*$ a timelike 1-form. Show: $|\omega X| \geq |\omega||X|$, and equality holds iff ω and aX are physically equivalent for some $a \in \mathbb{R}$. [The norm $|\omega|$ is taken with respect to the Lorentzian inner product $\hat{g}x$ on M_x^* (Exercise 0.0.15b).]

EXERCISE 1.1.11

Let (M, g) be a Lorentzian manifold and let $X \in M_x$ and $\omega \in M_x^*$ be physically equivalent. Show that the causal character assigned to X by gx is the same as the causal character assigned to ω by $\hat{g}x$, where $\hat{g}x$ is the Lorentzian inner product on M_x^* given in Exercise 0.0.15.

EXERCISE 1.1.12

Let (M, g) be a Lorentzian manifold, N a manifold, and $\phi: N \to M$ an immersion. ϕ^*g is called the *metric induced on N by* ϕ iff ϕ^*g is a metric on N. Show ϕ^*g is a metric on N iff ϕN is timelike or spacelike.

EXERCISE 1.1.13

Suppose $m \in (0, \infty)$. Define $\mathcal{T}_0^m \subset V$ by $\mathcal{T}_0^m = \{v \in V | g(v, v) = -m^2\}$. Assuming V is 4-dimensional and regarding (V, g) as a Lorentzian manifold, show $\mathcal{T}_0^m$ is a spacelike 3-submanifold.

EXERCISE 1.1.14

Show: the closure of $\mathcal{T}_0^+$ is the disjoint union of $\mathcal{T}_0^+$, $\mathcal{L}_0^+$, and the zero vector; the boundary of $\mathcal{T}_0^+$ is $\mathcal{L}_0^+ \cup \{0\}$.

> Some other algebraic properties of (V, g) are given in Optional exercises 8.1.

1.2 Time orientability

We now consider the concepts of "past" and "future." Let (M, g) be a connected Lorentzian manifold, TM be its tangent bundle and $\Pi: TM \to M$ be the projection. The *causal character* of $(x, X) \in TM$ is the causal character of $X \in M_x$.

Proposition 1.2.1. *The set $\mathcal{T} \subset TM$ of timelike points is an open submanifold. $\mathcal{T}$ has either one (connected) component or two.*

PROOF. Define $K: TM \to \mathbb{R}$ by $K(x, X) = g(X, X)$. Then K is C^∞. As the complete inverse image of $(-\infty, 0)$ under K, $\mathcal{T}$ is open. Let $\mathcal{A}$ be a component of $\mathcal{T}$. If $\psi: \mathcal{T} \to \mathcal{T}$ denotes the homeomorphism defined by $\psi(x, X) =$

$(x, -X)$, then $\psi\mathcal{A}$ is also a component of $\mathcal{T}$. We will show $\mathcal{T} = \mathcal{A} \cup \psi\mathcal{A}$. Let $\mathcal{B} = \mathcal{A} \cup \psi\mathcal{A}$, $\mathcal{C} = \mathcal{T} - \mathcal{B}$. $\mathcal{B}$ is open and closed in $\mathcal{T}$, and hence both $\mathcal{B}$ and $\mathcal{C}$ are open and closed in $\mathcal{T}$. It follows that $\mathcal{B}$ and $\mathcal{C}$ are open in TM.

We claim $\Pi\mathcal{B} \cap \Pi\mathcal{C} = \emptyset$. If not, then there exist $(x, Z) \in \mathcal{B}$ and $(x, Y) \in \mathcal{C}$ for some $x \in M$. Let $\mathcal{Y} \subset M_x$ be that one of the two components of $M_x \cap \mathcal{T}$ in which (x, Y) lies (Exercise 1.1.9a); then $\mathcal{C} \cap \mathcal{Y} \neq \emptyset$. Since $\mathcal{C}$ is a union of components of $\mathcal{T}$, this implies $\mathcal{Y} \subset \mathcal{C}$. Now either (x, Z) or $(x, -Z)$ is in $\mathcal{Y}$, while both are in $\mathcal{B}$ by definition of $\mathcal{B}$. Thus $\mathcal{B} \cap \mathcal{Y} \neq \emptyset$, and hence also $\mathcal{Y} \subset \mathcal{B}$ because $\mathcal{B}$ is a union of components. It follows that $\mathcal{B} \cap \mathcal{C} \neq \emptyset$, a contradiction.

We therefore have $\Pi\mathcal{B} \cap \Pi\mathcal{C} = \emptyset$ and $\Pi\mathcal{B} \cup \Pi\mathcal{C} = M$. Since M is connected, $\Pi\mathcal{C} = \emptyset$, $\Rightarrow \mathcal{C} = \emptyset$, $\Rightarrow \mathcal{T} = \mathcal{A} \cup \psi\mathcal{A}$. If $\mathcal{A} \cap \psi\mathcal{A} = \emptyset$, $\mathcal{T}$ has two components. Otherwise, $\mathcal{A} = \psi\mathcal{A}$. □

One can also give a proof of this proposition using the notion of parallel translation induced by the Levi–Civita connection of g. Indeed, the timelike vectors of each M_y split into two components $\mathcal{T}_y^+$ and $\mathcal{T}_y^-$ (Exercise 1.1.9b). Now fix an $x \in M$, and let $\mathcal{A}^+$ be the union of all the components of $\mathcal{T}_y$, $\forall y \in M$, which are the images of $\mathcal{T}_x^+$ under parallel translation along some curve from x to y. Similarly, define $\mathcal{A}^-$. Identifying $\mathcal{A}^+$ and $\mathcal{A}^-$ with subsets of $\mathcal{T} \subset TM$ in the obvious manner, we see that both $\mathcal{A}^+$ and $\mathcal{A}^-$ are connected and $\mathcal{T} = \mathcal{A}^+ \cup \mathcal{A}^-$. The details are left as an exercise (Optional exercise 8.2.3).

Definition 1.2.2. The connected Lorentzian manifold (M, g) is called *time orientable* iff $\mathcal{T}$ has two components.

EXAMPLE 1.2.3. Let $M = \mathcal{S}^1 \times \mathbb{R}$, the cylinder; we regard M as being obtained from $\mathbb{R}^2$ by identifying (u^1, u^2) with $(u^1, u^2 + \pi)$. We will consider two different Lorentzian metrics on M. First, define on $\mathbb{R}^2$ 1-forms $\omega = \cos(u^2)du^1 + \sin(u^2)du^2$ and $\chi = -\sin(u^2)du^1 + \cos(u^2)du^2$. Then $\hat{g} = \omega \otimes \omega - \chi \otimes \chi$ is a Lorentzian metric on $\mathbb{R}^2$ and the mapping $\mathbb{R}^2 \to \mathbb{R}^2$ defined by $(u^1, u^2) \to (u^1, u^2 + \pi)$ leaves $\hat{g}$ unchanged. Thus $\hat{g}$ determines a Lorentzian metric g on M. Then (M, g) is Lorentzian and orientable, but not time orientable, as the following figure indicates.

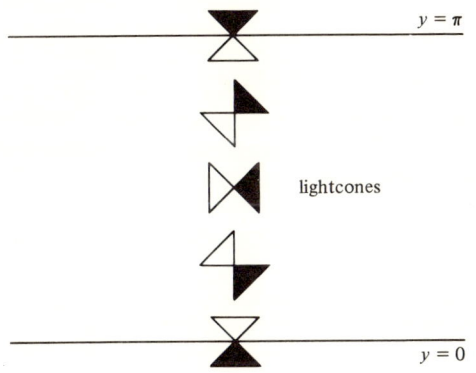

1 Spacetimes

On the other hand, the Lorentzian metric $\hat{g}_1 = du^1 \otimes du^1 - du^2 \otimes du^2$ on $\mathbb{R}^2$ is also invariant under the mapping $(u^1, u^2) \to (u^1, u^2 + \pi)$, so it too defines a Lorentzian metric, say g_1, on M. (M, g_1) is visibly time orientable.

> Thus time orientability involves the Lorentzian metric and not just the underlying C^∞ structure. However, if M is simply connected, then (M, g) is time orientable for all Lorentzian metrics g on M (cf. Section 8.2.3).

Suppose (M, g) is time orientable. (M, g) is *time oriented* iff one component of $\mathcal{T}$ is labelled $\mathcal{T}^+$ and called the *future*. The complement of $\mathcal{T}^+$ in $\mathcal{T}$, to be denoted by $\mathcal{T}^-$, is then called the *past*. Suppose there is a causal vector field X on (M, g). Then $(x, W) \to g(W, X)$ defines a C^∞ onto function $\psi: \mathcal{T} \to (-\infty, 0) \cup (0, \infty)$ (Exercise 1.1.9c). Thus $\mathcal{T}$ is not connected and (M, g) is time orientable. If we designate $\psi^{-1}(-\infty, 0)$ as $\mathcal{T}^+$, we say (M, g) is *time oriented by* X. In this case, the future is the component of $\mathcal{T}$ whose closure contains $Xx \ \forall x \in M$.

Let (M, g) be a time-oriented Lorentzian manifold with dimension $M \geq 3$. For each $x \in M$, define $\mathcal{T}_x^+ = \mathcal{T}^+ \cap \pi^{-1}x \subset M_x$. The corresponding component (Exercise 1.1.9) $\mathcal{L}_x^+$ of the lightcone in M_x is called *the future lightcone* $\mathcal{L}_x^+$ *in* M_x. $X \in M_x$ is called *future pointing* iff $X \in \mathcal{T}_x^+ \cup \mathcal{L}_x^+$. A vector field X defined on $\mathcal{U} \subset M$ is called *future pointing* iff each $Xx \in M_x$ is future pointing, $x \in \mathcal{U}$. A 1-form ω defined on $\mathcal{U} \subset M$ is *future pointing* iff the vector field physically equivalent to ω is future pointing. *Past lightcone* and *past pointing* are defined dually.

Exercise 1.2.4

Let (M, g) be a time-oriented Lorentzian manifold and suppose $W, V \in \mathcal{T}_x^+$. Show: (a) *convexity*, i.e., if $a \in [0, \infty)$ and $b \in (0, \infty)$, then $aV + bW \in \mathcal{T}_x^+$; (b) *wrong-way triangle inequality*—that is, $|V + W| \geq |V| + |W|$; (c) the inequality in (b) is equality iff span $V =$ span W; (d) that (a) remains valid for $V \in$ closure of $\mathcal{T}_x^+$ and that (b) remains valid if $V, W \in$ closure of $\mathcal{T}_x^+$ (cf. Exercise 1.1.14).

Exercise 1.2.5

Suppose (M, g) is a Lorentzian manifold. Show: (a) If $x \in M$, there is an open neighborhood $\mathcal{U}$ of x such that $(\mathcal{U}, g|_\mathcal{U})$ is time orientable. (b) If (M, g) is not time orientable, there is a double covering $\phi: \tilde{M} \to M$ (Bishop–Goldberg 3.C.3) such that $(\tilde{M}, \phi^*g)$ is time orientable.

Exercise 1.2.6

Show that "isometric by an orientation and time-orientation-preserving isometry" is an equivalence relation for oriented and time-oriented Lorentzian manifolds.

EXERCISE 1.2.7

Let (M, g) be a time-oriented Lorentzian manifold. (a) Show that a causal vector field on M must be either past pointing or future pointing. (b) Show that if $\gamma: \mathscr{E} \to M$ is a causal curve then either $\gamma_* u$ is future pointing $\forall u \in \mathscr{E}$ or $\gamma_* u$ is past pointing $\forall u \in \mathscr{E}$. Then γ and $\gamma\mathscr{E}$ are said to be *future pointing* or *past pointing*, as the case may be.

1.3 Spacetimes

Definition 1.3.1. A *spacetime* (M, g, D) is a connected 4-dimensional, oriented, and time-oriented Lorentzian manifold (M, g) together with the Levi–Civita connection D of g on M.

In context, we shall sometimes write (M, g) or just M for (M, g, D). A general relativistic *gravitational field* $[(M, g)]$ is an equivalence class of spacetimes where the equivalence is defined by orientation and time-orientation-preserving isometries (Exercise 1.2.6). Each $(M, g, D) \in [(M, g)]$ is a *representative* of $[(M, g)]$. Physically, all representatives of $[(M, g)]$ model the same situation. We shall normally work with one representative, but focus attention on properties shared by all representatives in the same gravitational field.

We discuss some motivations.

The spacetimes of significance in physics are all models of (a part of) the history of (some portion of) the universe. The dimension of a spacetime is intuitively accounted for by the three spatial dimensions of the known universe and an extra dimension of time. Since spacetimes model histories, "disconnected" would connote "always was, is, and always will be disconnected." Thus one assumes M connected. The requirement of time orientability is suggested by our knowledge of thermodynamical processes on the earth, now. The second law of thermodynamics implies that one can distinguish past directions from future directions on earth by measuring the increase in entropy. It seems somewhat reasonable to assume that thermodynamics will smoothly determine future directions in the whole universe. No one knows if this is true, but if we ever really met beings going the wrong way in time, trying to communicate with them would presumably be as confusing as trying to talk to some of the regents of the University of California. Orientability of M is also a plausible condition to impose because the nonconservation of parity is now established for a whole class of experiments (the so-called "weak interactions"). On earth, we can thus intrinsically distinguish between right-handed and left-handed coordinate systems in ordinary 3-space. Thus (M, g, D) can at least be oriented in the region surrounding the earth, now, in the following way: in each coordinate neighborhood, the 4-form $dx^1 \wedge dx^2 \wedge dx^2 \wedge dx^4$ is consistent with the orientation iff (a) each dx^1, dx^2, dx^3 is spacelike and $\{dx^1, dx^2, dx^3\}$ is dual to a right-handed spatial coordinate system of the tangent space at each point, and (b) dx^4 is future pointing and timelike. Again, the extrapolation of this property to other parts of the universe for all time involves some guesswork but is standard practice.

To a geometer, that M should be a C^∞ manifold is perhaps the most acceptable and the most obvious requirement. However, this is probably the most mystifying requirement on a deeper level. Why should all macroscopic physical phenomena—past, present and future—be regarded as occurring on a smooth structure? Offhand, one would think that nature might use something logically simpler—say, piecewise linear manifolds or more general topological spaces. Perhaps she does. The internal contradictions of present special relativistic quantum theory are severe. These contradictions may stem from trying to force a "jumpy" quantum world into a C^∞ manifold.

Many modifications of Definition 1.3.1 have been suggested. For example, one might use a metric connection with torsion in place of the Levi–Civita connection. There are perhaps a thousand such modifications of various kinds which have appeared in print. We shall not consider them here.

Although we have not done so, many physicists would include stable causality (Hawking–Ellis [1]) in the definition of a spacetime. On the other hand, a geometer approaching the same subject would most likely require M to be complete. This we have not done for the simple reason that even the weaker requirement of infinite extendibility of all non-spacelike geodesics would exclude most of the spacetimes of current interest (Section 1.4 and Chapters 6 and 7). For example, in the standard cosmological models particles enter the universe with a big bang (Chapter 7) and the history of such a particle is represented by an inextendible timelike geodesic whose parameter is bounded from below (compare Corollary 1.4.6 following). Whether incompleteness is a property of nature or a misleading feature of current models is a highly controversial question. We remark that infinite extendibility of spacelike geodesics has no direct physical interpretation.

Newtonian analogue. Let $\phi(\vec{x})$ be a time-independent Newtonian gravitational potential (Section 0.1.8). In our units (Section 0.1.4), $\max |\phi| \simeq 10^{-6}$ within the solar system. Whenever $\max |\phi| \ll 1$, Newtonian space, time, and gravitational potential can be replaced by a crude spacetime model as follows. Let $M = \mathbb{R}^4$. Define $\tilde{\phi}: M \to \mathbb{R}$ by $\tilde{\phi} x = \phi(u^1 x, u^2 x, u^3 x) \, \forall x \in M$. Let $g = (1 - 2\tilde{\phi}) \sum_{\mu=1}^{3} du^\mu \otimes du^\mu - (1 + 2\tilde{\phi}) du^4 \otimes du^4$. Take ∂_4 as future pointing and orient M by $du^1 \wedge \cdots \wedge du^4$. Then (M, g, D) is a spacetime. Using it for a general relativistic model, and following the rules of Chapters 2 and 3, gives results at worst as inaccurate as the corresponding Newtonian model (cf. Section 9.3). Roughly, g replaces ϕ and D replaces the Newtonian gravitational field $-\vec{\nabla}\phi$. However, even when $|\phi| \ll 1$ in Newtonian theory, more accurate general relativistic models are sometimes needed. Moreover, some spacetimes model situations altogether beyond the scope of Newtonian physics, such as black holes (Example 1.4.2 and Section 7.5) and gravitational waves (Section 7.6).

Let (M, g) and (N, h) be spacetimes. Define (N, h) to *contain* (M, g) iff M is an open submanifold of N, $h|_M = g$, and (M, g) has the induced orientation and time orientation. Define (M, g) as *maximal* iff each spacetime that contains (M, g) is (M, g). In physics, one prefers in principle to work with maximal spacetimes. However, one is sometimes too lazy to work out the properties of spacetime in regions where "matter" is present; moreover, one sometimes suspects that in some regions conditions may be so extreme that current physics cannot adequately describe them. Then one works with a spacetime that is not maximal. Compare Sections 7.3 to 7.5.

Proposition 1.3.2. *Suppose (N, h) contains (M, g) but, $\forall$ lightlike geodesic $\lambda: \mathscr{E} \to N$ such that $(\lambda \mathscr{E}) \cap M \neq \phi$, $\lambda \mathscr{E} \subset M$. Then $M = N$.*

Roughly, the proposition says that a spacetime is maximal iff one cannot see into it or out of it. Like many other results, it indicates the key role played by lightlike geodesics. The proof uses techniques more advanced than have been discussed here. The idea is to assume a point p on the boundary of M and show that to each point in a sufficiently small neighborhood of p there is a once-broken lightlike geodesic from p. We omit the details (but see Exercise 5.2.7).

EXERCISE 1.3.3

Show that a complete spacetime is maximal.

EXERCISE 1.3.4

Suppose $M = \mathbb{R}^2$, $g = du^1 \otimes du^1 - du^2 \otimes du^2$, $h = du^1 \otimes du^1 - (\exp u^2)du^2 \otimes du^2$. Show (M, g) is maximal and (M, h) is not.

Sections 8.2 to 8.4 outline some global properties of spacetimes.

1.4 Examples of spacetimes

The spacetimes most important in current physics are given in the next three examples. We define them mathematically now. They will be used to illustrate various mathematical and physical concepts as they arise. We will discuss in detail the physical applications of Schwarzschild spacetimes (Example 1.4.2) in Chapter 7, and of Einstein–de Sitter spacetime (Example 1.4.3) in Chapter 6.

EXAMPLE 1.4.1. MINKOWSKI SPACE. On $\mathbb{R}^4$ define $g = \sum_{\mu=1}^{3} du^\mu \otimes du^\mu - du^4 \otimes du^4$; time orient $(\mathbb{R}^4, g)$ by ∂_4 and orient $\mathbb{R}^4$ by $du^1 \wedge du^2 \wedge du^3 \wedge du^4$. The Levi–Civita connection of g is then uniquely determined by $D_{\partial_i}\partial_j = 0$ $\forall i, j = 1, \ldots, 4$ (Bishop–Goldberg 5.6). $(\mathbb{R}^4, g, D)$ is a spacetime; it is called *Minkowski space*. The gravitational field $[(M, g)]$, which contains Minkowski space, is the *trivial gravitational field*; (nonquantum) special relativity and (special relativistic) quantum theory use the trivial gravitational field. The trivial gravitational field $[(M, g)]$ is used iff gravity is negligible.

Special relativistic concepts are important even in analyzing nontrivial gravitational fields because any tangent space to any point in a spacetime has a structure isomorphic to that of Minkowski space. Many measuring devices are very small compared to the size of regions on which curvature becomes important. Such measuring devices are normally modelled as objects on a tangent space rather than objects on a manifold; special relativity is then used to analyze them. For example, a protractor is regarded as an object on a tangent space, not in a spacetime, when one is discussing nontrivial gravitational fields (cf. Section 2.1).

EXAMPLE 1.4.2. SCHWARZSCHILD SPACETIMES. The spacetimes most important in current physics, apart from Minkowski space, were found by Schwarzschild in 1916 when he was studying the gravitational field outside a spherically symmetric body.

Let $(\mathscr{S}^2, h, \zeta)$ be the unit 2-sphere (Section 0.0.9). Let $\mu \in (0, \infty)$ be given. Define $\mathscr{A} \subset \mathbb{R}^2$ by $\mathscr{A} = (u^1)^{-1}[(0, 2\mu) \cup (2\mu, \infty)]$; here $(u^1)^{-1}$ denotes the complete inverse image. Then $\mathscr{A}$ is open and has two connected components. Define $M = \mathscr{S}^2 \times \mathscr{A}$, with $P: M \to \mathscr{S}^2$ and $Q: M \to \mathscr{A}$ as the natural projections. Define $r = u^1 \circ Q: M \to (0, 2\mu) \cup (2\mu, \infty)$ and $t = u^2 \circ Q: M \to \mathbb{R}$. Thus $(1 - (2\mu/r))$ is a C^∞ function from M onto $(-\infty, 0) \cup (0, 1)$. We define a Lorentzian metric g on M by $g = (1 - (2\mu/r))^{-1} dr \otimes dr + r^2 P^* h - (1 - (2\mu/r)) dt \otimes dt$. Also define vector fields $\partial/\partial t$ and $\partial/\partial r$ on M by: $P_*(\partial/\partial t) = 0 = P_*(\partial/\partial r)$ and $Q_*(\partial/\partial t) = \partial_1$, $Q_*(\partial/\partial r) = \partial_2$.

Now define $N \subset M$ by $N = r^{-1}(2\mu, \infty)$. N is connected, and $g|_N$ is a Lorentzian metric. Relative to $g|_N$, $(\partial/\partial t)|_N$ is timelike. When $(N, g|_N)$ is time oriented by $(\partial/\partial t)|_N$ and oriented by $(dr \wedge P^*\zeta \wedge dt)|_N$, it is called the *normal Schwarzschild spacetime of active mass* $\bar{m} = 8\pi\mu$. We shall often omit the subscript N and write (N, g) for $(N, g|_N)$, and so on. As usual, the Levi-Civita connection is implied and the gravitational field that contains (N, g), rather than merely (N, g) itself, is the structure of interest.

A rough physical interpretation of N is the following. Consider a spherically symmetric stable star of radius r_0, with active mass $\bar{m}$, where $r_0 \gg \bar{m}$. Then the open submanifold $\mathscr{U}$ of N defined by $r > r_0$ is an excellent model for the complete history of the exterior of the star. However, the interior of the star is not modelled by any submanifold of (M, g).

$\mathscr{S}^2$ parametrizes angles; on $\mathscr{U}$, r is a kind of radius and t is a kind of time (Chapter 7). However, one must be very careful to interpret via intrinsic geometric properties and appropriate physical measurements; for example, on part of M, t is not in any sense whatsoever a time (cf. Exercise 1.4.8 and Chapter 7).

We next define $B \subset M$ by $B = r^{-1}(0, 2\mu)$. Then B is connected and $(\partial/\partial r)|_B$ is timelike relative to the Lorentzian metric $g|_B$ on B. $(B, g|_B)$, time oriented by $(\partial/\partial r)|_B$ and oriented by $(dr \wedge P^*\zeta \wedge dt)|_B$, is a spacetime called the *Schwarzschild black hole of active mass* $\bar{m} = 8\pi\mu$. To get a rough intuitive picture of $(B, g|_B)$, imagine that you notice gravity getting stronger and stronger; imagine that there is no escape; that no matter what you do you are forced to head for the future, where infinite curvature is lying in wait.

Roughly, $r|_B$ is a kind of radius, and $-r|_B$ is a kind of time, while $t|_B$ is neither. These for the moment rather confusing interpretations, models in which a black hole is "glued to" a normal Schwarzschild spacetime, and the collapsed stars whose exteriors are modelled by such spacetimes, will all be discussed in Chapter 7.

We sketch $PM \times rM$ below. Three points should be noted. First, we do not sketch M, merely a 3-dimensional slice. Second, M contains no point x such that $rx = 2\mu$ and nothing we have said so far justifies our putting $PN \times rN$ right next to $PB \times rB$, but we anticipate the results of Chapter 7 by so doing. Third, M contains no point x for which $rx = 0$; there is no valid reason to regard $r = 0$ as a point, and we sketch $r = 0$ as a sphere.

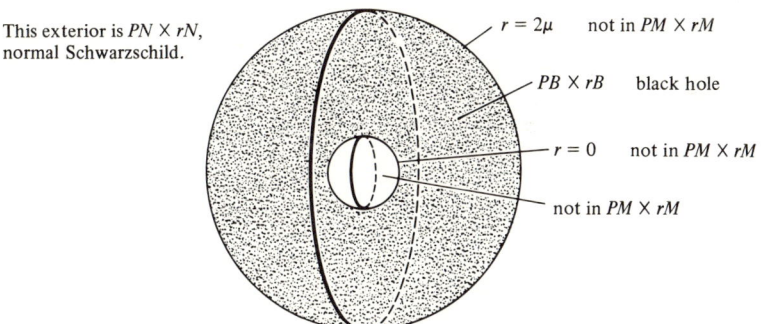

This exterior is $PN \times rN$, normal Schwarzschild.

$r = 2\mu$ not in $PM \times rM$

$PB \times rB$ black hole

$r = 0$ not in $PM \times rM$

not in $PM \times rM$

Schwarzschild spacetimes are geodesically incomplete, Ricci flat, but not flat (Chapter 7). A group theoretic characterization of Schwarzschild spacetimes is outlined in Optional exercises 8.4.

EXAMPLE 1.4.3. EINSTEIN–DE SITTER SPACETIME. Our final example is used more often in physics than any spacetime except those already discussed, is mathematically simple, and is just complicated enough to illustrate many of the concepts that will be introduced in Chapters 2 to 4.

Define M to be $M = \mathbb{R}^3 \times \mathscr{F}$, where $\mathscr{F}$ is an open interval in $\mathbb{R}$. Let $R: \mathscr{F} \to (0, \infty)$ be a function and define g on M by considering M as a subset of $\mathbb{R}^4$ and letting

$$g = \left\{ (R \circ u^4)^2 \sum_{\mu=1}^{3} du^\mu \otimes du^\mu \right\} - du^4 \otimes du^4.$$

Then g is a Lorentzian metric on M and ∂_4 is a timelike vector field on (M, g). (M, g, D) time oriented by ∂_4 and oriented by $du^1 \wedge \cdots \wedge du^4$ is a spacetime, called a *simple cosmological spacetime*. Such a spacetime is called *Einstein–de Sitter spacetime* iff $\mathscr{F} = (0, \infty)$ and R is given by $R(u) = u^{2/3}$, and this will be the case of main interest to us in our later discussion of

1 Spacetimes

cosmology. The gravitational field containing the Einstein–de Sitter spacetime is called the *Einstein–de Sitter gravitational field*. We shall now compute the curvature tensor of Einstein–de Sitter spacetime to show Einstein–de Sitter spacetime is not Ricci flat and is irremediably incomplete. Indeed, whenever someone simply hands you a spacetime, you should first compute the curvature tensor and some of the geodesics.

> Actually you should first find out whether you are dealing with some kind of a nut, as can happen in cosmology. On the contrary, the Einstein–de Sitter gravitational field was first found by Friedman and later singled out and rediscovered by Einstein and de Sitter. They used symmetry arguments and the Einstein field equation (Chapter 4). Roughly, if we assume a spacetime has the same symmetries as Euclidean space (Optional Exercises 8.4.4 and 8.4.5) and that the matter of the universe has a particular form, then the Einstein field equation leads to the Einstein–de Sitter gravitational field. In Chapter 6, we will see that current cosmological observations can be fitted by a model using the Einstein–de Sitter gravitational field, although the observations are not sufficiently precise to exclude other models. In the physics literature, the metric g above with an arbitrary R is called the $k = 0$ *Robertson–Walker* metric (cf. Exercise 6.2.14).
>
> *Newtonian analogue.* Imagine $\mathbb{R}^3$ filled uniformly with a gas. Suppose the gas is expanding in the sense that any two gas particles are running apart along the line joining them. Suppose this expansion rate is decreasing due to the gravity ϕ of the gas. Then (M, g) is like the complete history of $(\mathbb{R}^3, d\vec{x} \cdot d\vec{x}, \phi)$.

The following proposition gives an explicit formula for the Einstein tensor G of Einstein–de Sitter spacetime (compare Section 1.0.2). In case the reader has never computed a curvature tensor, we give some gory details. Hereafter, curvature computations will be left to the exercises. In the following, $(u^4)^{-1}$ means the real-valued function defined by $(u^4)^{-1}x = 1/u^4x$, and not the complete inverse image.

Proposition 1.4.4. *The Einstein tensor G of Einstein–de Sitter spacetime is*
$$G = (4/3)(u^4)^{-2}du^4 \otimes du^4.$$

PROOF. Let (M, g) be Einstein–de Sitter spacetime. The following are asserted or demanded $\forall i, j = 1, \ldots, 4$, and $\forall \mu, \nu = 1, 2, 3$. Define $\omega^4 = du^4$, $\omega^\mu = (u^4)^{2/3}du^\mu$. Then $\{\omega^i\}$ is an orthonormal basis of 1-forms on M. $d\omega^4 = 0$ and $d\omega^\mu = (2/3)(u^4)^{-1/3}du^4 \wedge du^\mu = (2/3)(u^4)^{-1}\omega^4 \wedge \omega^\mu$. Now define $\{\omega_j{}^i\}$ by $\omega_4{}^4 = 0 = \omega_\nu{}^\mu$, $\omega_\mu{}^4 = (2/3)(u^4)^{-1}\omega^\mu = \omega_4{}^\mu$. Then $\{\omega_j{}^i\}$ is the set of connection forms for $\{\omega^i\}$ since $\{\omega_j{}^i\}$ obey Equations 1.0.3d and e. We have $d\omega_4{}^4 = 0 = d\omega_\nu{}^\mu$, and

$$d\omega_\mu{}^4 = -(2/3)(u^4)^{-2}du^4 \wedge \omega^\mu + (2/3)(u^4)^{-1}d\omega^\mu$$
$$= -(2/9)(u^4)^{-2}\omega^4 \wedge \omega^\mu = d\omega_4{}^\mu.$$

1.4 Examples of spacetimes

Algebra now gives for the curvature forms (Equation 1.0.3b) $\Omega_4{}^4 = 0$, $\Omega_\mu{}^4 = -(4/9)(u^4)^{-2}\omega^4 \wedge \omega^\mu = \Omega_4{}^\mu$, and $\Omega_\nu{}^\mu = (8/9)\omega^\mu \wedge \omega^\nu = -\Omega_\mu{}^\nu$. Thus the curvature tensor 1.0.3e is

$$R = (4/9)(u^4)^{-2}\left\{\sum_{\rho,\sigma=1}^{3} 2X_\rho \otimes \omega^\sigma \otimes (\omega^\rho \wedge \omega^\sigma)\right.$$
$$\left. - \sum_{\rho=1}^{3}(X_\rho \otimes \omega^4 + X_4 \otimes \omega^\rho) \otimes (\omega^4 \wedge \omega^\rho)\right\},$$

where $\{X_i\}$ is the basis of vector fields dual to $\{\omega^i\}$. **Ric** can be obtained by contraction. For example, the relevant contraction of $X_4 \otimes \omega^\mu \otimes (\omega^4 \wedge \omega^\mu)$ $= (1/2)X_4 \otimes \omega^\mu \otimes (\omega^4 \otimes \omega^\mu - \omega^\mu \otimes \omega^4)$ is

$$(1/2)\{\omega^4(X_4)\omega^\mu \otimes \omega^\mu - \omega^\mu(X_4)\omega^\mu \otimes \omega^4\} = (1/2)\omega^\mu \otimes \omega^\mu.$$

Algebra gives **Ric** $= (2/3)(u^4)^{-2}\sum_{k=1}^{4}\omega^k \otimes \omega^k$. The (1, 1) tensor field physically equivalent to **Ric** is thus

$$(2/3)(u^4)^{-2}\left\{-X_4 \otimes \omega^4 + \sum_{\rho=1}^{3} X_\rho \otimes \omega^\rho\right\}.$$

Contracting gives for the scalar curvature $S = (4/3)(u^4)^{-2}$. Algebra now gives $G = (4/3)(u^4)^{-2}du^4 \otimes du^4$. □

Corollary 1.4.5. *Let Z be a future-pointing unit timelike vector field on Einstein–de Sitter spacetime which is an eigenvector of G in the following sense: $G(Z, \cdot) = fg(Z, \cdot)$ for some function f. Then $Z = \partial_4$.*

PROOF. Suppose $G(Z, \cdot) = fg(Z, \cdot)$. At each point, $du^4(Z)du^4 = ag(Z, \cdot)$ for some $a \in \mathbb{R}$. $du^4(Z) \neq 0$ since both du^4 and Z are timelike (Exercise 1.1.10b). Thus $du^4 = bg(Z, \cdot)$ for some $b \in \mathbb{R}$, $b \neq 0$. This implies $Z = e\partial_4$ for some $e \in \mathbb{R}$. Since Z is unit and future pointing, $Z = \partial_4$. □

Thus $Z = \partial_4$ is canonically distinguished in the sense that it can be defined solely in terms of g and the time orientation without referring to structures that $\mathbb{R}^4$ has but M does not; we shall call it the *comoving reference frame*.

> The concepts of "reference frame" and "comoving reference frame" are meaningful in a wider context. They will be defined in full generality, respectively, in Sections 2.3 and 3.13.

In the following corollary, we use the notation of Proposition 1.4.4 and Corollary 1.4.5; (M, g) is Einstein–de Sitter spacetime.

Corollary 1.4.6. *Let $(\tilde{M}, \tilde{g})$ be a spacetime and $\phi: M \to \tilde{M}$ an isometry of (M, g) onto $(\phi M, \tilde{g}|_{\phi M})$. Then $(\tilde{M}, \tilde{g})$ is incomplete. In particular, Einstein–de Sitter spacetime is incomplete.*

33

1 Spacetimes

PROOF. By Section 1.0.3 and the proof of Proposition 1.4.4,

$$D_Z Z = \sum_{i=1}^{4} Z(\omega^i Z) X_i + \sum_{i,j=1}^{4} \omega_j^i(Z)\omega^j(Z) X_i = (Z1)X_4 + 0 = 0.$$

Thus each integral curve of Z is a geodesic. Now $\gamma: (0, \infty) \to M$ defined by $\gamma u = (0, 0, 0, u) \in M$ is an integral curve of Z; moreover, the scalar curvature obeys $\lim_{u \to 0} S(\gamma u) = \infty$. If $(\tilde{M}, \tilde{g})$ were complete, then (Section 0.0.13) $\phi \circ \gamma: (0, \infty) \to \tilde{M}$ could be extended to a geodesic $\psi: (-\infty, \infty) \to \tilde{M}$. Then (Exercise 1.0.4) for $\tilde{S}$ the scalar curvature of $(\tilde{M}, \tilde{g})$,

$$\lim_{u \to 0^+} \tilde{S}(\psi u) = \lim_{u \to 0^+} (S \circ \phi^{-1})\psi u = \lim_{u \to 0} S(\gamma u) = \infty.$$

This is a contradiction and establishes the corollary. □

Actually a stronger result holds: Einstein–de Sitter spacetime is maximal. The proof is a straightforward application of Proposition 1.3.2 once the lightlike geodesics 5.2.2 have been computed. Both Proposition 1.4.4 and Corollary 1.4.6 show that the Einstein–de Sitter gravitational field is not the trivial one. According to the comments below Definition 1.3.1, Corollary 1.4.6 and its proof shows that undisturbed particles can enter Einstein–de Sitter spacetime.

Results like the proof of Corollary 1.4.6 can often be made more vivid by using spacetime diagrams. In a spacetime diagram, the lightcone in at least one tangent space is sketched with one dimension suppressed. A future pointing timelike direction in that tangent space is sketched as an arrow. The diagram itself represents M with one or two dimensions suppressed. Thus a spacetime diagram for Corollary 1.4.6 is the following.

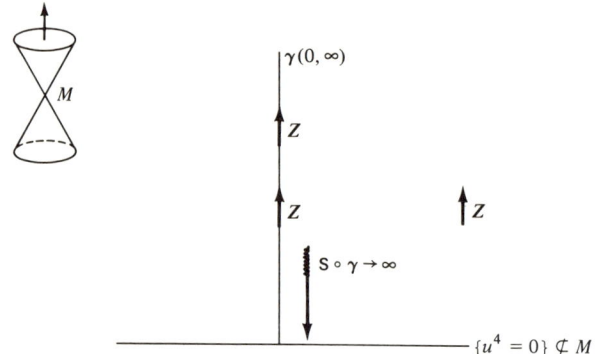

Spacetime diagrams have two advantages. First, they can be used to indicate properties of a gravitational field rather than merely properties of one of its representatives. Second, all of us unfortunately have a vivid sense of simultaneity. When we think of a physical process, we automatically think of a series of still-lives, representing the process "at one time," then "at a slightly later time," and so on. This habit is

1.4 Examples of spacetimes

utterly misleading for relativity. The only way to break this habit is to play with lots of spacetime diagrams.

Of course, we must also be aware of the limitations of a spacetime diagram: we cannot hope to faithfully portray a nonflat 4-dimensional Lorentzian manifold on a flat 2-dimensional Riemannian sheet of paper.

EXERCISE 1.4.7

Let $\tilde{G}$ be the (1, 1) tensor field physically equivalent to the Einstein tensor G on a spacetime. Show: (a) The scalar curvature S obeys S = −contraction $\tilde{G}$. (b) **Ric** = 0 iff G = 0.

EXERCISE 1.4.8

Let (M, g), N, B, r, and t be as in Example 1.4.2. Show that $dr|_N$ is spacelike and $dt|_N$ is timelike but that on B the opposite behavior holds.

EXERCISE 1.4.9

If x and $\hat{x}$ are points in Minkowski space (M, g), then there is a geodesic $\gamma: [a, b] \to M$ such that $\gamma a = x$, $\gamma b = \hat{x}$. Define $k(x, \hat{x}) = \int_a^b |\gamma_* s| \, ds$. (a) Show that $k(x, \hat{x})$ is the same for all such γ. (b) Find two geodesics $\psi: \mathbb{R} \to M$ and $\hat{\psi}: \mathbb{R} \to M$ such that $\psi\mathbb{R} \cap \hat{\psi}\mathbb{R} = \varnothing$, but $k(x, \hat{x}) = A \in \mathbb{R} \; \forall x \in \psi\mathbb{R}, \forall \hat{x} \in \hat{\psi}\mathbb{R}$, where A is independent of x and $\hat{x}$.

As a hint, we mention that ψ and $\hat{\psi}$ must be lightlike. This exercise and Optional exercise 8.1.12 indicate that many properties of Lorentzian manifolds are antiintuitive; often the most important cases for physics are those that are least like the results for Riemannian manifolds.

EXERCISE 1.4.10

The results of Bishop–Goldberg 5.6 and of our Optional exercises 8.4.1 show that a spacetime (M, g, D) is in the trivial gravitational field iff (M, g) is: (a) complete, (b) flat, and (c) simply connected. In the present exercise, you are asked to show that if any of the conditions (a) to (c) is dropped, (M, g) need not be isometric to Minkowski space.

Thus (a) to (c) are a sharp characterization of the trivial gravitational field. Corresponding characterizations for nontrivial gravitational fields rely heavily on group theory. Some are given in Optional exercises 8.4.

2 Observers

If you actually ask an astronomer for data, he will not normally give you just some numerical tables and photographic plates. Often he will present his observations in roughly the following form: "If Newtonian physics were valid, this would be the most obvious explanation of my results." For example, the statement that the earth is about 93 million miles from the sun implicitly assumes (an amazing amount of) Newtonian physics. Presented with observations in this form, the theorist must translate them into relativistic terms. The most convenient way to do this is to use the concept of an observer. This chapter is concerned with the precise relativistic formulations of observers, neighboring observers, and whether or not a frame of reference is rotating.

2.0 Mathematical preliminaries

In this section, M, N denote C^∞ manifolds.

2.0.1 Tensor fields over maps

Let $\gamma\colon \mathscr{E} \to M$ be a curve. In Section 0.0.6 we defined $\gamma_* u$ to be the tangent vector to γ at γu. One is tempted to say that the assignment $\gamma u \to \gamma_* u \; \forall u \in \mathscr{E}$ defines a vector field on $\gamma\mathscr{E}$, to be called the tangent vector field of γ. Unfortunately this will not be well-defined if, say, $\gamma u_1 = \gamma u_2$ for two distinct u_1 and u_2 and γ crosses itself transversally there; for in that case, $\gamma_* u_1 \neq \gamma_* u_2$ and the assignment $\gamma u \to \gamma_* u$ becomes (at least) double-valued at γu_1. The following mild extension of the concept of a vector field (and more generally a tensor field) remedies the situation.

Let $\phi\colon N \to M$ be a C^∞ map and let $T_s^r M$ be the bundle of (r, s) tensors over M with projection $P\colon T_s^r M \to M$ (Section 0.0.5). An (r, s) *tensor field* over ϕ is a map $A: N \to T_s^r M$ such that $P \circ A = \phi$. We also define a *function over ϕ* to be just a function $f: N \to \mathbb{R}$. The two cases of particular interest to us are: (a) $N = M$ and $\phi =$ identity; in this case, a tensor field over ϕ is just an ordinary tensor field (Section 0.0.5). (b) $\phi: N \to M$ is a curve $\gamma: \mathscr{E} \to M$; in this case, the assignment $u \to \gamma_* u \; \forall u \subset \mathscr{E}$ is by definition the *tangent vector field of γ*, and is denoted by γ_*.

For a map $\phi: N \to M$ as above, tensor fields over ϕ arise most naturally in the following ways. (a) If B is a contravariant tensor field on N, then the assignment $y \to \phi_*(By) \; \forall y \in N$ defines a tensor field over ϕ, which will be denoted by $\phi_* B$. In particular, given any vector field X on N, $\phi_* X$ is always a well-defined vector field over ϕ, even if there is no vector field in M ϕ-related to X. (b) If C is an arbitrary tensor field on M, then the assignment $y \to C(\phi y) \; \forall y \in N$ defines a tensor field over ϕ, which will be denoted by $C \circ \phi$. This is called the *restriction of C to ϕ*.

> If ϕ is an imbedding with closed image then, since M is paracompact, every tensor field over ϕ is the restriction of some tensor field on M to ϕ. This can be proved by using a partition of unity (Kobayashi–Nomizu [1], p. 58).

Suppose (M, g) is a Riemannian or semi-Riemannian manifold. For $\phi: N \to M$ as above, two tensor fields A and B over ϕ are *physically equivalent* iff $Ax \sim Bx \; \forall x \in N$ (see Section 1.0.1 for the definition of $\sim$).

A *derivation* of tensor fields over ϕ is an assignment of a tensor field DA over ϕ to each tensor field A over ϕ such that: (a) DA is of the same type as A; (b) $D(C \otimes A) = DC \otimes A + C \otimes DA \; \forall$ tensor field C over ϕ; (c) $D(aA + bB) = aDA + bDB \; \forall a, b \in \mathbb{R}$ and $\forall$ tensor field B over ϕ. Note that by our convention, D is C^∞ in the sense that DA is $C^\infty \; \forall C^\infty$ tensor field A over ϕ. This definition of derivation still makes sense if we require that, in the above notation, A and B should be vector fields or functions over ϕ, and C should be a function over ϕ; then we call such a D a *derivation of functions and vector fields over ϕ*. Suppose a derivation D of functions and vector fields over ϕ is given; then there is a unique extension of D to a derivation of tensor fields over ϕ which commutes with contraction (Bishop–Goldberg 3.6.8).

> We give an example to illustrate the meaning of "D commutes with contraction." Let X, V be vector fields over ϕ and ω a 1-form over ϕ. For a derivation D, we have:
> $$D(X \otimes V \otimes \omega) = DX \otimes V \otimes \omega + X \otimes DV \otimes \omega + X \otimes V \otimes D\omega.$$
> If D commutes with contraction, then
> $$D((\omega V)X) = (\omega V)DX + (\omega(DV))X + ((D\omega)V)X,$$
> and
> $$D((\omega X)V) = (\omega(DX))V + (\omega X)DV + ((D\omega)X)V.$$

2.0.2 Connections over maps

Let $\phi: N \to M$ be a C^∞ map. A *connection D over ϕ* is an assignment to each vector field X on N of a derivation D_X of tensor fields over ϕ such that: (a) If f is a function over ϕ, then $D_X f = Xf$; (b) D_X is linear in X with respect to

2 Observers

C^∞ functions on N; (c) D_X commutes with contraction. Again we are primarily interested in two special cases: (a) $N = M$ and $\phi =$ identity; in this case, a connection over ϕ is just a connection on M in the usual sense. (b) $\phi: N \to M$ is a curve $\gamma: \mathscr{E} \to M$. If $\bar{D}$ is a connection on M and $\phi: N \to M$ as above, then $\bar{D}$ gives rise to a connection over ϕ as follows: the *induced connection* $\phi^*\bar{D}$ is the unique connection over ϕ such that

$$(\phi^*\bar{D})_X(A \circ \phi) = \bar{D}_{\phi_*X}A$$

$\forall X \in N_x$, $\forall x \in N$, and $\forall$ tensor field A on M (Bishop–Goldberg 5.7.1).

Now let (M, g, D) be a spacetime and let $\gamma: \mathscr{E} \to M$ be a curve. Because the induced connection γ^*D in this special case appears so often in the remainder of the book, we want to introduce some *notational abbreviations*: we shall write

$$D_{\gamma_*}A \quad \text{for} \quad (\gamma^*D)_{d/du}(A \circ \gamma),$$

where A is a tensor field on M, and

$$D_{\gamma_*}X \quad \text{for} \quad (\gamma^*D)_{d/du}X,$$

where X is a vector field over γ. In the same context, we shall also write

$$g(\gamma_*, X) \quad \text{for} \quad (g \circ \gamma)(\gamma_*, X \circ \gamma)$$

and

$$|\gamma_*|^2 \quad \text{for} \quad |(g \circ \gamma)(\gamma_*, \gamma_*)|.$$

In this notation, the *acceleration* A_γ of γ is by definition the vector field over γ such that $A_\gamma = D_{\gamma_*}\gamma_*$. Then γ is a geodesic iff $A_\gamma = 0$.

> In the language of fibre bundles, tensor fields over a map $\phi: N \to M$ are sections of the induced tensor bundles $\{\phi^*T_s^rM\}$, and a connection over ϕ is a connection in the induced tangent bundle ϕ^*TM. The concepts of vector fields over a map and connections over a map are often needed in differential geometry proper whenever hand-waving is forbidden—for instance, in the computations of the first and second variations of arc-length in Riemannian geometry. However, if ϕ is an imbedding with closed image all computations may be regarded as taking place on M (Kobayashi–Nomizu [1], pp. 58, 67). This special case, which is conceptually simpler and computationally more manageable, usually suffices to yield the correct information even for the general case where ϕ is not an imbedding.

2.0.3 Lie parallel vector fields over integral curves

In physics, the intuitive concept of "infinitesimally nearby" observers is important. We will presently modify standard results on Lie derivatives (Bishop–Goldberg 3.6) to formalize this intuitive concept. Throughout this subsection, X is a vector field on M, $\{\mu_s\}$ is its flow, and $\gamma: \mathscr{E} \to M$ is an integral curve of X.

Let $e \in \mathscr{E}$ and $W \in M_{\gamma e}$ be given. Suppose first there exists a vector field V on M such that $V\gamma e = W$ and the Lie derivative $L_X V$ vanishes. Then, for each pair $s \in \mathbb{R}$ and $y \in M$ such that $\mu_s y$ is defined, $(\mu_s)_* V y = V \mu_s y$. In particular, if $y = \gamma u$ for $u \in \mathscr{E}$, we get: (a) $V\gamma(u + s) = (\mu_s)_* V\gamma u$ (cf. Bishop–Goldberg 3.6).

Now in general there need not exist such a V, much less a unique one (cf. Exercise 2.0.5 following). But (a) suggests the following definition: A vector field W over γ is *Lie parallel* (*with respect to* X) iff $W(u + s) = (\mu_s)_* W u$ whenever $u \in \mathscr{E}$ and $u + s \in \mathscr{E}$. Then, using the results of Bishop–Goldberg 5.8.1, one gets both existence and uniqueness: (b) For each pair $e \in \mathscr{E}$, $W \in M_{\gamma e}$ there is a unique Lie parallel vector field W over γ such that $We = W$.

More generally, one can introduce a Lie derivative L_X that acts on tensor fields over γ; then $L_X W = 0$ iff W is Lie parallel (Exercise 2.0.6). Moreover, (d) below remains valid when X has zeros. But we are here emphasizing the special cases we need later.

Suppose now X is nowhere zero. Then, locally, we have nothing new. In fact there exists a coordinate map $x : \mathscr{U} \to \mathbb{R}^N$ such that $X|_{\mathscr{U}} = \partial_1$ and $\gamma e \in \mathscr{U}$ (Bishop–Goldberg 3.5). A vector field V on $\mathscr{U}$ obeys $L_X V = 0$ iff $V = \sum_{A=1}^N f^A \partial_A$, where each f^A is a function on $\mathscr{U}$ such that $X f^A = 0$ (Bishop–Goldberg 3.6). Let $\{a^A\}$ be the constants for which $W = \sum_{A=1}^N a^A (\partial_A \gamma e)$ in (b) and define $V = \sum_{A=1}^N a^A \partial_A$. Thus $L_X V = 0$. Let $\mathscr{F} \subset \mathscr{E}$ be the largest connected interval such that $e \in \mathscr{F}$ and that $(\gamma \mathscr{F}) \cap \mathscr{U}$ is connected. Applying the uniqueness assertion in (b) to $\gamma|_{\mathscr{F}}$ and using (a) we get: (c) Suppose X is nowhere zero. Then $W : \mathscr{E} \to TM$ is Lie parallel with respect to X iff, for each $e \in \mathscr{E}$, there is a neighborhood $\mathscr{F}$ of e, a neighborhood $\mathscr{U}$ of γe, and a vector field V on $\mathscr{U}$ such that $L_X V = 0$ and $W = V \circ \gamma$ on $\mathscr{F}$.

Next we apply the standard geometric interpretation of Lie brackets (Bishop–Goldberg 3.8) to our case. The interpretation will be needed only for comparison with intuitive physics, not in proofs, so we will be somewhat informal.

Suppose X is nowhere zero. Let $x : \mathscr{U} \to \mathbb{R}^N$ and $W|_{\mathscr{F}} = \sum_{A=1}^N a^A (\partial_A \circ \gamma|_{\mathscr{F}})$ be as above. We may write $\mathscr{U} = \{(x^1, \ldots, x^N) \mid |x^A| < \varepsilon \, \forall A\}$ and assume $\gamma|_{\mathscr{F}}$ is given by $\gamma u = (u, 0, 0, \ldots)$, since $X|_{\mathscr{U}} = \partial_1$. There is a smooth one-parameter family of integral curves of X determined by

$$(u, t) \to (u + a^1 t, a^2 t, a^3 t, \ldots).$$

Here $t = 0$ gives $\gamma|_{\mathscr{F}}$; t equal to any appropriate constant gives another integral curve; and the interesting case is where $W|_{\mathscr{F}}$ and $X \circ \gamma|_{\mathscr{F}}$ are linearly independent so that different curves have different images. This family uniquely determines $W|_{\mathscr{F}}$ as its transversal vector field. Specifically, $(Wf)u = [(\partial/\partial t)\{f(u + a^1 t, a^2 t, a^3 t, \ldots)\}]_{t=0}$ for each $f : \mathscr{U} \to \mathbb{R}$ and each $u \in \mathscr{F}$. Conversely, given $W|_{\mathscr{F}}$ the family is determined by the stated condition up through first order in t, in the sense of a Taylor series expansion in t. In this sense: (d) A vector field W over an integral curve γ of X is Lie parallel with respect to X iff W is the linearized (or "infinitesimal") version of a one-parameter family of integral curves of X near γ.

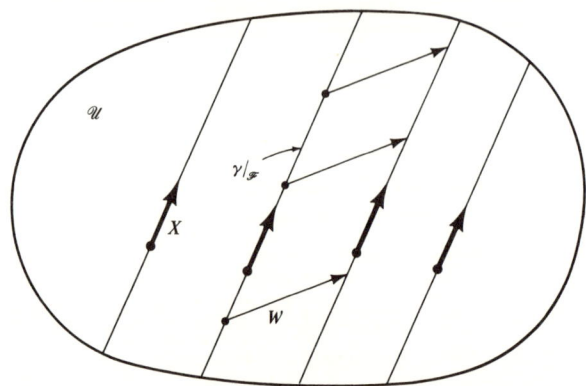

Finally, we need a result related to the Gauss Lemma of Riemannian geometry. Let X be a vector field on a spacetime (M, g). (e) Suppose: X is geodesic, $g(X, X)$ is constant on M, and W is a vector field, Lie parallel with respect to X, over an integral curve $\gamma: \mathscr{E} \to M$ of X. Then $g(\gamma_*, W) = $ constant and $g(D_{\gamma_*} W, \gamma_*) = 0 = g(D_{\gamma_*} D_{\gamma_*} W, \gamma_*)$.

Proof. Suppose $e \in \mathscr{E}$. Since X is geodesic, X is nowhere zero (Section 0.0.8) so we can take $\mathscr{U}$, $\mathscr{F}$ as in (c). Define $\tilde{\gamma} = \gamma|_{\mathscr{F}}$; let $\tilde{W}$ be a vector field on $\mathscr{U}$ such that $\tilde{W} \circ \tilde{\gamma} = W|_{\mathscr{F}}$ and $L_X \tilde{W} = 0$. Since D is symmetric (Section 0.0.8a), $D_X \tilde{W} - D_{\tilde{W}} X = L_X \tilde{W} = 0$. Moreover,

$$Xg(X, \tilde{W}) = g(D_X X, \tilde{W}) + g(X, D_X \tilde{W})$$
$$= g(X, D_X \tilde{W})$$
$$= g(X, D_{\tilde{W}} X)$$
$$= \tfrac{1}{2} \tilde{W} g(X, X) = 0,$$

where the last equality holds because $g(X, X)$ is constant by assumption. Restricting this equation to $\tilde{\gamma}$, we see that $(d/du) g(\tilde{\gamma}_*, W) = 0$. Since e is arbitrary and $\mathscr{E}$ is connected, $g(\gamma_*, W)$ is a constant. Furthermore, from the second line of the preceding string of equalities, we also have

$$g(X, D_X \tilde{W}) = 0.$$

Now using the definition of the induced connection $\gamma^* D$ (Section 2.0.2) and after restricting this equation to γ, we see that

$$g(\gamma_*, D_{\gamma_*} W) = 0.$$

Similarly $g(X, D_X D_X W) = XXg(X, \tilde{W}) = 0$. Restricting to γ gives $g(\gamma_*, D_{\gamma_*} D_{\gamma_*} W) = 0$. □

EXERCISE 2.0.4

Let X be a timelike geodesic vector field of constant norm in a spacetime M and let W be a vector field over one of its integral curves $\gamma: \mathscr{E} \to M$. Let pW denote the vector field over γ such that $\forall u \in \mathscr{E}$, $(pW)u$ = the orthogonal projection of

Wu into $(\gamma_* u)^\perp \subset M_{\gamma u}$. Show that if W is Lie parallel with respect to X, so is pW.

EXERCISE 2.0.5

Consider the cylinder $\mathbb{R} \times \mathscr{S}^1$ obtained by identifying (u^1, u^2) with $(u^1, u^2 + 1)$ on $\mathbb{R} \times [0, 1]$. Take $X = (\exp u^1)\partial_2$. (a) Show a vector field V on $\mathbb{R} \times \mathscr{S}^1$ obeys $L_X V = 0$ iff $V = aX$, $a \in \mathbb{R}$. (b) Take $\gamma: \mathbb{R} \to \mathbb{R} \times \mathscr{S}^1$ as $\gamma u = (0, u \bmod 1)$ and let $W = \partial_1 \gamma 0$. Show γ is an integral curve of X and find the W whose existence is asserted in Section 2.0.3b.

EXERCISE 2.0.6

Let X be a vector field on M, $\{\mu_s\}$ be its flow, and $\gamma: \mathscr{E} \to M$ be one of its integral curves. (a) Show there is a basis $\{W_A\}$ for vector fields over γ such that each W_A is a vector field over γ Lie parallel with respect to X. (b) Let f be any function over γ, and $W = \sum_{A=1}^N f^A W_A$ be any vector field over γ, where each f^A is a function over γ. Define $L_X f = f'$, the derivative, and define $L_X W = \sum f^{A'} W_A$. Show there is a unique derivation L_X of tensor fields over γ which obeys these conditions for each $\{W_A\}$ as in (a) and commutes with contractions. (c) Show $L_X W = 0$ iff W is Lie parallel with respect to X. (d) Show

$$(L_X W)u = \lim_{s \to 0} \frac{Wu - (\mu_s)_* W(u - s)}{s}$$

for each $u \in \mathscr{E}$.

2.1 Observers and instantaneous observers

Throughout this section, (M, g, D) is a spacetime.

We now define observers. Offhand, one might let the observer sit at a point of (M, g) and "observe" by using his exponential map. But a physical observer inexorably moves into the future and thus, intuitively speaking, consists of a continuum of points. Therefore, one models observers by curves. Formally, an *observer* in M is a future-pointing timelike curve $\gamma: \mathscr{E} \to M$ such that $|\gamma_*| = 1$.

The "timelike" requirement of this definition is a mathematical translation of the assumption that each observer travels slower than light (cf. Sections 0.2 and 2.1.3.). The normalization $|\gamma_*| = 1$ is imposed purely for convenience and may be compared with the classical theory of curves of Frenet and Serret where only unit speed curves need be used.

Let $\gamma: \mathscr{E} \to M$ be an observer. The image $\gamma\mathscr{E}$ may model his history from age six until he dies or becomes a consultant for the Pentagon. It is called his *world line*. $u \in \mathscr{E}$ is his *proper time*. It models time measured on any good clock whose history is $\gamma\mathscr{E}$ but has no simple relation to time measured on a clock with a different history. It can be computed as an arclength since

$$\int_a^u |\gamma_* s| ds = \int_a^u ds = u - a.$$

2 Observers

The proper time might be measured by a very small box full of cold dilute radium gas, with $N(u)$ radium atoms present at γu. Since radium atoms decay at a definite rate, u can be measured by counting and using $N(u) = N(u_0) \exp[-a(u - u_0)]$ for some universal constant a; a sets the scale of time.

γ_* is the observer's *world velocity* (or *4-velocity*) and A_γ ($= D_{\gamma_*}\gamma_*$) his or her *world acceleration* (Section 2.0.2). From $g(\gamma_*, \gamma_*) = -1$, we infer $g(A_\gamma, \gamma_*) = 0$. The observer is *freely falling* iff γ is a geodesic. The physical interpretation of a freely falling observer is that he experiences no external influence except perhaps gravity. The reader is not freely falling. The floor you are standing on or the chair you are sitting on exerts a non-gravitational force on you, thereby pushing you off a geodesic. However, a falling apple is, to good approximation, freely falling. (Compare the end of Section 0.2.)

EXAMPLE 2.1.1. THE TWIN "PARADOX." On Minkowski space ($\mathbb{R}^4$, g, D) (Example 1.4.1), take any three freely falling observers γ_1, γ_2, γ_3 with world lines A_1, A_2, A_3, respectively, as shown; here $\{x, y, z\} \subset \mathbb{R}^4$. Since Minkowski space is flat and each A_i is a geodesic, we may work vectorially. The wrong way triangle inequality (Exercise 1.2.4) gives:

$$|A_1| > |A_2| + |A_3|.$$

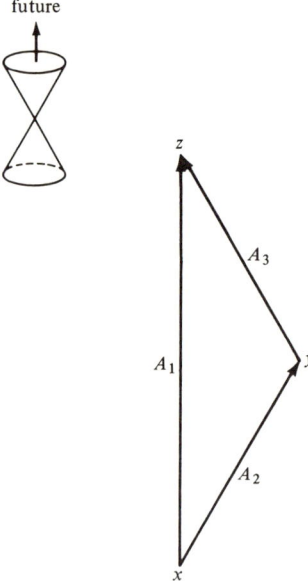

Recalling that proper time can be computed as arclength, we have (γ_1's proper time from x to z) > (γ_2's proper time from x to y) + (γ_3's proper time from y to z). Einstein dramatized the inequality by saying that γ_1 has aged more than a twin who first moves with γ_2 and then, by world accelerating near y, hops into γ_3's rocket.

There exist many such examples where the mathematics is simple, but the physical interpretation is rather striking. Some are found in Exercises 2.1.8 to 2.1.10; additional ones are given later.

> The effect of the twin paradox has been measured to high accuracy, though not on people. Contrary to popularizations: (a) The details of the world acceleration near y are not essential in the computation since the particular behavior of the twin near y can be made to approximate as closely as we please the behavior as depicted in the above figure. (b) The effect is correctly predicted by special relativity. One needs general relativity in the analysis only if the "turning near y" happens to be due to gravity. In that case our above discussion must be modified (cf. Section 8.3.9b).

For many purposes, such as Newtonian interpretation of relativistic concepts, it is not necessary to have the full force of the definition of an observer. Often the tangent vector to the observer at a point suffices. This leads to the following concept. An *instantaneous observer* is an ordered pair (z, Z), where $z \in M$ and Z is a future-pointing timelike unit vector in M_z. An approximate example is $(z, Z) = $ you-now. We give some examples of observers and instantaneous observers that arise naturally.

Let (N, g) be the normal Schwarzschild spacetime (Example 1.4.2). If W denotes the future-pointing timelike unit vector field everywhere proportional to $\partial/\partial t$, then an integral curve of W is called a *stationary observer*. Each (x, Wx) for $x \in N$ is then called an *instantaneous stationary observer*.

Let (M, g) be the Einstein–de Sitter spacetime. Then $Z = \partial_4$ is a future-pointing unit timelike vector field in M. An integral curve of Z is called a *comoving observer*, and each (x, Zx) for $x \in M$ is called an *instantaneous comoving observer*. Note that each comoving observer is (a subset of) the positive portion of a u^4-coordinate curve. According to Corollary 1.4.5, "comoving observer" and "instantaneous comoving observer" are concepts meaningful for the Einstein–de Sitter gravitational field.

> Both "stationary observer" and "comoving observer" are definable in a wider context. See Chapters 7 and 6, where the reader will also find a motivation for the terminology. For example, "stationary" refers to the fact that each element of the flow of $\partial/\partial t$ is an isometry. Assume the world line of the center of your room is an integral curve of W. The existence of the isometries makes it reasonable, or at least consistent, to hope that the room will be essentially unchanged after eight proper hours have gone by.

Throughout the rest of this section (z, Z) is an instantaneous observer. We shall indicate by some preliminary examples how and what (z, Z) observes.

2.1.2 Mathematics vs. physics; infinitesimals and negligible curvature

Now, whenever one discusses actual measurements a surprisingly large number of mathematically rather trivial but conceptually somewhat puzzling issues intrude. Some of these have already been touched on in the Guidelines for the Reader and in Section 0.1.1; a key one will be illustrated in a moment; one will be discussed in Section 2.1.6; the rest will be circumvented. Most are summarized by a comment of Einstein's, here quoted from Misner–Thorne–Wheeler [1]: "As far as the laws of mathematics refer to reality, they are not certain; and as far as they are certain, they do not refer to reality."

In particular, it is often convenient to interpret mathematically rigorous versions of "infinitesimal," such as a tangent space or a Lie transported vector field (Section 2.0.3), as applying to spacetime points which are at a "finite but very small separation." More specifically it often happens that our instantaneous observer (z, Z) can neglect curvature in analyzing his apparatus and can regard a small part of spacetime M as part of M_z; this parallels the prescription in Riemannian geometry for actually measuring, for example, the angle between two intersecting curves on the sphere. For example, suppose (M, g) is Einstein–de Sitter spacetime (Section 1.4) and $z \in M$ models here-now. Then it is appropriate to assume $u^4 z$ is about 10^{10} years $\simeq 3 \times 10^{17}$ seconds (Chapter 6). By Section 1.4 the scalar curvature Sz is about 10^{-35} seconds^{-2}. Now suppose we observe a distant-early galaxy through a telescope of intrinsic length $L = 10^{-8}$ seconds $\simeq 3$ m; suppose the actual observation requires $T = 1000$ seconds of our proper time. Then the dimensionless number $\varepsilon = LTSz$ is about 10^{-40}. But a hopping flea 100 miles away probably distorts an actual telescope by at least one part in 10^{40}. Thus $\varepsilon = 10^{-40}$ very strongly suggests one should model the telescope as an object on M_z, neglecting curvature. Since (M_z, gz), regarded as a spacetime, models the trivial gravitational field, the analysis of the telescope is then at least reduced to a problem in special relativity (Section 1.4). The influence of curvature on the light as it comes to z from the distant-early galaxy can be analyzed separately.

> Of course one can improve on the argument. In view of Einstein's comment and the existence of fleas, we do not wish to encourage this, but the following comments may be useful to a reader who disagrees. (a) Given a point in a Riemannian manifold, one can make precise the concept of a neighborhood so small the curvature tensor is negligible to order ε. (b) Given a point z in spacetime, (a) can fail, mainly because the set $\{X \in M_z \mid g(X, X) = -1\}$ is not compact. (c) Given (z, Z), (a) can be resurrected. Very roughly, this is the "principle of equivalence."

In the rest of this section we assume all measuring devices can be regarded as objects on a tangent space and analyzed special relativistically. But suppose

(z, Z) does not know special relativity. Of course he tries to hire a graduate student who does, and in principle this is the best solution. In practice (z, Z) usually takes advantage of the fact that $(Z^\perp, g|_{Z^\perp})$, regarded as a Riemannian manifold, is simply Euclidean 3-space (Sections 0.1.5 and 1.1). Roughly, (z, Z) then regards $Z^\perp$ as "(Euclidean 3-space)-now" and uses "almost Newtonian" concepts.

2.1.3 Newtonian velocity

For example, suppose $X \in M_z$ is future pointing causal; a special case would be when $\gamma: \mathcal{E} \to M$ is a (different) observer, $\gamma u = z$ and $\gamma_* u = X$. We have the unique orthogonal decomposition $X = eZ + p$, $e \in \mathbb{R}$, $p \in Z^\perp$. $e = -g(X, Z)$ and, by the results in Section 1.2, $e > 0$. p/e is by definition *the Newtonian velocity (of X or γ) observed by* (z, Z), our instantaneous observer. $|p|/e$ is by definition *the Newtonian speed observed by* (z, Z). The reader may check for himself that the Newtonian speed is never greater than 1 and that $\gamma_* u$ being timelike is equivalent to the Newtonian speed being less than the speed of light. The idea behind these definitions is that if (z, Z) proceeds naively, $|p|/e$ is like Euclidean distance divided by Newtonian time (cf. Sections 0.2 and 2.1.5 following).

We now set up some machinery appropriate for such interpretations, give another example, and add some warnings.

2.1.4 Terminology

(z, Z) is an instantaneous observer. Span Z is his *local time axis*, a 1-dimensional timelike subspace of M_z. $Z^\perp$ is his *local rest space*. The direct sum $M_z = (Z^\perp) \oplus (\text{span } Z)$ is his *associated orthogonal decomposition*. We will often write $R = Z^\perp$, $T = \text{span } Z$ and denote by $p: M_z \to R$ the orthogonal projection. If $\gamma: \mathcal{E} \to M$ is an observer, corresponding terms are used for the instantaneous observer $(\gamma u, \gamma_* u)$. For example $R_u = (\gamma_* u)^\perp$ is the observer's *local rest space at proper time* u. Thus the world acceleration $A_\gamma u$ lies in the local rest space R_u $\forall u \in \mathcal{E}$.

2.1.5 Projection tensor

Our next example concerns angles. An instantaneous observer (z, Z) determines a *projection tensor* h as follows: Let $p: M_z \to R$ be the orthogonal projection into the local rest space. Then by definition:

$$h(X, Y) = g(pX, pY),$$

$\forall X, Y \in M_z$. h is a symmetric bilinear form on M_z such that: (a) $h|_R = g|_R$; (b) $h(Z, \cdot) = 0$; (c) $h(X, \cdot) = g(X, \cdot) \Leftrightarrow g(X, Z) = 0$; (d) contraction $\tilde{h} = 3$, where $\tilde{h}$ is the (1,1) tensor physically equivalent to h; and (e) $\tilde{h} = p$.

Suppose $X \in M_z$, $Y \in M_z$; the case of main interest will be when neither X nor Y is spacelike. The *Newtonian angle between* X *and* Y *observed by*

2 Observers

(z, Z) is determined as follows: use $g|_R$ and ordinary Euclidean geometry to determine the angle θ between $pX \in R$ and $pY \in R$. Thus

$$\cos \theta = h(X, Y)/[h(X, X)h(Y, Y)]^{1/2}.$$

Note that θ depends on Z, not just on X and Y. Similarly, the *Newtonian length of X observed by* (z, Z) is defined to be $[h(X, X)]^{1/2}$.

> This rather Newtonian concept of the angle between two tangent vectors X and Y in a spacetime is often regarded as a fundamental interpretation rule for relativity. Physics books normally do not bother to state the above definition explicitly, overoptimistically assuming the idea is obvious. When X and Y are lightlike, the dependence of θ on Z corresponds to the astronomer's concept of "aberration" (cf. Exercise 5.1.4).

Anticipating more systematic later discussions, we mention that at a given point $z \in M$ a "particle of light" corresponds to a future-pointing lightlike vector $Y \in M_z$ (cf. the definition of a photon in Example 3.1.4). Regard yourself as an instantaneous observer (z, Z) and suppose the particle Y of light enters your eye. By algebra, as above, there is a unique $e > 0$ and unique unit vector $U \in Z^\perp$ such that $Y = e(Z - U)$. Intuitively speaking, U is the direction from which you see the light coming within your private version $Z^\perp$ of 3-space. For example, given two such particles Y, Y' the Newtonian angle θ you observe between them is given by $g(U, U') = \cos \theta$. e corresponds, among other things, to the color, as discussed in Chapter 5. Roughly, the bigger e the bluer the light. Note once again that someone whizzing past you at a measured Newtonian speed (Section 2.1.3) half that of light may see quite a different color, since $e = -g(Y, Z)$ depends on Z as well as on the intrinsic model Y. Now look up from the book for just one instant. You saw an enormous number of such Y, with various directions and colors. Seeing a collection $\{(e, U)\}$ is your main contact with the real world; this holds not only for people but also for astronomers and cosmologists.

2.1.6 Logic vs. history

We are systematically regarding nonquantum general relativistic concepts—for example, the world velocity γ_* of an observer—as primary; special relativity is treated as a special case; prerelativistic concepts—for example, Newtonian speed (Section 2.1.3)—are usually viewed as annoying anachronisms. This keeps the mathematics simple. But historically it is quite inaccurate; moreover, it often fails to take advantage of whatever physics background the reader may have; and the historically primary concepts are often very convenient (though never strictly necessary) when discussing actual measurements. Thus some Newtonian concepts will continue to crop up from time to time and the Optional exercises give details; the warnings in Section 0.1.5 then apply.

2.1 Observers and instantaneous observers

In his book, *Gravitation and Cosmology* [1], Weinberg argues for a more quantum theoretical approach to relativity. However, his opening paragraph applies also to the present case. "Physics is not a finished logical system. Rather, at any moment it spans a great confusion of ideas, some that survive like folk epics from the heroic periods of the past, and others that arise like utopian novels from our dim premonitions of a future grand synthesis. The author of a book on physics can impose order on this confusion by organizing his material in either of two ways: by recapitulating its history, or by following his own best guess as to the ultimate logical structure of physical law. Both methods are valuable; the great thing is not to confuse physics with history, or history with physics."

2.1.7 Spatial isotropy

We give one more example. Astronomers on earth have long observed that the distribution of galaxies in the sky is, roughly, uniform in all directions. Thus to a terrestrial observer, "the universe is isotropic." We now give the mathematical definition of spatial isotropy.

Let (z, Z) be an instantaneous observer. Define:

$$\mathcal{O}^3 = \{\psi: M_z \to M_z | \psi Z = Z, \text{ and } g(X, Y) = g(\psi X, \psi X) \forall X, Y \in M_z\}.$$

Thus each $\psi \in \mathcal{O}^3$ is linear (cf. O'Neill [1]). Let $\psi \in \mathcal{O}^3$, and let $M_z = R \oplus T$ be the associated orthogonal decomposition of M_z. Since $\psi|_T = $ identity, ψ is completely determined by $\psi|_R$. Thus $\mathcal{O}^3$ is isomorphic to the group of automorphisms of R which preserve the positive definite inner product $gz|_R$ of R. In other words, $\mathcal{O}^3$ is isomorphic to the rotation group of $\mathbb{R}^3$.

For each $\psi \in \mathcal{O}^3$, let $\psi_s{}^r: T_s{}^r(M_z) \to T_s{}^r(M_z)$ be the unique extension of ψ (Exercise 0.0.16). $T \in T_s{}^r(M_z)$ is called *spatially isotropic for* (z, Z) iff $\psi_s{}^r T = T \; \forall \psi \in \mathcal{O}^3$.

> Offhand, one might want to define a tensor T to be isotropic iff T is left invariant by the full group of automorphisms of M_z that preserve gz—that is, the Lorentz group (Section 8.4.2). However, the weaker restriction of spatial isotropy is the more important one in general relativity.
>
> In Chapter 4, we will see that the distribution of matter in the universe determines a (2, 0)-tensor field T. Therefore the astronomical observation of isotropy suggests the requirement that for $(z, Z) = $ us-now, $\psi_0{}^2 Tz = Tz \; \forall \psi$ as above.

Now let $\phi: M \to M$ be an isometry that is orientation and time orientation preserving, and such that $\phi z = z$, $\phi_* Z = Z$; then $\phi_* \in \mathcal{O}^3$. (M, g) is called *spatially isotropic for* (z, Z) iff given any two unit vectors $X_1, X_2 \in R$ [= local rest space of (z, Z)], there is an isometry $\phi: M \to M$ as above such that $\phi_* X_1 = X_2$. Minkowski space is spatially isotropic for every instantaneous observer and Einstein–de Sitter spacetime is spatially isotropic for every instantaneous comoving observer; the proofs of these facts use only linear algebra and will be left as an exercise (Exercise 2.1.11).

2 Observers

In the following list of exercises, the first three deal with some effects that were discussed in the early days of relativity. They are designed to familiarize the reader with physical interpretations of timelike and spacelike vectors.

Exercise 2.1.8 (Time Dilation)

In 2-dimensional Minkowski space (Section 0.2) let γ_1 and γ_2 be two freely falling observers such that $\gamma_1 0 = \gamma_2 0 = x$. For $s_1, s_2 > 0$, let $y = \gamma_1 s_1$, $z = \gamma_2 s_2$; assume that the geodesic joining y to z is orthogonal to the world line of γ_1 at y. (a) Show $s_1 > s_2$. (b) If v is the Newtonian velocity of γ_2 observed by $(\gamma_1)_* 0$, show that $s_1/s_2 = 1/(1 - |v|^2)^{1/2}$.

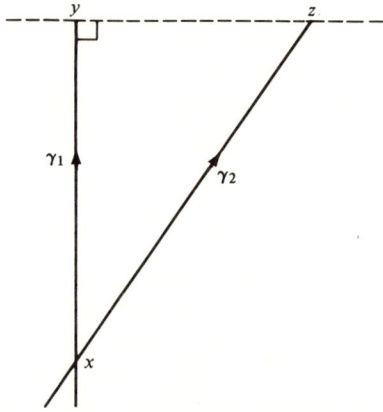

According to the discussion given after Exercise 2.1.13, we can interpret y and z as events taking place simultaneously as observed by $(y, (\gamma_1)_* s_1)$. The fact that there is a difference between the proper times of γ_1 and γ_2 (i.e., $s_1 > s_2$) was therefore regarded as a curious discrepancy in the early days of relativity.

Exercise 2.1.9

Let $\gamma_1, \gamma_2, \gamma_3$ be observers in 2-dimensional Minkowski space whose world lines intersect at $x = \gamma_1 0 = \gamma_2 0 = \gamma_3 0$. Let s_{ij} be the Newtonian speed of γ_i observed by $(x, (\gamma_j)_* 0)$. (a) Show $s_{ij} = s_{ji}$. (b) (Einstein addition law) Verify $s_{31} = (s_{21} + s_{32}) (1 + s_{21} s_{32})^{-1}$ or $(s_{21} - s_{32})(1 - s_{21} s_{32})^{-1}$. (c) Let (M, g) be a spacetime and let $\gamma_1, \gamma_2, \gamma_3$ be observers in (M, g). Retaining the preceding notation, show that $s_{ij} = s_{ji}$ and that when γ_1 observes a 90° Newtonian angle between the other two, $s_{32}^2 = s_{31}^2 + s_{21}^2 - s_{31}^2 s_{21}^2 < 1$.

Exercise 2.1.10 (Lorentz Contraction)

Let $\{e_1, e_2\}$ be an orthogonal basis of the Lorentzian vector space $\mathbb{R}^2$ and let an infinite strip $\mathscr{S}$ be defined by $\{a_1 e_1 + a_2 e_2 \mid |a_1| \leq L_1, a_2 \in \mathbb{R}\}$. Let γ_1 and γ_2 be two freely falling observers defined by: $\gamma_1 u = u e_1$, $\gamma_2 u = u(b_1 e_1 + b_2 e_2)$, where $u \in \mathbb{R}$, $b_1^2 - b_2^2 = -1$, and $b_1 \neq 0$. Let the geodesic orthogonal to γ_2 at the origin 0 meet the boundary of $\mathscr{S}$ at A and B, and let $L_2 =$ Lorentzian length of

2.1 Observers and instantaneous observers

AO or OB. (a) Show $L_2 < L_1$. (b) If $|v|$ is the Newtonian speed of γ_2 observed by $(0, (\gamma_1)_*0)$, compute L_1/L_2 in terms of $|v|$.

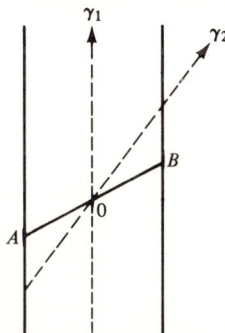

$\mathscr{S}$ represents the complete history of a rigid rod. γ_1 is an observer at rest with respect to the rod and γ_2 is moving relative to the rod. Then the length of the rod measured by γ_1 is strictly greater than that for γ_2.

EXERCISE 2.1.11

Show that Minkowski space is spatially isotropic for every instantaneous observer and that Einstein–de Sitter spacetime is spatially isotropic for every instantaneous comoving observer.

EXERCISE 2.1.12

Show that $T \in T_2^0(M_z)$ is spatially isotropic for an instantaneous observer (z, Z) iff $T = a(gz) + bh$ where $a, b \in \mathbb{R}$, and h is the projection tensor of Section 2.1.5.

EXERCISE 2.1.13

Let $W, V \subset M_z$ be 1-dimensional subspaces orthogonal to each other. Show four cases can occur: (a) W spacelike, V timelike; (b) W, V both spacelike; (c) W lightlike, V spacelike; (d) W, V both lightlike.

With the availability of instantaneous observers, the concept of orthogonality can now be interpreted. In each case the physical interpretation is different, as follows. (a) In general, an instantaneous observer is supposed to regard each point of a spacelike W as "simultaneous with" the origin of M_z iff $W \subset R$ [= local rest space of (z, Z)]. Thus assuming $V = T$ [= local time axis of (z, Z)], all the points of W are events taking place simultaneously as observed by (z, Z). (b) Consider a (z, Z) such that $W, V \subset R$. Then W, V are orthogonal in M_z iff they are orthogonal in R relative to $gz|_R$. Even measuring orthogonality with respect to $gz|_R$ involves unexpected devices, such as a "radar set" (Chapter 5), but at least orthogonality with respect to $gz|_R$ has the obvious analogue of Euclidean orthogonality. (c) Consider a (z, Z) such that $V \subset R$ and let $p: M_z \to R$ be the orthogonal projection. Then a short computation shows that pW and V are orthogonal in R with respect to $gz|_R$ iff V, W

2 Observers

are orthogonal. The projection pW can be interpreted. For example, if W is the spacetime propagation direction of an electromagnetic wave (compare the figure in the fine-print section of Example 1.1.6), then pW is interpreted by (z, Z) as the spatial direction the wave moves. (d) Two lightlike vectors are orthogonal iff they are proportional (Corollary 1.1.5). Thus, in this case, $W = V$. Rather artificial interpretations can be given, but let us simply regard this case as trivial.

EXERCISE 2.1.14

Two observers at a laboratory on the equator have exactly the same age. Observer 1 hops on a jet going 1000 miles per hour due west and returns to his laboratory after one circumnavigation (let us say that the earth rotates at a rate of 1000 miles per hour on the equator). Neglect the gravitational field of the earth and the sun; take the world line of the earth center as the u^4-axis of Minkowski space; take into account the earth's rotation about its axis. (a) Sketch the world lines of the earth center and of both observers. (b) How much older is 1 than 2 after the circumnavigation? (Answer: about 10^{-7} seconds. This experiment was recently performed.)

2.2 Gyroscope axes

To make observations, an observer γ must be able to decide when a unit vector X in his local rest space at proper time u has "the same spatial direction" as a unit vector $\hat{X}$ in his local rest space at another proper time $\hat{u}$. Put another way, how can an observer detect "rotation" in his local rest space? In prerelativistic physics, the problem has a simple solution: Newton pointed out that lack of rotation in the above sense is a local requirement that can be imposed operationally without observation of distant matter. A gyroscope axis is often used in practice to determine this absence of rotation.

> In Newtonian physics the characteristic property of an inertial reference frame is that Newton's laws hold in an inertial reference frame; the characteristic property of a gyroscope axis is that it makes a constant angle with each coordinate axis of each inertial reference frame. (A gyroscope is a rigid body, with axial symmetry, rotating around the axis of symmetry with one point on that axis constrained.) Thus suppose a Newtonian observer carries three gyroscopes whose axial directions are linearly independent. A given spatial direction then undergoes rotation iff the direction angles of this spatial direction with respect to the three gyroscope axes are not all constant.

In relativity, if the observer γ is freely falling, then one can formally define $X, \hat{X}$ above to have the same spatial direction iff X is a parallel transport of $\hat{X}$ along γ. If γ is world accelerated, then $X \in R_u$ no longer implies that the parallel transport of X is in the local rest space at $\hat{u}$. The following modification of parallel transport is then needed.

For the rest of this section, let (M, g, D) be a spacetime. Let $\gamma: \mathscr{E} \to M$ be an observer, and let $M_{\gamma u} = R_u \oplus T_u$ be the associated orthogonal decomposition of $(\gamma u, \gamma_* u)$. Let $p_u: M_{\gamma u} \to R_u$ and $q_u: M_{\gamma u} \to T_u$ be the orthogonal projections. If X is a vector field over γ (Section 2.0.1), then pX and qX are vector fields over γ which are defined by: $(pX)u = p_u(Xu)$, $(qX)u = q_u(Xu)$ (cf. Section 2.1.4 for the notation).

Proposition 2.2.1. *There is precisely one connection F over γ such that $F_Y X = [p(\gamma^* D)_Y p + q(\gamma^* D)_Y q] X$ $\forall$ vector field Y on $\mathscr{E}$ and $\forall$ vector field X over γ.*

PROOF. $F_Y X$ is $\mathbb{R}$-linear in X and linear (with respect to C^∞ functions over γ) in Y. Moreover, if f is a C^∞ function over γ, then $F_Y(fX) = fF_Y X + (Yf) \cdot (p^2 + q^2)X = fF_Y X + (Yf)X$. If we demand $F_Y f = Yf$, then there is precisely one derivation of tensor fields over γ that commutes with contractions and agrees with F_Y on vector fields and functions over γ (Section 2.0.1). □

F is called the *Fermi-Walker connection* over γ. We shall agree to write

$$F_{\gamma_*} \quad \text{for} \quad F_{d/du}.$$

(Compare the end of Section 2.0.2 for notation.) The main properties of F are summarized in Proposition 2.2.2; the proofs consist of straightforward computations.

Proposition 2.2.2. *Let X, Y be vector fields over γ. The Fermi–Walker connection F over γ satisfies the following:*

(a) $$F_{\gamma_*} X = D_{\gamma_*} X + g(\gamma_*, X) A_\gamma - g(A_\gamma, X) \gamma_*,$$

where A_γ is the world acceleration of γ. In particular, $F = \gamma^* D$ iff γ is freely falling.

(b) $$\frac{d}{du} g(X, Y) = g(F_{\gamma_*} X, Y) + g(X, F_{\gamma_*} Y).$$

(c) $$F_{\gamma_*} \gamma_* = 0.$$

(d) If $Xu \in R_u$, $Yu \in R_u$ $\forall u \in \mathscr{E}$,
then $F_{\gamma_*} X \in R_u$, $F_{\gamma_*} Y \in R_u$ $\forall u \in \mathscr{E}$,
and $g(F_{\gamma_*} X, Y) = g(D_{\gamma_*} X, Y)$.

Since γ is a curve, a standard property of connections tells us that if $V \subset M_{\gamma u}$ for some $u \in \mathscr{E}$, then there is a unique F-parallel vector field V over γ such that $Vu = V$ (Bishop–Goldberg 5.7). Thus, if $u_0 \in \mathscr{E}$ and $\{X_i\}$ is a basis of $M_{\gamma u_0}$, then there is precisely one set of F-parallel vector fields $\{X_i\}$ over γ such that $X_i u_0 = X_i$, for $i = 1, \ldots, 4$. In particular, if $\{X_i\}$ is orthonormal, then (b) of Proposition 2.2.2 implies that $\{X_i u\}$ is orthonormal $\forall u \in \mathscr{E}$. If furthermore $X_4 = \gamma_* u_0$, then $X_4 = \gamma_*$ by (c) of the same proposition. $\{X_\sigma u\}$ for $\sigma = 1, 2, 3$ then span R_u $\forall u \in \mathscr{E}$. $X_\sigma u$ is then interpreted as "the unit vector at proper time u along the axis of the σth gyroscope carried by γ." Two unit vectors $X_1 \in R_{u_1}$, $X_2 \in R_{u_2}$ are defined to have the

51

2 Observers

same spatial direction iff for some real numbers a_1, a_2, a_3, $X_1 = \sum_\sigma a_\sigma(\mathbf{X}_\sigma u_1)$, and $X_2 = \sum_\sigma a_\sigma(\mathbf{X}_\sigma u_2)$, or in other words, iff X_1 is the F-parallel transport of X_2.

> Let γ be the observer whose world line is the history of the center of a physical gyroscope. Then, though we shall not prove this here, the axis of the gyroscope traces out (with the passage of time) a vector field over γ that is F-parallel.
> We have thus defined the lack of rotation, mathematically, in terms of g and D. In general relativity, g and D are "influenced by distant matter" in the sense of the Einstein field equations (Chapter 4). Although there is no known precise sense in which g and D are "determined by distant matter," general relativity thus slightly modifies Newton's assumption of the strictly local character of the absence of rotation.

EXERCISE 2.2.3

Suppose $\omega \in (0, 1)$. In Minkowski space $(\mathbb{R}^4, g, D)$, let $\gamma: \mathbb{R} \to \mathbb{R}^4$ be defined by $\gamma u = (1 - \omega^2)^{-1/2}(\cos \omega u, \sin \omega u, 0, u)$. (a) Show that γ is an observer. (b) For $\omega \ll 1$, use the first two terms of the binomial expansion of γu to find the approximate solution V of the equations $F_{\gamma_*} V = 0$ and $g(\gamma_*, V) = 0$, where F is the Fermi–Walker connection on γ. (c) Use the isomorphism of Exercise 0.0.10 to compare $V(2\pi/\omega)$ with $V0$ and show that the two vectors differ slightly.

> *Physical interpretation.* γ is the complete history of an orbiting electron around the nucleus of a hydrogen atom, say. Since $\gamma(2\pi/\omega)$ and $\gamma 0$ describe the electron at the same point spatially, but at different times, the difference between $V(2\pi/\omega)$ and $V0$ is, from a Newtonian standpoint, a rotation of the electron's axis of spinning. The effect is (indirectly) observed and is called the *Thomas precession*. The experiment provides one check on the use of the Levi–Civita connection D and on the identification of Fermi–Walker transport with lack of rotation.

2.3 Reference frames

A single observer is so local that only cooperation between observers gives much information. In this section, we give the mathematical definition of a family of observers. Let (M, g, D) be a spacetime.

Definition 2.3.1. A *reference frame* Q on a spacetime M is a vector field each of whose integral curves is an observer.

> *Newtonian analogue.* A reference frame is like a cloud of Newtonian point particles that move without collisions but otherwise arbitrarily. The reader should not assume that a general reference frame is in any sense "rigid" or "irrotational"; compare the discussions given later in this section, especially Exercise 2.3.12.

Thus a vector field Q is a reference frame iff $g(Q, Q) = -1$ and Q is future pointing. In accordance with the terminology for a vector field introduced in Section 0.0.8, Q is a *geodesic reference frame* iff in addition $D_Q Q = 0$. Let Q be a reference frame in the rest of this section.

The integral curves of Q are called *observers in* Q. All observers in a geodesic reference frame are freely falling. Let ω be the 1-form physically equivalent to Q. If $\gamma: \mathscr{E} \to M$ is an observer in Q, then $du = -\gamma^*\omega$ since $(\gamma^*\omega)(d/du) = \omega(\gamma_*) = g(Q, \gamma_*) = g(\gamma_*, \gamma_*) = -1$. Thus geometric properties of ω can be related to the proper times of the observers in Q, as in the following discussion.

Q is called: *locally synchronizable* iff $\omega \wedge d\omega = 0$, *locally proper time synchronizable* iff $d\omega = 0$, *synchronizable* iff there are C^∞ functions h and t on M such that $h > 0$ and $\omega = -h\,dt$, and *proper time synchronizable* iff $\omega = -dt$. Since $-h\,dt \wedge d(-h\,dt) = 0$, a synchronizable reference frame is locally synchronizable; similarly $\omega = -dt \Rightarrow d\omega = 0$, so that a proper time synchronizable reference frame is locally proper time synchronizable. Conversely, if Q is locally synchronizable (respectively, locally proper time synchronizable) then the restriction of Q to every sufficiently small open set is a synchronizable (respectively, a proper time synchronizable) reference frame (Exercise 2.3.11). If Q is synchronizable or proper time synchronizable, respectively, any function t as above is called a *time function* or a *proper time function for Q*, respectively. If a time function exists, it is not unique. If a proper time function exists, it obeys $du = \gamma^* dt\ \forall$ observer γ in Q.

If Q is synchronizable, then all the level hypersurfaces of the time function t are orthogonal to Q, and hence also orthogonal to all the observers in Q. Conversely, if Q is an arbitrary reference frame on M and t is a function on M such that (1) dt is nowhere zero and (2) the level hypersurfaces of t are everywhere orthogonal to Q, then Q is synchronizable and $\pm t$ is a time function for Q.

From a mathematical point of view, the terminology used in the above definitions is easily understood: If Q is proper time synchronizable, let $\mathscr{N}_a$ be the level hypersurface of a proper time function t for Q defined by $t = a$. For simplicity, let us consider only the case where each observer in Q meets each $\mathscr{N}_a$ exactly once. Then every observer $\gamma: \mathscr{E} \to M$ in Q can adjust his "atomic clock" $\mathscr{E}$ so that his proper time is 0 when his world line intersects $\mathscr{N}_0$. Since $du = \gamma^* dt$, it follows that when the proper time of each observer in Q is a, his world line intersects $\mathscr{N}_a$. In this way, a proper time function for a proper time synchronizable Q achieves a uniform synchronization among all observers in Q. If Q is only synchronizable but not proper time synchronizable and t is a time function for Q, then $du = (h \circ \gamma)\gamma^* dt$ for each observer γ in Q. Since $h \circ \gamma$ is not identically equal to 1, $t \circ \gamma$ no longer equals u up to an additive constant, but is nevertheless explicitly expressible in terms of u by $t \circ \gamma = \int (du/h\gamma u)$. Thus if each observer agrees to use a modified time, cooperation among observers becomes comparatively convenient.

From a physical point of view, the reason for the above terminology is somewhat more profound. Using photons, which will be systematically discussed in Chapter 5, we shall show in Section 5.3 how the observers in a

2 Observers

synchronizable (respectively, proper time synchronizable) reference frame can experimentally correlate by "radar" to arrive at a compromise time (respectively, genuine synchronization of their proper times).

Synchronizability has no nontrivial Newtonian analogue.

We next formalize the concept of "neighboring observers" in a reference frame Q. Let $\gamma: \mathscr{E} \to M$ be an observer in Q and let p and q be the projection operators defined above Proposition 2.2.1.

Definition 2.3.2. A vector field W over γ is called a *neighbor of γ in Q* iff there exists a vector field W' over γ such that $pW' = W$ and $L_Q W' = 0$ (Section 2.0.3 and Exercise 2.0.6).

Let F be the Fermi–Walker connection over γ. $F_{\gamma_*} W$ is called the neighbor's *3-velocity relative to γ*, and $F_{\gamma_*}^2 W (\equiv F_{\gamma_*} F_{\gamma_*} W)$ is called the neighbor's *3-acceleration relative to γ*. Both $(F_{\gamma_*} W)u$ and $(F_{\gamma_*}^2 W)u$ lie in the local rest space of $(\gamma u, \gamma_* u) \forall u \in \mathscr{E}$ (which accounts for the "3" in these definitions) because, according to Proposition 2.2.2b and 2.2.2c, denoting the 3-velocity by V for the moment, $pW = W$ and $qW = 0 \Rightarrow V = pD_{\gamma_*} pW$ (Proposition 2.2.1) $\Rightarrow pV = V$, and similarly for the acceleration.

> *Newtonian analogue.* Let $\vec{x}(t)$ be the path of a point particle in Euclidean 3-space and $\vec{y}(t)$ be another such path with $\vec{n}(t) = \vec{y}(t) - \vec{x}(t)$ small $\forall t \in$ Newtonian time axis. Regard $\vec{n}(t)$ as an element of the tangent space at $\vec{x}(t)$. Then W, $F_{\gamma_*} W$, and $F_{\gamma_*}^2 W$ are, respectively, like $\vec{n}$, $d\vec{n}/dt$, $d^2\vec{n}/dt^2$.

In Definition 2.3.2, the key quantity is W'. We have chosen to call W (instead of W') a neighbor simply because for technical discussions, such as Newtonian interpretations, it is more convenient to have the property that a neighbor always lies in the local rest spaces of γ. Moreover, if Q is geodesic, then a neighbor W of γ in Q must himself satisfy $L_Q W = 0$ (Exercise 2.0.4). Conceptually, one thinks of a neighbor as an "infinitesimally nearby" observer in Q (compare Section 2.0.3, especially (d)).

> A neighbor replaces the more cumbersome concept of a one-parameter family of neighboring observers, in the same way that Jacobi fields replace a one-parameter family of geodesics in Riemannian geometry. Our next proposition in fact shows that a neighbor in a geodesic reference frame necessarily satisfies the Lorentzian version of the Jacobi equation. Compare the remarks in Section 2.1.2.

We need some notation for the next two propositions, which give mathematical interpretations of a neighbor's 3-acceleration and 3-velocity, respectively. Let (z, Z) be an instantaneous observer and let $M_z = R \oplus T$ be his associated orthogonal decomposition. Denoting the curvature tensor of (M, g, D) by R as usual, we define a linear transformation $\psi_z : R \to R$ by $\psi_z X \to R_{ZX} Z$, $\forall X \in R$. That, in fact, $\psi_z R \subset R$ follows from: $g(\psi_z X, Z) =$

$g(R_{ZX}Z, Z) = 0 \, \forall X \in R$, because R_{ZX} is skew-adjoint (Section 1.0.2 and Exercise 1.0.6). ψ_Z is self-adjoint with respect to $gz|_R$ because $\forall V, W \in R$, $g(\psi_Z V, W) = g(R_{ZV}Z, W) = \hat{R}(W, Z, Z, V) = \hat{R}(V, Z, Z, W) = g(R_{ZW}Z, V) = g(\psi_Z W, V)$, where $\hat{R}$ is the $(0, 4)$ tensor field physically equivalent to R (Section 1.0.2).

Proposition 2.3.3. *Let Q be a geodesic reference frame and let W be a neighbor of an observer $\gamma: \mathscr{E} \to M$ in Q. Then the 3-acceleration of W relative to γ satisfies:* $F_{\gamma_*}{}^2 W = \psi W$, *where* $(\psi W)u = \psi_{\gamma_* u}(Wu) \, \forall u \in \mathscr{E}$.

PROOF. By Exercise 2.0.4 $L_Q W = 0$. Fix a $u \in \mathscr{E}$, and let $\tilde{W}$ be a vector field defined in some neighborhood of γu such that $[\tilde{W}, Q] = 0$ and $\tilde{W} \circ \gamma = W$ (Section 2.0.3). Now $D_Q{}^2 \tilde{W} = D_Q D_Q \tilde{W} = D_Q(D_{\tilde{W}} Q + [Q, \tilde{W}]) = D_Q D_{\tilde{W}} Q = R_{Q\tilde{W}} Q + D_{\tilde{W}} D_Q Q + D_{[Q, \tilde{W}]} Q = R_{Q\tilde{W}} Q$; the last equality is because $D_Q Q = 0$ by assumption. Restricting to γ, we have $D_{\gamma_*}{}^2 W = R_{\gamma_* W} \gamma_*$. Since γ is a geodesic, the Fermi–Walker connection F coincides with $\gamma^* D$ (Proposition 2.2.2a). Thus the preceding equation is equivalent to $F_{\gamma_*}{}^2 W = R_{\gamma_* W} \gamma_* = \psi W$. □

Proposition 2.3.3 indicates the basic way to check, when inside a freely falling elevator, whether one is in the trivial gravitational field: Take two freely falling apples in the elevator; if they have nonvanishing relative 3-accelerations, then $R \neq 0$ and spacetime is not isometric to Minkowski space.

For Proposition 2.3.4, recall that given an observer $\gamma: \mathscr{E} \to M$, R_u denotes γ's local rest space at u.

Proposition 2.3.4. *Let Q be a reference frame and let $\gamma: \mathscr{E} \to M$ be an observer in Q. The assignment $X \to -D_X Q$ then defines a linear transformation $A_Q: R_u \to R_u$ which assigns to each neighbor of γ in Q the negative of his 3-velocity relative to γ.*

The negative sign in the definition $A_Q X = -D_X Q$ has no special significance; it is merely a convention universally adopted by differential geometers (cf. Sections 8.1.9 and 8.1.10 for more on A_Q).

Proof of 2.3.4. We first show that $A_Q R_u \subset R_u$. This is because $g(A_Q X, Q) = -g(D_X Q, Q) = -\frac{1}{2} X g(Q, Q) = \frac{1}{2} X 1 = 0, \forall X \in R_u$. Next let W be a neighbor of γ in Q. We have to show that $\forall u \in \mathscr{E}$, $F_{\gamma_* u} W = D_{W u} Q$. This is equivalent to showing $g(F_{\gamma_* u} W, V) = g(D_{Wu} Q, V) \forall V \in R_u$. By Proposition 2.2.2d, $g(F_{\gamma_* u} W, V) = g(D_{\gamma_* u} W, V)$. Thus it suffices to show:

$$g(D_{\gamma_* u} W, V) = g(D_{Wu} Q, V)$$

$\forall V \in R_u$. Let W' be a vector field over γ such that $pW' = W$ and $L_Q W' = 0$. Write $W' - W = f \gamma_*$ for some C^∞ function f over γ, and let $\tilde{W}'$ be a vector field defined in some neighborhood $\mathscr{U}$ of γu such that $\tilde{W}' \circ \gamma = W'$ and $[\tilde{W}', Q] = 0$ (Section 2.0.3). We may assume $\mathscr{U}$

is so small that there exists a C^∞ function $F: \mathcal{U} \to \mathbb{R}$ such that $F \circ \gamma = f$. Define a vector field $\tilde{W}$ in $\mathcal{U}$ by $\tilde{W} = \tilde{W}' - FQ$; then $\tilde{W} \circ \gamma = W$. Now,

$$\begin{aligned} D_{\tilde{W}} Q &= D_Q \tilde{W} + [\tilde{W}, Q] \\ &= D_Q \tilde{W} + [\tilde{W}' - FQ, Q] \\ &= D_Q \tilde{W} - [FQ, Q] \\ &= D_Q \tilde{W} + (QF)Q. \end{aligned}$$

At u, this becomes

$$D_{Wu} Q = D_{\gamma_* u} W + f'(u) \gamma_* u,$$

where f' denotes the derivative of f as usual. This immediately implies that

$$g(D_{\gamma_* u} W, V) = g(D_{Wu} Q, V) \ \forall V \in R_u, \quad \text{as desired.} \qquad \square$$

Q is called *irrotational at* $x = \gamma u$ iff A_Q is self-adjoint with respect to $gx|_{R_u}$, *rigid at* x iff A_Q is skew-adjoint with respect to $gx|_{R_u}$, and *irrotational* or *rigid* iff it is irrotational or rigid at every $x \in M$ (cf. Section 8.1.10). The terminology intentionally parallels that of Newtonian hydrodynamics and can be motivated mathematically, as follows. We have seen that a neighbor of an observer γ in Q corresponds to a one-parameter family of γ's neighboring observers in Q. Proposition 2.3.4 therefore implies that $A_Q: R_u \to R_u$ is the algebraic object that gives a complete infinitesimal description at $\gamma u = x$ of the behavior of γ's neighboring observers in Q relative to γ itself. Suppose Q is irrotational at x; then relative to a suitable orthonormal basis of R_u, the matrix of A_Q is a diagonal matrix. Thus in a small neighborhood of x, the observers in Q near γ exhibit no overall rotation. This absence of infinitesimal rotation accounts for the term "irrotational." On the other hand, let Q be rigid at x. Then relative to a suitable orthonormal basis $\{X_1, X_2, X_3\}$ of R_u, the matrix of A_Q is

$$\begin{bmatrix} 0 & r & 0 \\ -r & 0 & 0 \\ 0 & 0 & 0 \end{bmatrix}.$$

This is the infinitesimal generator of a rotation of R_u around the X_3-axis. Thus is a small neighborhood of x, the observers in Q near x revolve rigidly around γ, but do not approach or recede from γ. Hence the term "rigid."

Physical interpretation. In Newtonian physics, consider any point fixed on a rotating wheel. If the point obstinately regards itself as at rest, it sees infinitesimally neighboring points of the wheel revolving around itself. Proposition 2.3.4 makes this picture precise, makes it relativistic, and generalizes it to include "expansion" or "shearing" if the wheel is not rigid (cf. Section 8.1.10).

It would not be useful to consider only irrotational, only proper time synchronizable or, worst of all, only rigid reference frames. We live in a nearly rigid, slowly rotating earth and have relative speeds much less than the speed of light. Those facts influence our intuition in a very misleading way.

2.3 Reference frames

Newtonian analogue. Consider a cloud of gas particles in Euclidean 3-space. Suppose at Newtonian time t, the gas particle at $\vec{x}$ has Newtonian velocity $\vec{v}(\vec{x}, t)$. Let $v^i(\vec{x}, t)$ be the components of $\vec{v}$. Then A_Q is like $\{\partial v^i/\partial x^j | 1 \leq i, j \leq 3\}$. "Irrotational" and "rigid" are actually used in Newtonian hydrodynamics and aerodynamics to mean, respectively, the matrix $\{\partial v^i/\partial x^j\}$ is symmetric, respectively, antisymmetric.

High-precision modern astronomy is somewhat complicated by the fact that one cannot synchronize clocks on the rotating earth, as indicated by Proposition 2.3.5.

Proposition 2.3.5. *A reference frame is irrotational iff it is locally synchronizable.*

Proof. Let Q be the reference frame as usual and let ω be the 1-form physically equivalent to Q. In a sufficiently small neighborhood, let X, Y be vector fields such that $g(Q, X) = g(Q, Y) = 0$ everywhere. Then in this neighborhood $\omega \wedge d\omega = 0 \Leftrightarrow d\omega(X, Y) = 0$ for all such vector fields X and Y, by virtue of the definition of the exterior product, $\Leftrightarrow \omega([X, Y]) = 0$ for all such X and Y, by virtue of the formula for $d\omega$ (Section 0.0.5e), $\Leftrightarrow g(Q, [X, Y]) = 0$ for all such X and Y, by definition of ω, $\Leftrightarrow g(D_X Q, Y) = g(D_Y Q, X)$ for all such X and Y, because $D_X Y - D_Y X - [X, Y] = 0$ and $g(Q, X) = g(Q, Y) = 0$, $\Leftrightarrow g(A_Q X, Y) - g(X, A_Q Y) = 0$. □

EXAMPLE 2.3.6. Let (M, g, D) be Einstein–de Sitter spacetime (Example 1.4.3), and let $Z = \partial_4$ be the comoving reference frame (cf. the discussion after Corollary 1.4.5). Then Z is indeed a reference frame in the sense of Definition 2.3.1 (cf. Section 2.1).

Let γ be a comoving observer—that is, an observer in Z (Section 2.1). As an example of a neighbor, consider $W = (\sum_{\mu=1}^{3} a_\mu \partial_\mu) \circ \gamma$, where $a_1, a_2, a_3 \in \mathbb{R}$. We claim that W is a neighbor of γ in Z. Indeed, we have $g(\partial_\mu, Z) = 0 \; \forall \mu = 1, 2, 3$, so that $g(W, \gamma_*) = (g(\sum_\mu a_\mu \partial_\mu, Z)) \circ \gamma = 0$. In addition, $L_Z W = (L_Z(\sum_\mu a_\mu \partial_\mu)) \circ \gamma = (\sum_\mu a_\mu [\partial_4, \partial_\mu]) \circ \gamma = 0$, thereby proving our claim. Since γ is (part of) a u^4-coordinate curve, we may write $\gamma u = (b_1, b_2, b_3, u)$ for some $b_\mu \in \mathbb{R}$ ($\mu = 1, 2, 3$) and $u > 0$. In this case, we can exhibit a one-parameter family of comoving observers near γ corresponding to W (in the sense of Section 2.0.3): if t is the parameter ($t \in \mathbb{R}$), then the neighboring observers in question are given by $u \to (a_1 t + b_1, a_2 t + b_2, a_3 t + b_3, u)$. We leave as an exercise to show that in fact every neighbor of a comoving observer in Z is such a W for appropriate choices of $a_1, a_2, a_3 \in \mathbb{R}$ (Exercise 2.3.10a).

Some of the basic properties of Z are summarized in Proposition 2.3.7.

Proposition 2.3.7. *Let Z be the comoving reference frame, γ an observer in Z, and W a neighbor of γ in Z. Then: (a) Z is proper time synchronizable. (b) Z is geodesic. (c) W's 3-velocity relative to γ is $(2/(3u))W$. (d) W's 3-acceleration relative to γ is $(-2/(9u^2))W$.*

2 Observers

Proof. (a) The 1-form physically equivalent to Z is $-du^4$, so u^4 is a proper time function for Z. (b) was proved in Corollary 1.4.6. (c) Let $W = (\sum_{\mu=1}^{3} a_\mu \partial_\mu) \circ \gamma$ (Exercise 2.3.10a). Then (b) and Proposition 2.2.2a imply that: W's 3-velocity relative to $\gamma = D_{\gamma_*} W = (\sum_\mu a_\mu D_{\partial_4} \partial_\mu) \circ \gamma = (2/(3u))W$, where we have used the formulas given in the proof of Proposition 1.4.4; the details are left as an exercise (Exercise 2.3.10b). (d) As in (c), W's 3-acceleration relative to $\gamma = D_{\gamma_*}^2 W = D_{\gamma_*}\{(2/(3u))W\} = -(2/(3u^2))W + (2/(3u))D_{\gamma_*}W = (-2/(9u^2))W$. □

Note that by virtue of Proposition 2.3.5, (a) implies that Z is irrotational. However, (c) gives a much stronger result. For, according to Proposition 2.3.4, the linear operator A_Z is then a pure contraction:

$$A_Z X = (-2/(3u))X, \quad \forall X \in R_u \; [= \text{local rest space of } (\gamma u, \gamma_* u)].$$

According to the interpretation of a neighbor as a one-parameter family of neighboring observers, the fact that W's 3-velocity relative to γ is always a positive multiple of W means that the nearby observers in Z are receding from γ, and the fact that W's 3-acceleration relative to γ is always a negative multiple of W means that the recession is slowing down. Since this is true for each comoving observer γ, one can express this phenomenon as: "the comoving reference frame is expanding but the relative acceleration is inward." Roughly: "the universe expands."

EXERCISE 2.3.8

Suppose W is a neighbor of an observer γ in a reference frame Q and suppose W' and W'' are two vector fields over γ satisfying $pW' = pW'' = W$, and $L_Q W' = L_Q W'' = 0$, where p is the usual orthogonal projection into the local rest spaces of γ. (a) Show $W' - W'' = a\gamma_*$ for some $a \in \mathbb{R}$. (b) If $W \in R_u$ (= local rest space of γ at u) is given, show that there exists one and only one neighbor W of γ in Q such that $Wu = W$.

EXERCISE 2.3.9

Let γ be an observer in a geodesic reference frame Q. If W is a vector field over γ such that $L_Q W = 0$ and $g(Wu, \gamma_* u) = 0$ for one u, then W is a neighbor of γ in Q.

EXERCISE 2.3.10

Let γ be a comoving observer in Einstein–de Sitter spacetime and let W be a neighbor of γ in the comoving reference frame Z. (a) Show that for some $a_1, a_2, a_3 \in \mathbb{R}$, $W = (\sum_{\mu=1}^{3} a_\mu \partial_\mu) \circ \gamma$. (b) Give a detailed proof that the 3-velocity of W relative to γ equals $(2/(3u))W$.

EXERCISE 2.3.11

(a) Let Q be a locally synchronizable (respectively, locally proper time synchronizable) reference frame and suppose $x \in M$. Then the restriction of Q to a sufficiently small neighborhood of x is synchronizable (respectively, proper

2.3 Reference frames

time synchronizable). (b) Show that on a simply connected spacetime, every locally proper time synchronizable reference frame is proper time synchronizable.

EXERCISE 2.3.12

Recall that a tensor field S on M is *parallel* ($\equiv$ *covariant constant*) iff $D_X S = 0 \; \forall$ vector X. (a) Show ∂_4 and $(1/3)(5\partial_4 + 4\partial_1)$ are both parallel reference frames on Minkowski space. (b) Show from Definition 1.0.2 of the curvature tensor that if Z is a parallel reference frame on M, **Ric** $(Z, X) = 0 \; \forall X$. (c) Show there exists no parallel reference frame on Einstein–de Sitter spacetime. (d) (difficult) Show that in general there is no rigid reference frame defined in any neighborhood of a spacetime point.

> Reference frames as in (a) are called "inertial" in physics texts. (c) Shows it would not be useful to introduce this concept in general: when gravity is not negligible there is usually no reference frame that has all the properties "inertial" suggests. (d) Is one reason why a "rigid meter stick" is not a very useful concept in general relativity.

EXERCISE 2.3.13

Let Q be a reference frame and ω be the physically equivalent 1-form. (a) Show $d\omega = 0$ iff Q is geodesic and irrotational (iff Q is locally proper time synchronizable, by definition). (b) On a normal Schwarzschild spacetime (Example 1.4.2) find a rigid, irrotational reference frame that is not geodesic.

EXERCISE 2.3.14

Let Q be a geodesic reference frame on M and let $\{\partial_i\}$ be coordinate vector fields defined in some neighborhood $\mathcal{U}$ in M. Let $Q = \sum_i Q^i \partial_i$ in $\mathcal{U}$, and for fixed i, j, define a function f_{ij} on $\mathcal{U}$ by $f_{ij} = g(D_{\partial_i} Q, \partial_j) - g(\partial_i, D_{\partial_j} Q)$. (a) Show that for an observer $\gamma: \mathscr{E} \to \mathcal{U}$ in Q,

$$(f_{ij} \circ \gamma)' + \sum_k \{(\partial_i Q^k) \circ \gamma\}(f_{kj} \circ \gamma) + \sum_k \{(\partial_j Q^k) \circ \gamma\}(f_{ik} \circ \gamma) = 0,$$

where the prime denotes differentiation. (b) Observing that Q is irrotational at γu_0 iff $(f_{ij} \circ \gamma) u_0 = 0 \; \forall i, j$, deduce from (a): if Q is irrotational at γu_0, $u_0 \in \mathscr{E}$, then it is irrotational at $\gamma u \; \forall u \in \mathscr{E}$. (c) Show that if $X \in M_x$ is any given timelike unit vector, then there is an open neighborhood $\mathcal{U}$ of x such that there is a geodesic irrotational reference frame Q on $\mathcal{U}$ with $Qx = \pm X$.

EXERCISE 2.3.15

On a neighborhood of the origin in Minkowski space, define $Q = [1 - (u^1)^2]^{-1/2} \cdot [\partial_4 + u^1 \partial_2]$. Show: (a) Q is a reference frame. (b) Q is not irrotational and thus not locally synchronizable.

3 Electromagnetism and matter

In this chapter, we shall analyze electromagnetism, matter, their mutual influences, and the influence of spacetime on each. Chapter 4 will close the web by analyzing the influence of electromagnetism and matter on spacetime.

Mathematicians notoriously find discussions of matter difficult. We have thus split this chapter into three parts. Part One emphasizes basic definitions, simple examples, and intuition. Part Two, on interactions, attempts to summarize all the underlying laws normally covered in a six-year physics curriculum, except those of quantum and statistical physics. Part Three is merely technical. Mathematically, the material is rather routine. But there are serious conceptual problems. Some that have already been mentioned—for example, in Section 2.1—will recur here. In addition, we will be plagued throughout by the fact that one needs "matter" as a general concept but has no honest formal definition available.

The concept of a spacetime is simpler than the Newtonian concepts it replaces, primarily because g unifies the structure. Similarly the relativistic laws of electromagnetism, discussed in Sections 3.4 and 3.7, are far simpler than their Newtonian counterparts. Unfortunately, however, relativistic matter models are no simpler than Newtonian ones. In any case, we have no matter model from which all other useful matter models can be deduced. Presumably, this rather sad fact just reflects our ignorance. Perhaps quantum theory will eventually provide a precise matter model applicable to all the rich structures one can see if one looks around. But at present we must work with many intuitively interrelated yet mathematically independent models. Some mathematical presentations circumvent the resulting mess by flatly refusing to discuss matter in any detail. The gain in elegance is then large, but the loss in physical relevance is larger. In Section 3.1, and following, we attempt to take the bull by the horns. The use of intuitively interrelated, mathematically independent models is puzzling, so we now discuss an example and some generalizations.

Suppose we have many small identical bodies and want to describe where they are "now." We can use two Newtonian models. We can say that at Newtonian time $t = 0$, a point particle is at $\vec{x}_1 \in \mathbb{R}^3$, another is at $\vec{x}_2 \in \mathbb{R}^3$, and so on. Or we can give a smooth function $n: \mathbb{R}^3 \to [0, \infty)$ with $n(\vec{x})$ interpreted as the number of bodies per unit volume at $\vec{x} \in \mathbb{R}^3$ when $t = 0$. Neither model is physically precise; for example, both neglect quantum effects. Intuitively, the second model is obtained from the first by "averaging," by "adding up and smoothing out in position space." Sometimes one can regard the first model as a special case of the second by allowing n to be an appropriate distribution of slow growth (generalized function in the sense of L. Schwartz). But in general neither model can be regarded as a consequence of the other. On the other hand, one must remember the intuitive interrelation, since both models are supposed to approximate a single situation.

When similar model pairs occur in general relativity, there is a superficial problem and a serious one. The superficial problem is that tensors above different tangent spaces cannot directly be added, so one sometimes has to do a little extra work to assign even a vague meaning to "averaging." Normally one can just use past light cones in an appropriate way. The serious problem is the same as in Newtonian theory: Attempts to make one model a mathematical consequence of the other are usually shipwrecked by the physical inaccuracies of both models.

In Section 3.2, and following, we make the formal independence of our matter models explicit, indicating their heuristic interrelation by phrases such as those in quotes above. We shall define particles, consistent particle sets, and models obtained by "adding up contributions from many particles and smoothing out in M." In Chapter 5, we consider more sophisticated models obtained by "adding up and smoothing out in TM—in position and in energy-momentum space."

Throughout this chapter, (M, g, D) is a spacetime.

PART ONE: BASIC CONCEPTS

3.0 Review and notation

3.0.1 Integration

If N is an oriented Riemannian or semi-Riemannian manifold, then there is a distinguished volume element, called the metric volume element (Bishop–Goldberg 4.6). The *metric volume element* Ω on M is the unique 4-form on M such that if $\{X_1, \ldots, X_4\}$ is a consistently oriented orthonormal basis of

M_x, then $\Omega(X_1, \ldots, X_4) = (4!)^{-1}$. The factorial arises from the Bishop–Goldberg conventions.

For the rest of the book, Ω will always denote the metric volume element, and any basis of vector fields $\{X_1, \ldots, X_4\}$ on $\mathcal{U} \subset M$ will always be assumed to be consistent with the orientation determined by Ω.

Let $\{X_1, \ldots, X_4\}$ be a basis of vector fields in $\mathcal{U} \subset M$ and let $\{\omega^1, \ldots, \omega^4\}$ be the dual basis. Denoting $g(X_i, X_j)$ by g_{ij}, we let $g = |\det\{g_{ij}\}|$ (det = determinant). It follows directly from the definition of Ω and an algebraic manipulation that in $\mathcal{U}$,

(a) $$\Omega = \sqrt{g}\,\omega^1 \wedge \cdots \wedge \omega^4.$$

In particular:

(b) If $\{\omega^1, \ldots, \omega^4\}$ is an orthonormal basis of 1-forms in $\mathcal{U} \subset M$ (consistent with the orientation of M), then $\Omega = \omega^1 \wedge \cdots \wedge \omega^4$.

Being a nowhere zero top degree form, Ω enjoys a useful algebraic property. Recall first that if X is a vector field, the interior product operator $i(X)$ is defined by: if ω is a q-form, $q > 0$, then $i(X)\omega$ is the $(q-1)$-form such that

$$i(X)\omega(Z_1, \ldots, Z_{q-1}) = q\omega(X, Z_1, \ldots, Z_{q-1}),$$

$\forall$ vector fields $Z_1, \ldots, Z_{q-1}$ (Bishop–Goldberg 4.4). Now algebra gives:

(c) If two vector fields X and W defined in $\mathcal{U} \subset M$ satisfy $i(X)\Omega = i(W)\Omega$, then $X = W$ in $\mathcal{U}$.

The metric volume element enables us to integrate functions: if f is a continuous function on M and $K \subset M$ is an open set with compact closure K^-, then $\int_K f\Omega$ is defined.

Let ω be a p-form on M and let $\mathcal{N}$ be an imbedded, compact, oriented $(p+1)$-dimensional topological manifold with boundary $\partial\mathcal{N}$ (Bishop–Goldberg). For the purpose of integration, it suffices to assume that $\mathcal{N}$ and $\partial\mathcal{N}$ are *piecewise* C^∞ in the sense that $\mathcal{N}$ (respectively, $\partial\mathcal{N}$) is a C^∞ manifold outside a compact subset which is a finite union of C^∞ manifolds of dimension $\leq p$ (respectively, $\leq p-1$). Since $\mathcal{N}$ is oriented, $\partial\mathcal{N}$ has an induced orientation (Section 0.0.5). *Stokes' theorem* then states that, assumptions as above,

$$\int_\mathcal{N} d\omega = \int_{\partial\mathcal{N}} \omega.$$

Using Stokes' theorem, standard arguments show that if $\int_{\partial\mathcal{N}} \omega = 0$ for every such $\mathcal{N}$, then ω is closed. However, one can draw the same conclusion while restricting attention to a special class of such $\mathcal{N}$, as described below.

First recall that for an arbitrary $\mathscr{A} \subset M$, an arbitrary mapping $\phi: \mathscr{A} \to N$ into another manifold N is said to be C^∞ iff ϕ can be extended to a C^∞ mapping $\tilde{\phi}: \tilde{\mathscr{A}} \to N$ where $\tilde{\mathscr{A}}$ is an open set containing $\mathscr{A}$. Now consider the *standard cylinder* $\mathcal{N}_0$ in Minkowski space $(\mathbb{R}^4, g, D)$ which is defined to be $\mathcal{N}_0 = \{(u^1, \ldots, u^4) \mid \sum_{\nu=1}^{3} u^\nu u^\nu \leq 1, |u^4| \leq 1\}$. The *boundary manifolds* of $\mathcal{N}_0$ are the following five imbedded submanifolds of $\mathbb{R}^4$:

$$\mathcal{B}_1 = \left\{(u^1, u^2, u^3, -1) \,\middle|\, \sum_\nu u^\nu u^\nu < 1\right\},$$

$$\mathcal{B}_2 = \left\{(u^1, u^2, u^3, 1) \,\middle|\, \sum_\nu u^\nu u^\nu < 1\right\},$$

$$\mathcal{B}_3 = \left\{(u^1, u^2, u^3, u^4) \,\middle|\, \sum_\nu u^\nu u^\nu = 1, |u^4| < 1\right\},$$

$$\mathcal{C}_1 = \left\{(u^1, u^2, u^3, -1) \,\middle|\, \sum_\nu u^\nu u^\nu = 1\right\}.$$

$$\mathcal{C}_2 = \left\{(u^1, u^2, u^3, 1) \,\middle|\, \sum_\nu u^\nu u^\nu = 1\right\}.$$

$\mathcal{B}_3$ is timelike; the rest are spacelike. A *causal box* in M is then defined to be a C^∞ imbedding $\phi: \mathcal{N}_0 \to M$ which preserves orientation, time orientation, and the causal character of each of the boundary manifolds of $\mathcal{N}_0$. By the usual abuse of language, we also refer to $\mathcal{N} \equiv \phi\mathcal{N}_0$ as a causal box. Using the picture as shown, we give an explicit description of the relevant data of $\mathcal{N}$. We denote the interior of $\mathcal{N}$ by $\mathcal{A}$ and denote $\phi\mathcal{B}_1$, $\phi\mathcal{B}_2$, and so on by $\mathcal{B}_1$, $\mathcal{B}_2$, and so on.

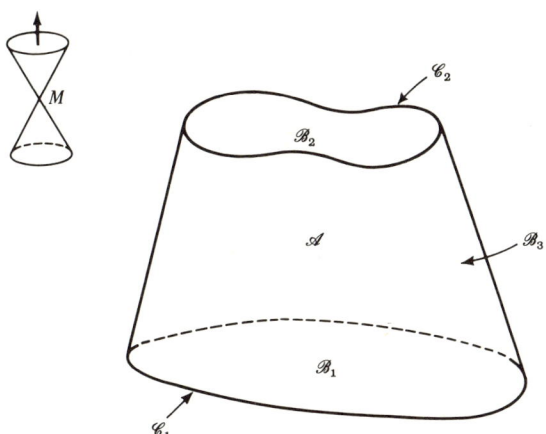

Submanifold	Topology	Orientation	Causal character	Closure
$\mathcal{A}$	$\mathbb{R}^4$	from M	timelike	$\mathcal{A}^- = \mathcal{A} \bigcup_{j=1}^{3} \mathcal{B}_j \bigcup_{i=1}^{2} \mathcal{C}_i$
$\mathcal{B}_i \, (i = 1, 2)$	$\mathbb{R}^3$	from $\mathcal{A}$	spacelike	$\mathcal{B}_i^- = \mathcal{B}_i \cup \mathcal{C}_i$
$\mathcal{B}_3$	$\mathbb{R} \times \mathscr{S}^2$	from $\mathcal{A}$	timelike	$\mathcal{B}_3^- = \mathcal{B}_3 \cup \mathcal{C}_1 \cup \mathcal{C}_2$
$\mathcal{C}_i \, (i = 1, 2)$	$\mathscr{S}^2$	from $\mathcal{B}_i$ (not $\mathcal{B}_3$)	spacelike	$\mathcal{C}_i^- = \mathcal{C}_i$

3 Electromagnetism and matter

Newtonian analogues. $\mathscr{B}_1$ is like an open solid ball in Euclidean 3-space at one Newtonian instant. However, one can also imagine that two distinct points in $\mathscr{B}_1$ are at slightly different Newtonian times provided the Newtonian time difference is so small light cannot reach either from the other quickly enough. In general, when M is not flat it is neither possible nor necessary to distinguish between these two intuitive pictures (cf. Exercise 2.3.12). $\mathscr{A}$ is like a history of such a ball, assuming the ball wobbles, expands or contracts so slowly each point in it moves at less than the speed of light. In the simplest cases, one may imagine the ball rigid during the history but in general this is neither possible nor, when possible, particularly useful. Analogous interpretations hold for the other submanifolds (cf. Section 9.1).

Let $\mathscr{D}$ denote the oriented manifold-with-boundary $\mathscr{B}_2^- = \mathscr{B}_2 \cup \mathscr{C}_2$, where the orientation is the one specified above. We shall call such a $\mathscr{D}$ a *space-section*. Note that due to the difference in orientation, $\mathscr{B}_1^-$ is not a space-section.

(d) A 3-form ω is closed iff $\int_{\partial \mathscr{N}} \omega = 0$ for all causal boxes $\mathscr{N}$; a 2-form ψ is closed iff $\int_{\partial \mathscr{D}} \psi = 0$ for all space-sections $\mathscr{D}$.

We indicate the idea of the proof by considering the case of a 0-form, i.e., a C^∞ function $f: M \to \mathbb{R}$. Then for $\gamma: [a, b] \to M$ a curve we have:

$$\int_a^b df(\gamma_* u) du = f\gamma b - f\gamma a.$$

This corresponds to Stokes' theorem stated above.

By continuity, we have $f\gamma b - f\gamma a = 0 \;\forall$ such curve γ iff $df(W) = 0 \;\forall$ vector W, i.e., iff $df = 0$. Now suppose we restrict the causal character by considering only timelike curves (the spacelike case is similar). Then $f\gamma b = f\gamma a \;\forall$ timelike γ iff $df(W) = 0 \;\forall$ timelike W. Suppose $df(W) = 0 \;\forall$ timelike W, take $x \in M$, $W \in M_x$ timelike and $X \in M_x$. Since the set of timelike vectors in M_x is open, there exists an $a \in (0, \infty)$ such that $W + aX$ is timelike. Then $0 = df(W + aX) = a \cdot df(X)$. Since (x, X) was arbitrary, we have $df = 0$. Thus $f\gamma b - f\gamma a = 0 \;\forall$ timelike γ iff $df = 0$, corresponding to assertion (d).

The proof of (d) itself is similar (cf. Bishop–Goldberg 4.8 and 4.9).

3.0.2 Conservation laws for vector fields

Let X be a vector field on spacetime M. Then $d[i(X)\Omega]$ is a 4-form. This implies that there is a unique C^∞ function, which we denote by div $X: M \to \mathbb{R}$ and define as the *divergence* of X, such that $d[i(X)\Omega] = (\text{div } X)\Omega$. Let L_X denote the Lie derivative; then $L_X = i(X)d + di(X)$ (Bishop–Goldberg p. 172). $d\Omega = 0$ since the only 5-form on M is the zero 5-form. Thus we also have $L_X\Omega = (\text{div } X)\Omega$.

A vector field X is said to obey a *differential conservation law* iff div $X = 0$; it is said to obey an *integral conservation law* iff $\int_{\partial \mathscr{N}} i(X)\Omega = 0$ for all causal boxes $\mathscr{N}$. The definition of div X and Section 3.0.1 imply: X obeys a differential conservation law iff it obeys an integral conservation law.

There are many Newtonian analogues; the following is typical. Suppose water molecules and water ions are present in a moving fluid. Let $n(\vec{x}, t)$ be the number of water molecules per unit volume at $\vec{x} \in \mathbb{R}^3$ and Newtonian time t; let $\vec{v}(\vec{x}, t)$ be the Newtonian fluid velocity. Take Minkowski space $(\mathbb{R}^4, g, D)$. n and $\vec{v}$ may be regarded as defined on $(\mathbb{R}^4, g)$. Now define a vector field $X = n\partial_4 + n \sum_{\mu=1}^{3} v^\mu \partial_\mu$ on $(\mathbb{R}^4, g)$ where $\vec{v} = (v^1, v^2, v^3)$. A computation shows that div $X = \partial_4 n + \sum_{\mu=1}^{3} \partial_\mu(nv^\mu)$, which in Newtonian notation becomes $\partial n/\partial t + \vec{\nabla} \cdot (n\vec{v})$ (Section 9.0.1). Thus div $X = 0$ iff $\partial n/\partial t + \vec{\nabla} \cdot (n\vec{v}) = 0$. This last equation (the so-called equation of continuity) holds in Newtonian physics iff no net creation of water molecules out of water ions or vice versa occurs. Thus a differential conservation law for X corresponds to the conservation of matter in physics.

Integral conservation laws are physically relevant because integral quantities, such as total electric charge, are often more directly related to observation. We will give a direct interpretation of $\int_{\partial \mathcal{N}} i(X)\Omega = 0$ by using the standard cylinder $\mathcal{N}_0$ in Minkowski space $(\mathbb{R}^4, g, D)$ and by using the above vector field X. We will use the notation associated with $\mathcal{N}_0$ (Section 3.0.1) without comment. Thus $\mathcal{N}_0$ is the history of the closed unit ball $\mathcal{B}$ in $\mathbb{R}^3$ from time -1 to time 1. Similarly, $\mathcal{B}_3$ is the history of the boundary $\partial \mathcal{B}$ of $\mathcal{B}$ from time -1 to 1.

Since $\Omega = du^1 \wedge \cdots \wedge du^4$, a computation gives:

$$\int_{\partial \mathcal{N}_0} i(X)\Omega = -\int_{\mathcal{B}_1} n\Omega_1 + \int_{\mathcal{B}_2} n\Omega_2 + \int_{\mathcal{B}_3} \psi,$$

where $\Omega_1 = du^1 \wedge du^2 \wedge du^3$ (respectively, $\Omega_2 = du^2 \wedge du^1 \wedge du^3$) is the metric volume element of $\mathcal{B}_1$ (respectively, $\mathcal{B}_2$) in the induced orientation and

$$\psi = \{nv^1 du^2 \wedge du^3 \wedge du^4 - nv^2 du^1 \wedge du^3 \wedge du^4 + nv^3 du^1 \wedge du^2 \wedge du^4\}.$$

Now $\int_{\mathcal{B}_1} n\Omega_1 = \int_{\mathcal{B}} n(\vec{x}, -1) dV$, where dV is the ordinary volume element of $\mathbb{R}^3$. Thus $\int_{\mathcal{B}_1} n\Omega_1 =$ total amount of water in $\mathcal{B}$ at time -1. Similarly, $\int_{\mathcal{B}_2} n\Omega_2 =$ total amount of water in $\mathcal{B}$ at time 1. Also, by the definition of ψ in terms of $n\vec{v}(\vec{x}, t)$ and the interpretation of u^4 and $\mathcal{B}_3$ as Newtonian time and the history of $\partial \mathcal{B}$, respectively, $\int_{\mathcal{B}_3} \psi =$ total amount of water which escapes through $\partial \mathcal{B}$ from time -1 to 1. Thus, an integral conservation law $\int_{\partial \mathcal{N}_0} i(X)\Omega = 0$ in this case means: (amount of water in $\mathcal{B}$ at time -1) = (amount of water in $\mathcal{B}$ at time 1) + (amount of water escaping through $\partial \mathcal{B}$ from time -1 to 1). This then directly justifies the use of the term "conservation law."

Note once more that a relativistic expression $\int i(X)\Omega$ can have different Newtonian analogues, depending on the causal character.

3.0.3 Certain tensor fields

We shall frequently need tensor fields physically equivalent to a $(0, 2)$-tensor field and shall set up a few conventions. Let S be a $(0, 2)$-tensor field on spacetime M; suppose $x \in M$, $X, Y \in M_x$, χ (respectively ψ) $\in M_x^*$ is physically equivalent to X (respectively Y), $(X_1, \ldots, X_4)$ is a basis for M_x and $(\omega^1, \ldots, \omega^4)$ is the dual basis. Thus the physically equivalent $(1, 1)$-tensor field $\tilde{S}$ obeys $\tilde{S}(\chi, Y) = S(X, Y)$ (note the order). If Z is a vector field on M we will denote by $\tilde{S}Z$ that vector field for which $\chi(\tilde{S}Z) = \tilde{S}(\chi, Z)$ $\forall \chi \in M_x^*$, $\forall x \in M$. The (2, 0) tensor field $\hat{S}$ physically equivalent to S and $\tilde{S}$ obeys $\hat{S}(\chi, \psi) = S(X, Y)$. For example, $\hat{S}$ is symmetric iff S is and similarly for skew-symmetry. We define *traces* as follows: trace S = trace $\hat{S}$ = trace $\tilde{S}$ = contraction $\tilde{S}$. Thus (trace $\tilde{S}$)$(x) = \sum_{i=1}^{4} \tilde{S}(\omega^i, X_i)$; moreover, if $(X^1, \ldots, X^4)$ is orthonormal, (trace S)$(x) = \sum_{\rho=1}^{3} S(X^\rho, X^\rho) - S(X^4, X^4)$. The above notation, including the way in which "~" and "^" are used and the obvious analogues for an S defined only on a subset of M, will be used rather often.

EXERCISE 3.0.4

Show that the metric volume element Ω satisfies $D\Omega = 0$.

EXERCISE 3.0.5

Let ($\mathbb{R}^2$, $du^1 \otimes du^1 - du^2 \otimes du^2$) be the 2-dimensional version (Section 0.2) of Minkowski space. Thus $\omega = du^1 \wedge du^2$ is the metric area element. For $a \in [-1, 1]$, $P = \partial_2 + a\partial_1$ is a future-pointing vector field. Now suppose a finite number of inextendible integral curves of P are "occupied"—for example, represent the histories of small bodies actually present, as sketched. Suppose so many are occupied that you prefer to "smooth out" by introducing the "number density" of occupied curves. (a) Show that the concept "number of curves crossing an orthogonal curve per unit length of the orthogonal curve" would not be appropriate for $a = \pm 1$. (b) Argue that the following is appropriate. There is a C^∞ world-density $\eta: \mathbb{R}^2 \to [0, \infty)$ of curves, interpreted by saying that $\forall(x, X) \in T\mathbb{R}^2, (\eta x)|\omega(P, X)|$ is (approximately) the number of occupied P curves which cross a curve of tangent X per unit parameter interval of the latter. The key point is that one can take ηx to be independent of the direction of X.

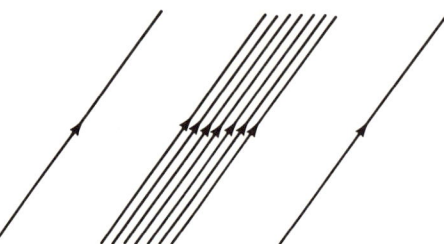

EXERCISE 3.0.6

Let Ω be the metric volume element, X be a vector field, ω is a 1-form. Show $\omega \wedge [i(X)\Omega] = \omega(X)\Omega$.

3.1 Particles

To model the history of a small object we need a curve. Suppose $m \in [0, \infty)$ is given.

Definition 3.1.1. A *particle of rest-mass m* is a future-pointing curve $\gamma\colon \mathscr{E} \to M$ such that $g(\gamma_*, \gamma_*) = -m^2$ (cf. Section 0.2).

Here m is the analogue of a Newtonian inertial-mass (Section 0.1.6). The main difference is that $m = 0$ is allowed in the relativistic case. For $m \neq 0$ the term "rest-mass" refers to one particular way m can be measured (cf. Feynman [1]). For $m = 0$, the term "rest-mass" is misleading. We sometimes write (γ, m) for a particle γ of rest-mass m.

> Atoms are often modelled as $m \neq 0$ particles; so are galaxies. We do not use the concept of a "mass that increases as the particle's speed increases," which is not basis-free and is thus obsolete.

If γ is a particle of nonzero rest-mass, there exists a positive affine reparametrization γ' such that γ' is an observer (Sections 0.0.6 and 2.1). The terminology for observers is thus taken over for such particles; for example, the arc-length of γ is also referred to as *proper time*. Similarly, let γ be any particle. γ is *freely falling* iff γ is geodesic, as interpreted in Section 0.2.

> In the simplest cases, the Newtonian equation $\ddot{\vec{r}} = -\nabla\phi$ of Section 0.1.7 for motion in a gravitational potential is an approximation to the geodesic condition $D_\gamma \gamma_* = 0$; see Section 9.3 for an example. However, the geodesic condition is applicable also to situations where Newtonian concepts are useless.

The tangent γ_* is defined as the *energy-momentum of γ*. Historically speaking, energy-momentum unifies and replaces two different concepts— namely, energy and momentum. To make this explicit we now give some auxiliary definitions and results. Actually, all the results merely follow from the basic equation $g(\gamma_*, \gamma_*) = -m^2$ (cf. Section 2.1.6).

3.1.2 Auxiliary concepts

Let $\gamma\colon \mathscr{E} \to M$ be a particle of rest-mass m. Let (z, Z) be an instantaneous observer with $z = \gamma u$, $u \in \mathscr{E}$. Then we have the orthogonal decomposition $\gamma_* u = \mathrm{e} Z + \mathrm{p}$ of the energy-momentum, with $\mathrm{e} = -g(\gamma_* u, Z) \in \mathbb{R}$ and $\mathrm{p} \in Z^\perp$. e is defined as the *energy (z, Z) measures for γ at u*. By Section 1.1, this measured energy is positive, even when $m = 0$. p is the *3-momentum (z, Z) measures for γ at u*. Denote the Newtonian velocity (z, Z) observes for $\gamma_* u$ by $\vec{v}$, where the arrow is added merely for vividness; thus $\vec{v} = \mathrm{p}/\mathrm{e} \in Z^\perp$ (2.1). From Exercise 0.2.4 and Sections 1.1 and 2.1 the reader can check the following. (a) $0 \leq |\vec{v}| \leq 1$, as usual. (b) Suppose $m \neq 0$; then $\mathrm{e} = m/(1 - |\vec{v}|^2)^{1/2}$, $\mathrm{p} = m\vec{v}/(1 - |\vec{v}|^2)^{1/2}$. (c) Suppose $|\vec{v}| \ll 1$. Then $m \neq 0$ and $\mathrm{e} \simeq mc^2 + \tfrac{1}{2}m|\vec{v}|^2$, $\mathrm{p} \simeq m\vec{v}$, where $c = 1$ is the speed of light. In this sense e includes rest-mass energy mc^2 and Newtonian kinetic energy $\tfrac{1}{2}m|\vec{v}|^2$ while p replaces Newtonian momentum $m\vec{v}$ (cf. Section 0.2). (d) In any case, $\mathrm{e}^2 = m^2c^4 + |\mathrm{p}|^2c^2$; thus Einstein's famous $\mathrm{e} = mc^2$ here holds iff $\mathrm{p} = 0$ iff $\vec{v} = 0$.

3 Electromagnetism and matter

But within relativity, an energy-momentum γ_* is interpreted directly, via the collision conservation law of Section 3.8, following. Suppose no observer is actually observing. Then it is only $\gamma_* u$, not energy e, nor 3-momentum p, nor even the pair (p, e), nor $m\vec{v}$, and so on, which models something present in nature.

3.1.3 Electric charge

Each particle in Definition 3.1.1 is assigned an *electric charge* $e \in \mathbb{R}$, measured essentially as in Newtonian physics. Empirically, one finds $e = 0$ whenever $m = 0$. This can be partially explained by quantum theory but will here be assumed *ad hoc*. We shall sometimes write (γ, m, e) for a particle γ of rest-mass m and electric charge e with $(m, e) \in [0, \infty) \times \mathbb{R}$ the *type of* γ.

EXAMPLE 3.1.4. PHOTONS. Two kinds of rest-mass zero particles have been observed: photons ("particles of light") and neutrinos; in all likelihood a third kind, gravitons, exists. However, we shall not really need the latter two here and thus formally define a *photon* as a particle of zero rest-mass. For example the electric charge of a photon is zero (Section 3.1.3). All of Chapter 5 concerns photons.

> To really pin down a particle type one needs some microphysical, quantum theoretic parameters: baryon number, electron-lepton number, mu-lepton number, and spin. For example neutrinos are distinguished from photons by having different lepton numbers and a different spin. We shall not need such quantum theoretical parameters in this book, but we briefly outline the significance of baryon and lepton numbers.
>
> As far as is known, every body in nature could be in principle be built by using a few basic constituents. We need a source or sink for baryon number, say a box with protons and antiprotons in it. (The protons in the box should be kept away from the antiprotons.) Two similar boxes, say one with electrons and antielectrons and one with mu-mesons and anti-mu-mesons can act as sources or sinks for lepton numbers. A box of pi-minus mesons could act as a source or sink for electric charge. Finally, a box of photons could be a source or sink of energy and angular momentum. If we used these constituents to build, say, a neutron or a baseball, we could proceed in various ways. But no matter in which order we proceeded, the total number of protons used minus the total number of antiprotons used would always be the same. This difference is the baryon number of the constructed particle: 1 for the neutron, about 10^{25} for the baseball. The two lepton numbers have a similar significance, as does the electric charge.

EXERCISE 3.1.5

Let $\gamma: \mathscr{E} \to M$ be a particle. Show $m \neq 0$ iff γ is timelike, iff one instantaneous observer observes a Newtonian speed less than 1 for γ, iff all instantaneous observers on $\gamma\mathscr{E}$ observe a Newtonian speed less than 1 ("when $m \neq 0$ even an observer who runs head on at the particle as fast as he can still measures a collision speed less than the speed of light").

EXERCISE 3.1.6

(a) Suppose $x \in M$, X, $Y \in M_x$. Show that any one of the following three conditions implies $X = Y$: $\omega(X) = \omega(Y) \,\forall \omega \in M_x^*$; $\omega(X) = \omega(Y) \,\forall$ timelike $\omega \in M_x^*$; $g(X, Z) = g(Y, Z) \,\forall$ instantaneous observer (x, Z). (b) In practice, it can happen that an instantaneous observer has available a device—for example, a set of Geiger counters—for measuring the energy of particles but not one for measuring 3-momentum. Thus in Section 3.1.2, suppose $z = \gamma u$ and we have instantaneous observers $(z, Z), (z, Z'), \ldots$ at z. Let e, e', ..., respectively, be the energy each measures for γ. Show from (a) that given sufficiently many such measured values there exists at most one corresponding energy-momentum $\gamma_* u \in M_z$; that in practice there exists exactly one can be regarded as a law of macrophysics.

EXERCISE 3.1.7

Suppose g' is a Lorentzian inner product on M_x, $x \in M$. Show: (a) If $g'(X, X) = g(X, X) \,\forall$ causal $X \in M_x$, then $g' = gx$. (b) $g'(X, X) = 0 \,\forall$ lightlike $X \in M_x$ iff $g' = agx$ for some $a \in (0, \infty)$.

3.2 Particle flows

Without yet attempting to describe matter models in general we give an example. Suppose we have an enormous number of particles, each having the same rest-mass $m \in [0, \infty)$. Suppose, intuitively speaking, there is no "random motion" so that all the energy-momenta near any one point are nearly equal. Examples are a very cold gas streaming smoothly in space ($m \neq 0$) or a laser beam ($m = 0$). Then the following idealization is often useful.

Definition 3.2.1. A *particle flow* (P, η) *of rest-mass m* is a function $\eta: M \to [0, \infty)$, called the *world density*, and an *energy-momentum* vector field $P: M \to TM$ such that each integral curve of P is a particle of rest-mass m.

Thus P is future pointing and $g(P, P) = -m^2$. The integral curves of P are called *particles* (*potentially*) *in the particle flow*. The main idea is that the world density η specifies, in a "smoothed-out" way, how many particles are actually present (Exercise 3.0.5). To make this explicit, suppose $\mathscr{D}$ is a space-section (Section 3.0.1). Then $N = \int_\mathscr{D} i(\eta P)\Omega$ is defined as the *total number of particles in $\mathscr{D}$*. In the cases of interest $N \gg 1 \,\forall \mathscr{D}$ of interest so one does not demand N be an integer.

Suppose, intuitively speaking, no particles in (P, η) are being created—for example, from other kinds of particles—or destroyed, for example, by radioactive decay. Then it is appropriate to demand the integral conservation law $\int_{\partial \mathcal{N}} i(\eta P)\Omega = 0$ for all causal boxes $\mathcal{N}$ or, equivalently, the differential conservation law div $(\eta P) = 0$ (Section 3.0.2). One says the particle world density η is *conserved* iff div $(\eta P) = 0$.

3 Electromagnetism and matter

EXAMPLE 3.2.2. Let $(\mathbb{R}^4, g)$ be Minkowski space, $\eta: \mathbb{R}^4 \to (0, \infty)$ be C^∞, and suppose $m \in (0, \infty)$. Then $(m\partial_4, \eta) = (P, \eta)$ is a particle flow of rest-mass m. The total number of particles in the space-section $\mathcal{D}$ given by

$$\left\{ u^4 = a, \sum_{\mu=1}^{3} (u^\mu)^2 \leq 1 \right\}$$

is $m \int_\mathcal{D} \eta(u^1, u^2, u^3, a) du^1 du^2 du^3$. η is conserved iff div $(\eta \partial_4) = 0$ iff $\partial_4 \eta = 0$; in this case the above total number is independent of a.

For purposes of Newtonian analogues one might regard $m\eta$, rather than η, as number density, as the example suggests. But for $m = 0$ this does not work. $e\eta$ would work but depends on a gratuitously introduced instantaneous observer who measures e. η is best.

Generally, and roughly, speaking the equations of physics are supposed to have the property that in some sense "the present determines the future." As an example we analyze div $(\eta P) = 0$. Suppose (M, g) and P are given, with P a future-pointing vector field and $g(P, P) = -m^2$, $m \in [0, \infty)$. Suppose $\phi: N \to M$ is a spacelike imbedding, with N a 3-manifold. Suppose each inextendible integral curve $\gamma: \mathscr{E} \to M$ of P intersects ϕN for exactly one $u \in \mathscr{E}$. Suppose we are given, as initial data, a C^∞ function $\eta_0: N \to [0, \infty)$.

Proposition 3.2.3. *There is at most one particle flow* (P, η) *such that* div $(\eta P) = 0$ *and* $\eta \circ \phi = \eta_0$.

PROOF. Suppose $x \in M$. There is exactly one γ as above such that: (a) $\gamma 0 \in \phi N$, and (b) $\gamma u^1 = x$ for at least one $u^1 \in \mathscr{E}$. In fact, u^1 is then unique: since γ is an integral curve of a vector field, $\gamma u^1 = \gamma u^2$ implies $\gamma(u^2 - u^1) = \gamma 0 \in \phi N$, which in turn implies $u^2 - u^1 = 0$ by our hypotheses.

Now suppose η exists and define functions $f = \eta \circ \gamma: \mathscr{E} \to \mathbb{R}$ and $j = (\text{div } P) \circ \gamma: \mathscr{E} \to \mathbb{R}$. We have div $(\eta P)\Omega = d[i(\eta P)\Omega] = d[\eta i(P)\Omega] = d\eta \wedge [i(P)\Omega] + \eta(\text{div } P)\Omega = [d\eta(P) + \eta \text{ div } P]\Omega$ (Exercise 3.0.6). Hence $0 = \text{div}(\eta P) = d\eta(P) + \eta \text{ div } P$; restricting to γ gives $f' + fj = 0$ and we also have $f(0) = \eta_0 \phi^{-1} \gamma 0$, say $f(0) = a \in [0, \infty)$. Since $\mathscr{E}$ is connected we can simply integrate the ordinary differential equation $f' + fj = 0$ to get

$$\eta x = fu^1 = a \exp\left[-\int_0^{u_1} j(u) du \right].$$

Thus if an η exists its value at each $x \in M$ is uniquely determined by (M, g), P, and η_0. □

In fact an η does exist, but showing this requires, among other things, a C^∞ proof here omitted; in later examples we shall give existence proofs.

In a physical context, it may not be kosher to assume (M, g), much less P, given since (P, η) may "influence spacetime" (Chapter 4). For the moment we are sticking to the simplest formulation.

Given (M, g) and P there need not exist an imbedding $\phi: N \to M$

meeting our conditions. However, in the simplest special cases—for example, when (M, g) is globally hyperbolic as defined in Section 8.3—ϕ always exists. For example, if M is Minkowski space or Einstein–de Sitter spacetime, choosing ϕN as the level surface $u^4 = 1$ works for any P.

3.2.4 Electric charge and photon beams

If (P, η) is a particle flow of rest-mass m we shall henceforth assume each particle (potentially) in (P, η) has the same electric charge $e \in \mathbb{R}$. We sometimes write (m, e, P, η), with e the *electric charge of* and (m, e) the *type of the particle flow*. In particular we define a *photon beam* as a particle flow of type $(0, 0)$.

EXERCISE 3.2.5

Let (M, g) be a simple cosmological spacetime (Example 1.4.3) with $\mathscr{F} = (0, \infty)$, $n: (0, \infty) \to (0, \infty)$ be a C^∞ function; and suppose $m \in (0, \infty)$. Set $P = m\partial_4$, $\eta = (1/m)n \circ u^4$. (a) Show (P, η) is a particle flow of rest-mass m. (b) Show η is conserved iff there is an $a \in (0, \infty)$ such that $n = aR^{-3}$. (c) In the Einstein–de Sitter case—that is, $Ru = u^{2/3}$—interpret $n = aR^{-3}$ physically in terms of integrals over appropriate space-sections (Section 3.0.1) and the "expansion of the universe" (Proposition 2.3.7). (An answer: "no particles appear or disappear so their world density must decrease as the comoving observers run apart".)

3.3 Stress-energy tensors

We now introduce a conceptually difficult, mathematically easy concept which will dominate the later chapters. The conventions of Section 3.0.3 are in force; for example, if $\hat{E}$ is a (2, 0)-tensor field on M, the physically equivalent (0, 2)-tensor field is denoted by E.

3.3.1 Formal definition

Formally, a *stress-energy tensor* on spacetime M is merely a symmetric (2, 0)-tensor field $\hat{E}$ on M such that $\hat{E}(\omega, \omega) \geq 0$ $\forall$ causal 1-form $\omega \in M_x$, $\forall x \in M$. Physically, more is involved, as we now discuss.

3.3.2 Earlier concepts

A stress-energy tensor replaces and unifies the following prerelativistic concepts. Energy of electromagnetism and/or matter, including rest-mass contributions as in Section 3.1.2, per unit $\mathbb{R}^3$ volume; momentum per unit $\mathbb{R}^3$ volume; energy flux; and momentum flux, which corresponds to stress, in fact to the pair (pressure, anisotropic stress). Hence the term "stress-energy" (or in some references "energy-momentum") as an abbreviation. The prerelativistic quantities were found independently. They are observer-dependent and quite messy even in simple situations, as illustrated in Optional exercises 9.1.9 and 9.1.10. But around 1905, physicists realized, with glee, that if one interrelates the measurements made by observers in relative motion, a single concept suffices, as follows.

3.3.3 Measured energy density

Suppose any instantaneous observer (z, Z) actually measures the energy in any unit 3-volume of his local rest space $Z^\perp$ (the method is indicated more explicitly in Definition 3.3.5 below). He is supposed to get $E(Z, Z)$, where E is physically equivalent to a stress-energy tensor (Section 3.3.1) which is the same for all instantaneous observers. Thus when $\hat{E}$ is a stress-energy tensor, $E(Z, Z)$ is defined as the *energy density (z, Z) measures for $\hat{E}$*. That the observations actually correspond to $E(Z, Z)$ can be regarded as a basic law of macrophysics; *a priori* one might have found, for example, measured energy density $= S(Z, Z, Z)$ with S a $(0, 3)$-tensor on M (cf. Exercise 3.3.9); but one does not. In particular, each measured energy density is nonnegative, corresponding to $\hat{E}(\omega, \omega) \geq 0$ in Section 3.3.1. The measurements determine $\hat{E}$ uniquely in the following sense.

Proposition 3.3.4. *Let E, E' be symmetric $(0, 2)$ tensor fields on M and suppose $E(Z, Z) = E'(Z, Z) \forall$ instantaneous observer (z, Z). Then $\hat{E} = \hat{E}'$.*

PROOF. Fix an instantaneous observer (z, Z). $\forall$ future pointing timelike $X \in M_z$, $E(X, X) = E'(X, X)$ since $(z, X/|X|)$ is an instantaneous observer at z. Now the set $\mathcal{T}_x^+ \subset M_x$ of all such X is open and contains Z (Section 1.1). Thus $\forall W \in M_x$, there exists $b \in (0, \infty)$ such that $Z + uW \in \mathcal{T}_x^+ \forall u \in [0, b)$ and then $E(Z + uW, Z + uW) = E'(Z + uW, Z + uW)$. Differentiating twice with respect to u we find $E(W, W) = E'(W, W) \forall W \in M_z$. By symmetry (Section 3.3.1) $Ez = E'z$. But z was arbitrary. □

Thus $E(Z, Z) =$ measured energy density $\forall (z, Z)$ can serve as an operational definition of a stress-energy tensor $\hat{E}$ (cf. Exercise 3.1.6). Now this operational definition, though as general and precise as anything else in macrophysics, refers to actual measurements so it cannot be used in mathematical proofs (Section 2.1.2). But as soon as a mathematically precise matter model on M is given, the operational definition leads to a natural way to a mathematically precise formulation. We now illustrate, using the only example as yet available to us.

> Within physics Definition 3.3.5 and the motivation below would count as a proposition and a proof, rather than a definition and a motivation, so we use the corresponding format (cf. Section 0.1.1). But we shall need to regard part of a tangent space as part of physical spacetime, and no amount of formal machinery can really bridge this gap (Section 2.1.2).

Definition 3.3.5. *Let (P, η) be a particle flow on spacetime M. The stress-energy tensor of (P, η) is $\hat{T} = \eta P \otimes P$.*

MOTIVATION. $\hat{T} = \eta P \otimes P$ is symmetric and C^∞; $\forall$ 1-form $\omega \in M_x$ $\hat{T}(\omega, \omega) = \eta x[\omega(P)]^2 \geq 0$; thus $\hat{T}$ is a stress-energy tensor (Section 3.3.1). We must show in what sense the measured energy density is $T(Z, Z) \forall$ instantaneous observer Z.

Let (z, Z) be an instantaneous observer, $X_1, X_2, X_3 \in Z^\perp$ be linearly independent, and $K \subset Z^\perp$ be the parallelipiped defined by X_1, X_2, X_3. By the definition of the matric volume element Ω (and by Section 2.1.2), the 3-volume of K is $V = |\Omega(X_1, X_2, X_3, Z)| > 0$. By the definition of total particle number in Section 3.2 (and by Section 2.1.2), the number of particles in K is $N = (\eta z)|\Omega(X_1, X_2, X_3, P)| \geq 0$. The energy (z, Z) measures for each particle is $e = -g(P, Z) > 0$ (Sections 3.1 and 3.2). The measured energy density is thus Ne/V (cf. Definition 3.8.4a for the motivation for simply adding all the energies). But $Pz = eZ + p$, where $p \in \text{span}(X_1, X_2, X_3) = Z^\perp$. Since Ω is antisymmetric this gives $Ne/V = (\eta z)e^2 = \eta z[g(P, Z)]^2$ for the measured energy density, cleverly independent of the particular parallelipiped chosen.

On the other hand $\hat{T} = \eta P \otimes P$ also implies $T(Z, Z) = (\eta z)[g(P, Z)]^2$. The argument holds $\forall$ instantaneous observer (z, Z) so by the uniqueness Proposition 3.3.4, $\hat{T} = \eta P \otimes P$ as was to be motivated.

Once we have a formal definition like 3.3.5 available we are back in business mathematically. We give an example.

Proposition 3.3.6. *Let $\hat{T}$ be the stress-energy tensor of a particle flow and X be a causal vector field. Then $T(X, X) - \frac{1}{2}(\text{trace } T)g(X, X) \geq 0$.*

PROOF. $T(X, X) - \frac{1}{2}(\text{trace } T)g(X, X) = \eta\{[g(X, P)]^2 - \frac{1}{2}g(P, P)g(X, X)\}$. η is nonnegative by definition and the factor in curly brackets is nonnegative by the wrong-way Schwarz inequality (Section 1.1). □

> As we shall discuss, the algebraic property Proposition 3.3.6 holds very generally for the stress-energy tensors that arise in nonquantum physics and its geometric and physical consequences are very far reaching (cf. Exercise 4.3.7 and Section 6.2).

EXERCISE 3.3.7

Let E be a symmetric $(0, 2)$-tensor field on M such that $E(Z, Z)$ is nonnegative $\forall$ instantaneous observer (z, Z). Show $\hat{E}$ is a stress-energy tensor (Definition 3.3.1) on M.

EXERCISE 3.3.8

(a) Let J and J' be vector fields on M such that $g(J, Z) = g(J', Z) \forall$ instantaneous observer (z, Z). As in Exercise 3.1.6 and Proposition 3.3.4, show $J = J'$. (b) Let F, F' be 2-forms on M. Thus, in the notation of Section 3.0.3, $\tilde{F}X \in M_x \forall X \in M_x$. Suppose $\tilde{F}Z = \tilde{F}'Z \forall$ instantaneous observer (z, Z). Show, much as in (a), that $F = F'$.

EXERCISE 3.3.9

Let $(\mathbb{R}^4, g)$ be Minkowski space and suppose that at $z \in \mathbb{R}^4$ we are given four instantaneous observers determined by $Z = \partial_4 z$, $Z^\pm = (1/3)(5\partial_4 z \pm 4\partial_1 z)$, and $Z' = (1/4) \times (5\partial_4 z + 3\partial_1 z)$. Suppose each actually measures an energy density and the measured values are, respectively, $U = 1$, $U^+ = (5/3)^2 = U^-$, and $U' = 3$. Assuming all four competent, show general relativity (and thus current

3 Electromagnetism and matter

macrophysics) is wrong—that is, no stress-energy tensor which corresponds to these measurements exists.

EXERCISE 3.3.10

Use the definitions in Section 3.1 and the methods of Definition 3.3.5 to motivate in detail the following definition. Let $\hat{T}$ be the stress-energy tensor of a particle flow, (z, Z) be an instantaneous observer. $\hat{T}Z \in M_z$ is the *energy-momentum density* (z, Z) measures. The same definition is used for an arbitrary stress-energy tensor.

3.4 Electromagnetism

Nonquantum relativistic electromagnetic theory is perhaps the most elegant part of physics. Though it is formally simple, its applications are well nigh endless: on a human scale electromagnetism is the most important of the four known interactions (Section 0.1.2). But in this book we need the theory primarily for background. In our applications the electromagnetic field will normally be zero, and we shall not analyze the practical applications. This section merely gives two basic definitions and a little history.

3.4.1 Charge-current density

Formally, a *charge-current density* on spacetime M is just a vector field J on M. But, historically speaking, J replaces and unifies two prerelativistic concepts: electric charge per unit $\mathbb{R}^3$ volume, and electric current density (charge flux; for details on these two, see Section 9.1). Moreover, much as in Section 3.3.3, we need a prescription for measuring. Suppose any instantaneous observer (z, Z) actually measures electric charge (Section 3.1.3) per unit 3-volume of $Z^\perp$. He is supposed to get $-g(Z, J)$, where J is a charge-current density on M. The appropriate uniqueness result then holds (Exercise 3.3.8a). So we can regard this prescription both as an operational definition and as a physical law (cf. Exercise 3.3.9). Almost exactly as in Definition 3.3.5 one can motivate the following definition. Let (m, e, P, η) be a particle flow; the *charge-current density* of the particle flow is $J = e\eta P$. For example, suppose the world density η is conserved; then J obeys the differential conservation law div $J = 0$ ("charge doesn't get lost").

> In the above, and throughout Chapters 3 and 4, it is useful to keep the following heuristic analogy in mind. J is "a source" for electromagnetism; a stress-energy tensor is "a source" for gravity. The formal similarities will be very strong.

Definition 3.4.2. Formally, an *electromagnetic field* on spacetime M is a 2-form on M.

In the end, only this formal definition is essential. But an electromagnetic field replaces and unifies two prerelativistic quantities: an electric field $\vec{E}$ and a magnetic field $\vec{B}$. Indeed, the simplification thereby achieved was one of

3.4 Electromagnetism

the main original motivations for introducing spacetimes. Reversing the historical order, we now indicate the sense of the unification. Let F be an electromagnetic field on M, and (z, Z) be an instantaneous observer; we use the conventions in Section 3.0.3.

Since F is antisymmetric, $\tilde{F}Z \in Z^\perp$. $E = \tilde{F}Z$ is defined as *the electric vector (z, Z) measures for F*. It is kosher to imagine (z, Z) measuring the electric vector in $Z^\perp$ by essentially Newtonian methods—for example, using a "test charge" (Alonso-Finn [2]). Now by exterior algebra there is a unique vector $B \in Z^\perp$ such that $4! \, \Omega(X, Y, B, Z) = F(X, Y) \, \forall X, Y \in Z^\perp$, where Ω is the metric volume element (cf. the discussion of Hodge duality in Bishop-Goldberg). B is defined as the *magnetic vector (z, Z) measures for F*.

EXAMPLE 3.4.3. Let $(\mathbb{R}^4, g)$ be Minkowski space. Thus there exist parallel reference frames—for example, ∂_4; this facilitates the interpretations in this special case (cf. Exercise 2.3.12). Suppose E and B are C^∞ functions $\mathbb{R}^4 \to \mathbb{R}$. Then $F = 2E du^1 \wedge du^4 + 2B du^3 \wedge du^1$ is an electromagnetic field on Minkowski space. Unravelling the definitions, we find that $\forall z \in \mathbb{R}^4$ the electric vector $(z, \partial_4 z)$ measures for F is $E \partial_1 z$ while his magnetic vector is $B \partial_2 z$. Thus the vector fields $E \partial_1$ and $B \partial_2$ are, respectively, called the *electric field* and *magnetic field in the reference frame ∂_4*. Now suppose E = 0 and B is nowhere zero. Then the electric field in the reference frame ∂_4 is identically zero. But $\forall z \in \mathbb{R}^4$, $(z, [1/3][5\partial_4 z + 4\partial_1 z])$ is an instantaneous observer who measures a nonzero electric vector. One thus often says: "An observer moving in a magnetic field observes an electric vector"; or "electric and magnetic fields are merely two aspects of the same thing"; or "electric vectors are not physically well-defined in the sense that two different observers at the same point may not even be able to agree on whether the electric vector is zero, but electromagnetic fields are intrinsic."

> Though thus a little unphysical, and considerably more complicated than the relativistic theory, the 19th-century theory of electric and magnetic fields has considerable elegance in its own right. Section 9.1 outlines it, and its relation to the relativistic theory, rather systematically. But we suggest the reader not already familiar with upper-division physics courses on electromagnetism focus attention on the simpler relativistic version.

Returning now to the general case, we note that the appropriate uniqueness result holds for the measured electric vector $\tilde{F}Z$ (Exercise 3.3.8b). For example if z is given and $\tilde{F}Z = 0$ for every instantaneous observer (z, Z) at z, then $Fz = 0$, and each such observer measures zero magnetic vector as well. Thus our definition of measured electric vectors can serve as an operational definition of an electromagnetic field F on spacetime M.

> A fully intrinsic, relativistic interpretation of F, which dispenses with instantaneous observers, is given by the Lorentz world-force law of Section 3.8.

3 Electromagnetism and matter

EXERCISE 3.4.4

Let F be an electromagnetic field on M, and (z, Z) be an instantaneous observer. (a) Show that the measured electric vector $\tilde{F}Z$ is zero iff $F(Z, X) = 0 \,\forall X \in Z^\perp$. (b) Show $Fz = 0$ iff $F(Z, X) = 0 = F(X, Y)\,\forall X, Y \in Z^\perp$. (c) Show that if both the electric vector and the magnetic vector (z, Z) measures are zero then $Fz = 0$, and vice versa. (d) Show that one could also regard measuring magnetic vectors as the primary operation—for example, that if every instantaneous observer (z, X) at $z \in M$ measures zero magnetic vector, then $Fz = 0$.

EXERCISE 3.4.5

On Minkowski space, suppose

$$F = 2 \sum_{\alpha=1}^{3} E^\alpha du^\alpha \wedge du^4 + 2[B^1 du^2 \wedge du^3 + B^2 du^3 \wedge du^1 + B^3 du^1 \wedge du^2],$$

where $\forall \alpha \in (1, 2, 3)$, E^α and B^α are functions. Show that $\forall z \in \mathbb{R}^4$, $(z, \partial_4 z)$ measures electric and magnetic vectors $E = \sum_{\alpha=1}^{3} E^\alpha \partial_\alpha z$ and $B = \sum_{\alpha=1}^{3} B^\alpha \partial_\alpha z$, respectively.

3.5 Matter and relativistic models

The basic object of interest in mathematical general relativity is a *relativistic model*, defined as a triple $(M, \mathcal{M}, F)$, where M is a spacetime (Definition 1.3.1), F is an electromagnetic field (Definition 3.4.2) on M, and $\mathcal{M}$ is a matter model on M.

But what on earth is a matter model on M? Since, as already indicated, there exists no precise, universal, overriding model for matter even in current microphysics, let alone in macrophysics, no fully satisfactory answer can be given. Willy nilly we shall have to proceed by enumeration. To make this less vague we now give a matter model that can be treated as generic in most arguments.

It merely consists of a collection $\mathcal{M} = \{(m_A, e_A, P_A, \eta_A) | A = 1, \ldots, N\}$ of N particle flows (Section 3.2.4) on M where N is a nonnegative integer. For example, suppose we have two theatre spotlight beams that cross; a pair $\mathcal{M}$ of photons beams is a reasonable model. By making N sufficiently large, one can model almost any form of matter. Moreover, all our other matter models—for example, a perfect fluid—will be obtained from such a collection by abstraction. Thus we will systematically use such a collection $\mathcal{M}$ to illustrate properties shared by all matter models used in this book and by most used in general relativity. We now give two examples of such properties.

> $\mathcal{M}$ similarly leads to more sophisticated models than we shall need: imperfect fluids, $m \neq 0$ gases, plasmas, solids, and so on. Sometimes the physics involved is then quite tricky. For example, to get any reasonably detailed, fundamental model, general relativistic or not, of an every day object, such as your chair, is difficult. Fortunately, the very large objects of main interest in gravitational theory are often less tricky. Semiempirical models apart, we (probably) know far more about the inner core of the star Sirius than we do about the cement in your driveway.

3.5 Matter and relativistic models

In the last few years, serious and fascinating attacks on the problem of combining quantum matter models with general relativity have been initiated. These models cannot be obtained from $\mathcal{M}$ above. It will be a long time before such models are brought into a form a mathematician would consider reasonably precise—probably this cannot be done until one can go whole hog and in some sense quantize spacetime itself.

Finally, various other matter models are sometimes used merely for convenience or mathematical elegance. Most of these cannot be motivated by $\mathcal{M}$, or by any other physically plausible argument either. We ignore them.

3.5.1 Matter stress-energy tensors and charge-current densities

Let $(M, \mathcal{M}, F)$ be a relativistic model. Then one always gets a matter stress-energy tensor and a charge-current density for $\mathcal{M}$. For example, suppose $\mathcal{M}$ is a finite collection $\{(m_A, e_A, \boldsymbol{P}_A, \eta_A) | A = 1, \ldots, N\}$ of particle flows on M. Then $\hat{T} = \sum_{A=1}^{N} \eta_A \boldsymbol{P}_A \otimes \boldsymbol{P}_A$ is a stress-energy tensor (Section 3.3.1) on M; by definition $\hat{T}$ is the *stress-energy tensor of* $\mathcal{M}$; the motivation is as in Definition 3.3.5. Similarly, $\boldsymbol{J} = \sum_{A=1}^{N} e_A \eta_A \boldsymbol{P}_A$ is the *charge-current density of* $\mathcal{M}$.

Proposition 3.5.2. *Let $(M, \mathcal{M}, F)$ be a relativistic model with $\mathcal{M}$ a finite collection of particle flows, $\hat{T}$ be the stress-energy tensor of $\mathcal{M}$. Then $T(Z, Z) = 0$ for one instantaneous observer (z, Z) iff $\hat{T}z = 0$ iff $\eta_A z = 0$ holds for the world density η_A of each particle flow in $\mathcal{M}$.*

PROOF. $T(Z, Z) = \sum (\eta_A z)[g(\boldsymbol{P}_A, Z)]^2$. $\forall A$, $\eta_A z$ is nonnegative by definition, and $e_A = -g(\boldsymbol{P}_A, Z) > 0$ as usual. $\square$

Suppose $N \neq 0$. The condition $\eta_A z = 0 \, \forall A$ is interpreted as no particles at z (Section 3.2). Generalizing, suppose $(M, \mathcal{M}, F)$ is any relativistic model and $\hat{T}$ is the stress-energy tensor of $\mathcal{M}$. We henceforth interpret $\hat{T}z = 0$ to mean matter-vacuum at z: "no matter at z except perhaps 'test matter' which responds but does not influence."

EXERCISE 3.5.3

Let $(M, \mathcal{M}, F)$ be a relativistic model with $\mathcal{M}$ a particle flow $(m, e, \boldsymbol{P}, \eta)$. Suppose $(\overline{M}, \bar{g}) \in [(M, g)]$, the spacetime equivalence class; thus there is an isometry $\psi: \overline{M} \to M$ that preserves orientation and time orientation. Let $\overline{F} = \psi^* F$, $\bar{\eta} = \eta \circ \psi$, $\overline{\boldsymbol{P}} = \psi_*^{-1} \boldsymbol{P}$, $\overline{\mathcal{M}} = (m, e, \overline{\boldsymbol{P}}, \bar{\eta})$. Show $(\overline{M}, \overline{\mathcal{M}}, \overline{F})$ is a relativistic model. It models the same physical situation as $(M, \mathcal{M}, F)$; more generally, relativistic models are regarded as defined only up to isomorphism.

PART TWO: INTERACTIONS

Having introduced the basic concepts, we now consider how electromagnetism and matter interact and how they respond to spacetime. Henceforth we shall take the following attitude toward a relativistic model $(M, \mathcal{M}, F)$. In macrophysics, the spacetime M, matter model $\mathcal{M}$, and electromagnetic field F are equal partners, each influencing the others.

3.6 Some mathematical methods

Although we work exclusively in a spacetime (M, g) in this section, most of the results and their proofs are valid for arbitrary semi-Riemannian manifolds.

3.6.1 Indices and the summation convention

In the previous chapters we have managed to avoid the use of indices. However, in general relativity, the emphasis is often on very detailed properties of spacetimes rather than general theorems about manifolds. The amount of computation required sometimes make indices indispensable, and the Einstein summation convention (Bishop–Goldberg 2.3) then becomes a convenient tool.

Hereafter, if a lower-case Latin index, say i, appears precisely once in each additive term of an equation (except perhaps the term "0"), the equation holds for every $i = 1, \ldots, 4$. If such an index appears twice in any additive term, once as a subscript and once as a superscript, a sum from 1 to 4 is implied. We give some examples and make some simple observations.

(a) $\qquad f_i \omega^i = f_1 \omega^1 + f_2 \omega^2 + f_3 \omega^3 + f_4 \omega^4 = f_j \omega^j = f_k \omega^k.$

In the remainder of Section 3.6.1, let $\{X_i\}$ be a basis of vector fields on an open set $\mathcal{U} \subset M$, and let $\{\omega^i\}$ be the dual basis.

(b) Let $g_{ij} = g(X_i, X_j)$; $\forall i, j = 1, \ldots, 4$, g_{ij} is a C^∞ function on $\mathcal{U}$. Let $\hat{g}$ be the $(2, 0)$ tensor physically equivalent to g, and let $g^{ij} = \hat{g}(\omega^i, \omega^j)$. Then each g^{ij} is also a C^∞ function on $\mathcal{U}$ and $g_{ij} g^{jk} = \delta_i^{\ k}$ (cf. Exercise 0.0.15e). Moreover, $g|_\mathcal{U} = g_{ij} \omega^i \otimes \omega^j$ and $\hat{g}|_\mathcal{U} = g^{ij} X_i \otimes X_j$. The symmetry of g is equivalent to $g_{ij} = g_{ji}$ or $g^{ij} = g^{ji}$ $\forall i, j$. Suppose $\{X_i\}$ is orthonormal. Then $\forall \mu, \nu = 1, 2, 3$, $g_{\mu\nu} = \delta_\nu^{\ \mu} = g^{\mu\nu}$, $g_{\mu 4} = 0 = g^{\mu 4}$, and $g_{44} = -1 = g^{44}$.

(c) Any tensor field C of type (p, q) can be expressed in terms of $\{X_i\}$ and $\{\omega^j\}$ locally in $\mathcal{U}$. For example,
$$C = C^{i_1 \ldots i_p}{}_{j_1 \ldots j_q} X_{i_1} \otimes \cdots \otimes X_{i_p} \otimes \omega^{j_1} \otimes \cdots \otimes \omega^{j_q}.$$
$\{C^{i_1 \ldots i_p}{}_{j_1 \ldots j_q}\}$ will be called *the components of C relative to $\{X_i\}$*. Let $\mathbf{R}$ be the curvature tensor and **Ric** be the Ricci tensor. Then on $\mathcal{U}$, $\mathbf{R} = R^i{}_{jkl} X_i \otimes \omega^j \otimes \omega^k \otimes \omega^l$ and **Ric** $= R_{ij} \omega^i \otimes \omega^j$, where $R_{ij} = R^l{}_{ilj}$. Moreover, with g^{ij} as in (b), the scalar curvature on $\mathcal{U}$ is $\mathrm{S} = g^{ij} R_{ij}$.

(d) We introduce a notation for the components of the covariant derivatives of a tensor. If a tensor $C = C^i{}_{jk} X_i \otimes \omega^j \otimes \omega^k$ (say), then we write its covariant differential as $DC = C^i{}_{jk|m} X_i \otimes \omega^j \otimes \omega^k \otimes \omega^m$, so that $D_{X_m} C =$

$C^i_{jk|m} X_i \otimes \omega^j \otimes \omega^k$. Thus the *second Bianchi identity* of the curvature tensor can be written as

$$R^i_{jkl|m} + R^i_{jlm|k} + R^i_{jmk|l} = 0$$

(Bishop–Goldberg p. 235).

(e) Let $\{\omega_j{}^i\}$ ge the connection forms for $\{\omega^i\}$ (Section 1.0.3). Then it is customary to write

$$\omega_j{}^i = \Gamma^i{}_{kj}\omega^k,$$

where each $\Gamma_{kj}{}^i$ is a C^∞ function on $\mathcal{U}$. In this notation, we have

$$DX_k = \Gamma^i{}_{jk} X_i \otimes \omega^j,$$

and using the fact that each D_{X_i} is a derivation commuting with contraction (Section 2.0.2c), we have

$$D\omega^i = -\Gamma^i{}_{jk}\omega^j \otimes \omega^k.$$

As an example, we see that for $C = C^i{}_{jk} X_i \otimes \omega^j \otimes \omega^k$ as in (d),

$$C^i{}_{jk|l} = X_l C^i{}_{jk} + C^m{}_{jk}\Gamma^i{}_{lm} - C^i{}_{mk}\Gamma^m{}_{jl} - C^i{}_{jm}\Gamma^m{}_{kl}$$

(cf. Bishop–Goldberg 5.9).

(f) Components of physically equivalent tensors (Section 1.0.1) are obtained from each other by "raising" or "lowering" indices. For instance, given a (1,3) tensor field $P = P^i{}_{jkl} X_i \otimes \omega^j \otimes \omega^k \otimes \omega^l$, then when using components the (0, 4) tensor field $\hat{P}$ physically equivalent to it will be written as $\hat{P} = P_{ijkl}\omega^i \otimes \omega^j \otimes \omega^k \otimes \omega^l$, where $P_{ijkl} = g_{im}P^m{}_{jkl}$. As another example, the (1, 1) tensor field $\tilde{G}$ physically equivalent to the Einstein tensor G (Section 1.0.2) is $G^i{}_j X_i \otimes \omega^j$, where $G^i{}_j = g^{im}R_{mj} - \frac{1}{2}\delta^i{}_j S = R^i{}_j - \frac{1}{2}\delta^i{}_j S$.

The reader should note the *notational convention* implicitly adopted in the above: the components of $\hat{g}$ are g^{ij} and not $\hat{g}^{ij}$, those of $\hat{P}$ are P_{ijkl} and not $\hat{P}_{ijkl}$, those of $\tilde{G}$ are $G^i{}_j$ and not $\tilde{G}^i{}_j$, and so on.

(g) Let T be a (2,0) tensor field on M. Then (Section 3.0.3)

$$(\text{trace } \hat{T})|_{\mathcal{U}} = T^{ij}g_{ij} \equiv T^i{}_i \equiv T^j{}_j.$$

A basis of vector fields $\{X_i\}$ is said to be *normal* at $x \in M$ iff $\{X_i x\}$ is orthonormal and $(DX_i)x = 0$ $\forall i$. By (e), this is equivalent to requiring $\{\omega^i x\}$ orthonormal and $(D\omega^i)x = 0$ $\forall i$. Given any $x \in M$, locally there exist coordinate vector fields $\{\partial_i\}$ which are normal at x (cf. Bishop–Goldberg, Proposition 5.13.1, and Problem 5.13.1).

(h) Let $\{\partial_i\}$ be coordinate vector fields around $x \in M$ and let $g_{ij} = g(\partial_i, \partial_j)$. Then $\{\partial_i\}$ is normal at x iff $\forall \mu, \nu \in (1, 2, 3)$ $g_{\mu\nu}x = \delta_{\mu\nu}$, $0 = g_{\mu 4}x = g_{4\mu}x$, $g_{44}x = -1$, and $(\partial_k g_{ij})x = 0$ $\forall i, j, k$.

Proof. The algebraic part is straightforward. In addition, we only have to show: $(\partial_k g_{ij})x = 0$ is equivalent to $(D\partial_i)x = 0$. If $(D\partial_i)x = 0$ $\forall i$, then $(\partial_k g_{ij})x = [\partial_k g(\partial_i, \partial_j)]x = [g(D_{\partial_k}\partial_i, \partial_j) + g(\partial_i, D_{\partial_k}\partial_j)]x = 0$.

3   Electromagnetism and matter

Conversely, suppose $(\partial_k g_{ij})x = 0 \; \forall i, j, k$. For $\Gamma_{jk}{}^i$ defined by $D_{\partial_k}\partial_k = \Gamma^i_{jk}\partial_i$, it is classical that $\Gamma_{jk}{}^i = \frac{1}{2}g^{il}(\partial_k g_{jl} + \partial_j g_{kl} - \partial_l g_{kj})$ (cf. Bishop–Goldberg, formula 5.11.5). Thus $(\partial_k g_{ij})x = 0 \; \forall i, j, k \Rightarrow \Gamma^i_{jk}x = 0 \; \forall i, j, k \Rightarrow (D\partial_i)x = 0 \; \forall i$.   □

(i) If $\{X_i\}$ is a basis of vector fields, $\{\omega^j\}$ its dual basis, and η is a p-form, then
$$d\eta = \omega^i \wedge D_{X_i}\eta.$$

Proof. Algebra shows that the right side is independent of the choice of $\{X_i\}$, and is hence globally defined on M. Thus both sides are tensor fields on M and it suffices to prove equality at each $x \in M$. We may then assume that $\{X_i\}$ is a basis of coordinate vector fields normal at x. In this case, both sides equal
$$(\partial_i \eta_{j_1\ldots j_p})dx^i \wedge dx^{j_1} \wedge \cdots \wedge dx^{j_p}$$
at the point x.   □

(j) Let η be a 1-form, 2-form, and 3-form successively. Then
$$(d\eta)_{ij} = \tfrac{1}{2}(\eta_{j|i} - \eta_{i|j}),$$
$$(d\eta)_{ijk} = \tfrac{1}{3}(\eta_{ij|k} + \eta_{ki|j} + \eta_{jk|i}),$$
$$(d\eta)_{ijkl} = \tfrac{1}{4}(\eta_{ijk|l} - \eta_{ijl|k} - \eta_{ilk|j} - \eta_{ljk|i}).$$
This is just a specialization of (i).

3.6.2 Divergence

Let C be a skew-symmetric $(p, 0)$-tensor field on M, where we regard each function ($p = 0$) and each vector field ($p = 1$) as skew-symmetric. For $p \geq 1$ we generalize Section 3.0.1b, defining the *interior product operator* $i(C)$ for C as follows. Let ω be a q-form, $q \geq p$, and let $Z_1, \ldots, Z_p$ be arbitrary vector fields. Then by definition, $i(Z_1 \wedge \cdots \wedge Z_p)\omega$ is the $(q - p)$-form such that
$$[i(Z_1 \wedge \cdots \wedge Z_p)\omega](W_1, \ldots, W_{q-p})$$
$$= \frac{q!}{p!(q-p)!}\omega(Z_1, \ldots, Z_p, W_1, \ldots, W_{q-p}),$$
for all vector fields $W_1, \ldots, W_{q-p}$. $i(C)\omega$ is then defined for a general C by forcing $i(C)\omega$ to be linear in C with respect to C^∞ functions. For $p = 0$, that is, C is a function—we define $i(C)\omega = C\omega$.

Let C be a $(p, 0)$-tensor field, $p \geq 1$. The *divergence of* C, denoted by div C, is the $(p - 1, 0)$ tensor field obtained from C by contracting the last two variables of DC. Equivalently, if $\{X_i\}$ is a basis of vector fields in $\mathcal{U} \subset M$ and $\{\omega^i\}$ is the dual basis, then
$$\operatorname{div} C(\psi^1, \ldots, \psi^{p-1}) = DC(\psi^1, \ldots, \psi^{p-1}, \omega^j, X_j),$$
for all 1-forms $\psi^1, \ldots, \psi^{p-1}$ in $\mathcal{U}$. From the definition of div C, it follows that if C is skew-symmetric, so is div C.

Since M happens to be orientable, there is an alternate description of div C available when C is skew-symmetric (cf. Section 3.0.2). Let Ω be the metric volume element. We claim: If C is a skew-symmetric $(p, 0)$ tensor field, $p \geq 1$, then $d(i(C)\Omega) = i(\operatorname{div} C)\Omega$.

Proof. We will verify this equality at an arbitrary $x \in M$. Choose local coordinates $\{x^1, \ldots, x^4\}$ so that $\{\partial_i\}$ is normal at x. Since both sides of the inequality are linear in C over $\mathbb{R}$, it suffices to check the equality for $C = f\partial_{i_1} \wedge \cdots \wedge \partial_{i_p}$ where $i_1 < \cdots < i_p$ and f is some C^∞ function in this coordinate neighborhood.

From this point on, the proof is a silly computation involving the combinatorial properties of permutations on the indices $i_1, \ldots, i_p$. It would therefore be more enlightening if we illustrate this procedure by a special case. Thus let $p = 2$ and assume for definiteness that $C = f\partial_1 \wedge \partial_2$. By definition of exterior product (cf. Section 0.0.5d and Bishop–Goldberg 2.18), we have

$$C = \tfrac{1}{2} f(\partial_1 \otimes \partial_2 - \partial_2 \otimes \partial_1).$$

In view of the normality of $\{\partial_i\}$ at x and the formulas in Section 3.0.1e,

$$(\text{div } C)x = [\tfrac{1}{2}(\partial_2 f)\partial_1 - \tfrac{1}{2}(\partial_1 f)\partial_2]x.$$

Now $\Omega x = (dx^1 \wedge \cdots \wedge dx^4)x$, so

$$i(\text{div } C)\Omega = \tfrac{1}{2}[(\partial_2 f)dx^2 \wedge dx^3 \wedge dx^4 + (\partial_1 f)dx^1 \wedge dx^3 \wedge dx^4]x.$$

On the other hand,

$$[i(C)\Omega]x = \tfrac{1}{2}(fdx^3 \wedge dx^4)x$$

so that

$$d(i(C)\Omega)x = \tfrac{1}{2}([(\partial_i f)dx^i] \wedge dx^3 \wedge dx^4)x$$
$$= \tfrac{1}{2}[(\partial_1 f)dx^1 \wedge dx^3 \wedge dx^4 + (\partial_2 f)dx^2 \wedge dx^3 \wedge dx^4]x. \quad \square$$

3.6.3 Killing vector fields

A vector field X on M is a *Killing vector field* iff $L_X g = 0$.

(a) X is Killing iff each member of its flow is locally an isometry (Bishop–Goldberg 5.11.5). In particular, a one-parameter group of isometries always induces a Killing vector field (Bishop–Goldberg 3.5).

(b) X is Killing iff $g([X, W], Z) + g(W, [X, Z]) = Xg(W, Z)$ for all vector fields W, Z.

This is an immediate consequence of the fact that $L_X g = 0$ and the fact that L_X is a derivation of tensor fields commuting with contraction.

(c) X is Killing iff $g(D_W X, Z) + g(W, D_Z X) = 0 \ \forall W, Z \in M_x, \forall x \in M$.

This follows from (b) and the fact that if W, Z are vector fields which extend W, Z to a neighborhood, then $D_X W - D_W X - [X, W] = 0 = D_X Z - D_Z X - [X, Z]$ and $Xg(W, Z) = g(D_X W, Z) + g(W, D_X Z)$.

(d) X parallel $\Rightarrow X$ Killing. This follows from (c) (cf. Exercise 2.3.12).

(e) If $\gamma: \mathscr{E} \to M$ is a geodesic and X is Killing, then $g(\gamma_*, X)$ is a constant.

Indeed, according to Proposition 2.2.2, $\forall u_0 \in \mathscr{E}$, $[(d/du)g(\gamma_*, X)]u_0 = g(D_{\gamma_* u_0}\gamma_*, X) + g(\gamma_*, D_{\gamma_* u_0}X) = g(\gamma_*, D_{\gamma_* u_0}X) = 0$, where the last equality is by (c).

3 Electromagnetism and matter

EXERCISE 3.6.4

(a) Let T be a (p, q)-tensor field on M with $p > 0$. Let the components of T be $T^{i_1\cdots i_p}{}_{j_1\ldots j_q}$. We define div T to be the $(p - 1, q)$-tensor field given in components by

$$(\text{div } T)^{i_1\cdots i_{p-1}}{}_{j_1\ldots j_q} = T^{i_1\cdots i_{p-1}k}{}_{j_1\ldots j_q | k}.$$

Show that for $q = 0$, this coincides with the definition given in Section 3.6.2.

(b) Show from the second Bianchi identity (Section 3.0.1d) that if $\hat{G}$ is the $(2, 0)$ tensor field physically equivalent to the Einstein tensor G, then div $\hat{G} = 0$.

(c) Let C be a $(p, 0)$-skew-symmetric tensor field. If $p \geq 2$, show div div $C = 0$.

Suppose ω is a p-form, $p \geq 1$. Let $\tilde{\omega}$ be the $(1, p - 1)$-tensor field physically equivalent to ω. Then div $\tilde{\omega}$ is equal to $\delta\omega$ (up to a universal constant), where δ is the co-differential of Hodge theory which lowers the degree of forms by 1.

(d) Suppose that $\forall A \in (1, 2)$, η_A is a function $M \to \mathbb{R}^4$ and P_A is a vector field. Use the definitions, the $\mathbb{R}$ linearity and Leibnitz properties of a covariant derivative and (a) to show:

$$\text{div } (\eta_1 P_1 \otimes P_1 + \eta_2 P_2 \otimes P_2)$$
$$= [\text{div } (\eta_1 P_1)]P_1 + [\text{div } (\eta_2 P_2)]P_2 + \eta_1 D_{P_1} P_1 + \eta_2 D_{P_2} P_2.$$

EXERCISE 3.6.5

Let $Z = Z^i X_i$ be a vector field on $\mathcal{U} \subset M$, where $\{X_i\}$ is a basis of vector fields on $\mathcal{U}$; thus (Section 3.6.1f) $Z_i = g_{ij} Z^j$. Show Z is Killing iff $Z_{i|j} + Z_{j|i} = 0$.

EXERCISE 3.6.6

Let X be a Killing vector field. Show that if X vanishes on an open set, it vanishes identically.

EXERCISE 3.6.7

Let $\gamma: \mathscr{E} \to M$ be a geodesic and let X be a Killing vector field on M. Writing $W = X \circ \gamma$, prove that $D_{\gamma_*}{}^2 W = R_{\gamma_* W} \gamma_*$, where R is the curvature tensor of M (cf. Proposition 2.3.3.).

EXERCISE 3.6.8

Show: (a) In local coordinates ∂_1 is Killing iff $\partial_1 g_{ij} = 0$. (b) On Minkowski space ∂_i is Killing $\forall i$. (c) On Einstein–de Sitter spacetime ∂_μ is Killing $\forall \mu \in (1, 2, 3)$. (d) On a normal Schwarzschild spacetime or Schwarzschild black hole, $\partial/\partial t$ is Killing; moreover, $\forall$ Killing vector field on the ordinary 2-sphere (Section 0.0.9), there exists a "corresponding" spacetime Killing vector field. (e) In (b) and (c) there are Killing vector fields that are $\mathbb{R}$-linearly independent of the ones mentioned (hint: consider rotations). (f) (Hard) In (d) every Killing vector field is $\mathbb{R}$-linearly dependent on those mentioned.

3.7 Maxwell's equations

EXERCISE 3.6.9

Show that if X and Y are Killing vector fields, then so is $[X, Y]$.

3.7 Maxwell's equations

We can now analyze the influence of spacetime M and of matter on electromagnetism. Thus let $(M, \mathcal{M}, F)$ be a relativistic model (Section 3.5), J be the charge-current density 3.5.1 of the matter model $\mathcal{M}$. We use the notation of Section 3.0.3 for the electromagnetic field F. The basic laws for the influence of M and $\mathcal{M}$ on F are as follows.

Definition 3.7.1. $(M, \mathcal{M}, F)$ *obeys Maxwell's equations* iff: (a) F is closed—that is, $dF = 0$; and (b) div $\hat{F} = 4\pi J$. One then calls J the *source of F* and also says the triple (M, F, J) *obeys Maxwell's equations*.

Historically speaking Definitions 3.7.1a–b replace and unify the four classical Maxwell equations: the Biot–Savart–Ampere–Maxwell law $\vec{\nabla} \times \vec{B} = -\vec{j} + (1/c)(\partial \vec{E}/\partial t)$ for the magnetic field $\vec{B}$ generated by currents and displacement currents and three further equations. Readers already familiar with the classical versions may find it useful to see, from Section 9.1, how they relate to the simpler equations in Definition 3.7.1.

Sometimes one can regard M and J as given *ab initio*, by neglecting the influence of F on M and on $\mathcal{M}$. Then Maxwell's equations become conditions that help determine F. We give two examples.

EXAMPLE 3.7.2. A CONSTANT MAGNETIC FIELD. In Example 3.4.3 set E = 0. Thus $F = 2B du^3 \wedge du^1$ is an electromagnetic field on Minkowski space $(\mathbb{R}^4, g)$, B is a function on $\mathbb{R}^4$, and the electric field in the covariant constant (parallel, "inertial") reference frame ∂_4 is everywhere zero. $dF = 0$ iff $d\text{B} \wedge du^3 \wedge du^1 = 0$ iff $\partial_4 \text{B} = 0 = \partial_2 \text{B}$. Suppose $dF = 0$. Then (M, F, J) obeys Maxwell's equation for zero source J iff div $\hat{F} = 0$ iff $d[i(\hat{F})\Omega] = 0$ iff $d(\text{B} du^2 \wedge du^4) = 0$ iff $\partial_3 \text{B} = 0 = \partial_1 \text{B}$ iff there is a constant $B_0 \in \mathbb{R}$ such that $F = B_0 du^3 \wedge du^1$ iff F is parallel iff the magnetic field $= \text{B} \partial_2$ in the reference frame ∂_4 is covariant constant, specifically $\boldsymbol{b} = B_0 \partial_2$.

In physics notation, $\vec{B} = B_0 \vec{e}_y$ = constant (Section 9.1).

Such constant magnetic fields are very useful in physics. Roughly, the reasons are the following. Suppose we have any electromagnetic F on any spacetime M. Near $z \in M$ we can often approximate M by Minkowski spacetime, and approximate F by a parallel (covariant constant) tensor, in the sense of Taylor series expansion. Moreover the observed electric vector is often negligible since opposite electric charges have a strong tendency to cancel out each other's electric effects by coming close together. When all these idealizations are appropriate

$$F = 2[B^3 du^1 \wedge du^2 + B^1 du^2 \wedge du^3 + B^2 du^3 \wedge du^1], \qquad B^\alpha \in \mathbb{R}$$

(cf. Section 9.1). Choosing appropriate "spatial" axes now leads to $F = B_0 du^3 \wedge du^1$ as in the example.

3 Electromagnetism and matter

EXAMPLE 3.7.3. WAVES. Let $(\mathbb{R}^4, g)$ be Minkowski space. Suppose we have a situation that observers in the (parallel) reference frame ∂_4 would describe, intuitively, as follows. Near the origin of 3-space are some electric charges that move back and forth in the ∂_1 direction of 3-space. An electromagnetic field is generated. At points on or near the u^3 axis of 3-space, very far from the charges, where the wavefronts are nearly plane, the electromagnetic field is observed. We now give an electromagnetic field appropriate in the observation region ("wave zone"); it is not appropriate at or near the charges. Let $f: \mathbb{R} \to \mathbb{R}$ be C^∞ and not identically zero. Define $\phi = (u^3 - u^4): \mathbb{R}^4 \to \mathbb{R}$ and $F = 2(f \circ \phi)d\phi \wedge du^1$. Thus F is an electromagnetic field (Section 3.4). We now show $(\mathbb{R}^4, F, 0)$ obeys Maxwell's equations and indicate why F is called a *plane, linearly polarized electromagnetic wave* on Minkowski space.

More generally, this term is applied to a 2-form F' on Minkowski space iff there is an isometry $\psi: \mathbb{R}^4 \to \mathbb{R}^4$ such that $\psi^* F' = F$, F as above.

A special case, which may be familiar to some readers, is $fu = A \sin \omega u$, $A, \omega \in (0, \infty)$. Introducing the electric and magnetic fields (Section 3.4) and using physics notation (Section 9.0) this gives $\vec{E} = A\vec{e}_x \sin \omega(z - ct)$, $\vec{B} = A\vec{e}_y \sin \omega(z - ct)$. Then A is called the *amplitude*, and ω the *angular frequency* of the wave, with the parallel (inertial) reference frame ∂_4 understood.

In fact we have $dF = (f' \circ \phi)d\phi \wedge d\phi \wedge du^1 = 0$. Moreover, $\hat{F} = 2(f \circ \phi) \times (\partial_3 + \partial_4) \wedge \partial_1$. Our definition 3.6 of div thus gives div $\hat{F} = \partial_1(f \circ \phi) - (\partial_3 + \partial_4)(f \circ \phi) = 0 - 0 = 0$. Thus $(\mathbb{R}^4, F, J)$ obeys Maxwell's equations with source $J = 0$.

By the definitions in Example 3.4.3, the electric and magnetic fields in the reference frame ∂_4 are, respectively, $e = f(u^3 - u^4)\partial_1$ and $b = f(u^3 - u^4)\partial_2$. Neither depends on u^1 or u^2; hence the term "plane." e is everywhere proportional to the covariant constant vector field ∂_1; hence the term "linearly polarized." f is an arbitrary function on $\mathbb{R}$—it might be sinusoidal, for example—hence the term "wave." Note that ∂_3, e, and b are pairwise orthogonal and $|e| = |b|$, a behavior sketched in elementary physics texts.

The observers in ∂_4 regard the wave as travelling in the ∂_3 direction. To see roughly why, suppose that $u_0 \in \mathbb{R}$, that $fu_0 > 0$, and that f has a local maximum at u_0. Then $|e| = |f \circ \phi|$ will have a local maximum at the points for which $u^3 = u^4 + u_0$. In the (u^3, u^4) plane these points form a line along which $du^3/du^4 = 1 =$ speed of light. The same argument applies to any other identifiable feature of f—say a point of inflection. In this sense the observers in ∂_4 see the wave pattern moving at the speed of light in the ∂_3 direction. If f is sinusoidal, the pattern is sinusoidal in u^4 for fixed u^3 and sinusoidal in u^3 for fixed u^4.

By a nasty calculation the reader can check that analogous results are obtained by using any other parallel reference frame on $(\mathbb{R}^4, g, D)$. The reference-frame-independent structures of main interest are: the span of the lightlike vector field $Y = \partial_3 + \partial_4$; the lightlike level

3.7 Maxwell's equations

hypersurfaces of ϕ; and the "spacetime polarization plane" $Y \wedge \partial_1$, which determines a two-dimensional lightlike subspace in each tangent space.

We now introduce the *stress-energy tensor* E *of an electromagnetic field* F on M. E is by definition the $(0, 2)$-tensor field on M whose components are given by $E_{ij} = (1/4\pi)[F_{im}F_j{}^m - (1/4)g_{ij}F^{mn}F_{mn}]$ (Section 3.6.1a,b,f). Once we have investigated the influence of electromagnetism on matter we shall show how E helps one keep track, roughly speaking, of the way energy flows back and forth between matter and F. For the moment we deal only with the formal properties of E. Notation is as in Sections 3.0.3 and 3.6.1.

Proposition 3.7.4. (a) $\hat{E}$ *is symmetric and trace* $\hat{E} = 0$. (b) $\hat{E}(\omega, \omega) \geq 0$ *for every causal 1-form* ω. (c) *If* (M, F, J) *obeys Maxwell's equations, then* $\text{div } \hat{E} = -\tilde{F}J$.

PROOF. (a) is straightforward. For (b), we first show that $\hat{E}(\omega, \omega) \geq 0 \; \forall$ timelike 1-form ω. Replacing ω by $\omega/|\omega|$ if necessary, we may assume that $|\omega| = 1$. Locally, choose an orthonormal basis of 1-forms $\{\omega^i\}$ so that $\omega^4 = \omega$. Observe that relative to $\{\omega^i\}$, $g_{\alpha\beta} = g^{\alpha\beta} = \delta_{\alpha\beta}$, $g_{\alpha 4} = g^{\alpha 4} = 0$ for $1 \leq \alpha, \beta \leq 3$, and $g_{44} = g^{44} = -1$ (Section 3.6.1b). Hence $F^{\alpha\beta} = F_{\alpha\beta} = F^\alpha{}_\beta = F_\alpha{}^\beta$ for $1 \leq \alpha, \beta \leq 3$, while $F^{4i} = -F^i{}_4$ and $F_{4i} = -F^4{}_i$. In the following argument, we let α, β, γ run from 1 to 3 and such a repeated Greek index (even if both are superscripts or subscripts) will imply summation from 1 to 3. The skew-symmetry of F will be used without comment.

$$4\pi \hat{E}(\omega, \omega) = 4\pi E^{44} = F^{4n}F^4{}_n + \tfrac{1}{4}F^{mn}F_{mn}$$
$$= F^{4\alpha}F^4{}_\alpha + \tfrac{1}{4}(F^{4n}F_{4n} + F^{\alpha n}F_{\alpha n})$$
$$= F^{4\alpha}F^{4\alpha} + \tfrac{1}{4}(F^{4\alpha}F_{4\alpha} + F^{n\alpha}F_{n\alpha})$$
$$= F^{4\alpha}F^{4\alpha} + \tfrac{1}{4}(-F^{4\alpha}F^{4\alpha} + F^{4\alpha}F_{4\alpha} + F^{\beta\alpha}F_{\beta\alpha})$$
$$= F^{4\alpha}F^{4\alpha} + \tfrac{1}{4}(-F^{4\alpha}F^{4\alpha} - F^{4\alpha}F^{4\alpha} + F^{\beta\alpha}F^{\beta\alpha})$$
$$= \tfrac{1}{2}F^{4\alpha}F^{4\alpha} + \tfrac{1}{4}F^{\beta\alpha}F^{\beta\alpha} \geq 0.$$

Thus (b) is proved if ω is timelike. To prove (b) in general, it suffices to show: if $\omega \in M_x^*$ and ω is lightlike, then $\hat{E}(\omega, \omega) \geq 0$. There exists a sequence of timelike ω_i in M_x^* such that ω_i converges to ω as $i \to \infty$ (compare Section 1.2). We already know that $\forall i$, $\hat{E}(\omega_i, \omega_i) \geq 0$. Thus

$$\hat{E}(\omega, \omega) = \lim_{i \to \infty} \hat{E}(\omega_i, \omega_i) \geq 0.$$

Finally, we prove (c). We have to show $E^{ij}{}_{|j} = -F^j{}_m J^m$. By Section 3.6.1j, $dF = 0 \Leftrightarrow F_{ij|k} + F_{jk|i} + F_{ki|j} = 0$. Moreover, $\text{div } \hat{F} = 4\pi J \Leftrightarrow F^j{}_{m|j} = -4\pi J_m$ (Exercise 3.6.4a). Thus, using $g^{ij}{}_{|k} = 0$, we have:

$$4\pi E^{ij}{}_{|j} = (F^j{}_m F^{im})_{|j} - \tfrac{1}{4}(g^{ij}F^{mn}F_{mn})_{|j}$$
$$= F^j{}_{m|j}F^{im} + F^j{}_m F^{im}{}_{|j} - \tfrac{1}{4}g^{ij}F^{mn}{}_{|j}F_{mn} - \tfrac{1}{4}g^{ij}F^{mn}F_{mn|j}$$
$$= -4\pi J_m F^{im} + F^{im}{}_{|j}F^j{}_m - \tfrac{1}{2}g^{ij}F^{mn}F_{mn|j}$$
$$= -4\pi F^{im}J_m + g^{ik}F_{km|j}F^{jm} - \tfrac{1}{2}g^{ik}F_{mj|k}F^{mj}.$$

3 Electromagnetism and matter

Now, $g^{ik}F_{km|j}F^{jm} = \frac{1}{2}g^{ik}F_{km|j}(F^{jm} - F^{mj}) = -\frac{1}{2}g^{ik}F^{mj}(F_{km|j} - F_{kj|m})$.
Thus,

$$E^{ij}{}_{|j} = \frac{1}{4\pi}\{-4\pi F^{im}J_m - \frac{1}{2}g^{ik}F^{mj}(F_{km|j} - F_{kj|m} + F_{mj|k})\} = -F^i{}_m J^m. \quad \Box$$

By Proposition 3.7.4a,b, $\hat{E}$ is a stress-energy tensor (Section 3.3.1) on M. It is defined as the *stress-energy tensor of* the electromagnetic field F.

> The reader may check from Section 3.4 and the proof just given that ∀ instantaneous observer (z, Z) the measured energy density $E(Z, Z)$ is $(1/8\pi)(|E|^2 + |B|^2)$, where E and B are the electric and magnetic vectors (z, Z) measures. In fact E unifies and replaces the classical energy density $(1/8\pi)(\vec{E}^2 + \vec{B}^2)$, Poynting vector $(1/4\pi)\vec{E} \times \vec{B}$ and Maxwell stress tensor (Section 9.1).

There are elegant integral versions of Maxwell's equations, as follows. Let J be a vector field on M, Ω be the metric volume element on M (Section 3.0.2), and F be a 2-form on M.

Proposition 3.7.5. (M, F, J) *obeys Maxwell's equations iff:* (a) $\int_{\partial \mathscr{D}} i(\hat{F})\Omega = 4\pi \int_{\mathscr{D}} i(J)\Omega$ *for every space-section* $\mathscr{D}$; *and* (b) $\int_{\partial \mathscr{D}} F = 0$ *for every space-section* $\mathscr{D}$. *In that case:* (c) div $J = 0$ *and, equivalently,* $\sum_{\mu=1}^{3} \int_{\mathscr{B}_\mu} i(J)\Omega = 0$ *for every causal box (Section 3.0.1).*

PROOF. div $\hat{F} = 4\pi J$ iff $d[i(\hat{F})\Omega] = 4\pi i(J)\Omega$ iff (a) holds; $dF = 0$ iff (b) holds (Sections 3.0.1 and 3.6.2). Moreover, if div $\hat{F} = 4\pi J$ then

$$\text{div } J = (1/4\pi) \text{ div div } F = 0$$

(Exercise 3.6.4c); by Section 3.0 div $J = 0$ is equivalent to the integral conservation law stated. $\quad \Box$

When J is any charge-current density and $\mathscr{D}$ is a space-section $\int_{\mathscr{D}} i(J)\Omega$ is defined as the *electric charge in* $\mathscr{D}$ for J (cf. Section 3.2). (a) Says that when Maxwell's equations hold, the electromagnetic field cleverly registers on the boundary of $\mathscr{D}$ the electric charge in $\mathscr{D}$. The differential and integral conservation laws in (c) are interpreted as conservation of electric charge.

> One pre-relativistic analogue of (a) is Gauss' law of electric flux $Q = \oint \vec{E} \cdot d\vec{A}$ ("measure electric charge in a volume by counting electric flux lines through the boundary 2-surface"); one such analogue of (c) states "the electric charge in a volume at a later time is that at an earlier time minus what has been carried through the surface"; see Section 9.1 for these and other analogues of a–c. Because of the similarity between (a) and (b), $dF = 0$ is often referred to as the hypothesis that there are no magnetic monopoles. Recent experiments suggest this hypothesis may possibly need modification.

3.7 Maxwell's equations

EXERCISE 3.7.6

Let (M, g) be Minkowski space with the u^4-axis deleted. Let $r: M \to (0, \infty)$ be defined by $r = (\sum_{\nu=1}^{3} u^\nu u^\nu)^{1/2}$, and, for $e \in \mathbb{R}$, define $F = (4\pi)^{-1} e r^{-2} dr \wedge du^4$. (a) Show that $(M, F, 0)$ obeys Maxwell's equations. (b) Let $\mathscr{C} \subset M$ be an imbedded submanifold homeomorphic to the 2-sphere. Show that if $\mathscr{C}$ is contractible, $\int_\mathscr{C} i(\hat{F})\Omega = 0$. (c) Give an example of a $\mathscr{C}$ such that $\int_\mathscr{C} i(\hat{F})\Omega = e$. (d) Give an example of an instantaneous observer who observes a nonzero magnetic vector. (e) Show that no instantaneous observer observes zero electric vector.

F in Exercise 3.7.6 is the electric field of a point charge at rest; the deleted u^4-axis corresponds to the history of the point charge. (d) Shows that "an observer moving in an electric field observes a magnetic field."

EXERCISE 3.7.7

(a) Show that for the plane wave F of Example 3.7.3, $\hat{E} = \frac{1}{4\pi}(f \circ \phi)^2 Y \otimes Y$, where $Y = \partial_3 + \partial_4$.

Note from Section 3.3 that the stress-energy tensor of a photon beam likewise has the form $\hat{T} = \eta Y \otimes Y$, where η is nonnegative and Y is lightlike. Electromagnetic waves and photon beams are intuitively related, mathematically independent models for light. However, the intuitive interrelation involves quantum theory, so we shall not discuss it in detail.

EXERCISE 3.7.8

Let (M, g) and (M, g') be spacetimes. They are *conformal* iff $g' = fg$ for some C^∞ function $f: M \to (0, \infty)$. (a) Suppose $((M, g), F, 0)$ obeys Maxwell's equations and (M, g') is conformal to (M, g). Show $((M, g'), F, 0)$ obeys Maxwell's equations. (b) Let (M, g) be Einstein–de Sitter spacetime. By considering $u = 3(u^4)^{1/3}$, show (M, g) is conformal to the open submanifold $\mathbb{R}^3 \times (0, \infty)$ of Minkowski space with $f = (u^4)^{4/3}$. (c) Use (a), (b), and Example 3.7.3 to find an F on Einstein–de Sitter spacetime (M, g) such that $(M, F, 0)$ obeys Maxwell's equations. Show that then the energy density $E(Z, Z)$ a comoving instantaneous observer (z, Z) measures is small (for a given value of f in Example 3.7.3) when the "cosmological time" $u^4 z$ is large.

Such waves have apparently been wandering around the universe since very early times, and observing them is our main empirical handle on conditions near a big bang. (c) says, roughly, that the wave gets weaker as the universe expands. In our detailed discussion (Chapter 6) of these effects we will use photons, rather than waves, but a wave model gives equivalent results; compare the preceding fine print comment.

EXERCISE 3.7.9

Suppose $\hat{E}(\omega, \omega) = 0 \, \forall$ causal $\omega \in M_z$. Show $\hat{E}z = 0 = Fz$.

3.8 Particle dynamics

In nonquantum relativity there are three basic laws for particles: the Lorentz world-force law and two conservation laws. We now discuss them.

Throughout the section F is a given electromagnetic field on spacetime M and the conventions in Section 3.0.3 are used. Thus if $\gamma\colon \mathscr{E} \to M$ is a particle (Section 3.1), $\tilde{F}\gamma_*$ is a vector field over γ with $\tilde{F}\gamma_* u \in (\gamma_* u)^\perp$ $\forall u \in \mathscr{E}$. Since then $g(\gamma_*, \gamma_*) = -m^2$, we also have $(D_{\gamma_*}\gamma_*)(u) \in (\gamma_* u)^\perp$. The influence of (M, g, D) and F on γ is given by the following replacement for Newton's $\vec{F} = m\vec{a}$.

Definition 3.8.1. A particle (γ, m, e) on M *obeys the Lorentz world-force law with respect to* F iff $e\tilde{F}\gamma_* = D_{\gamma_*}\gamma_*$.

The simplest, and physically most fundamental, way to interpret the law is by using it directly, as in Proposition 3.8.2, Theorem 3.8.3, and so on. But, historically speaking, the law incorporates the effects of electric, magnetic, and gravitational forces. To see roughly what was involved, suppose γ obeys the law and assume for brevity the rest-mass of γ is one. Then γ is an observer (Section 2.1) with world-acceleration $D_{\gamma_*}\gamma_*$. Moreover, $\forall u \in \mathscr{E}$, the electric vector the instantaneous observer $(\gamma u, \gamma_* u)$ measures is $E = \tilde{F}\gamma_* u$; thus $e\tilde{F}\gamma_*$ in Definition 3.8.1 corresponds to the electric Coulomb force eE of elementary physics. Now one can (but need not) rewrite this Coulomb force eE in terms of the electric and magnetic vectors some other instantaneous observer $(\gamma u, Z)$ at γu measures. It then takes one a more complicated form, first pointed out by H. A. Lorentz in 1897 (and written out explicitly in Section 9.2.4). Thus the law reads: (Coulomb–Lorentz force) = (unit mass) (world-acceleration). Hence the term "Lorentz world-force law." Finally, recall that gravity is built in automatically, mainly via D. For example, if $e = 0$ the particle is freely falling.

EXAMPLE 3.8.2. The following special case of the Lorentz world-force law is used in discussing nuclear magnetic moments, radio signals from quasars, and many other processes. It is thus relevant from nuclear physics to cosmology—from lengths of 10^{-24} seconds to those of 10^{18} seconds (42 powers of 10).

Suppose $(\mathbb{R}^4, g)$ is Minkowski space, $B_0 \in (0, \infty)$ and $F = 2B_0 du^1 \wedge du^2$. Then $(M, F, 0)$ is a key example of an (M, F, J), which obeys Maxwell's equation (Example 3.7.2). Let (γ, m, e) be a particle on $\mathbb{R}^4$ with nonzero rest-mass m. Define $\omega = (eB_0/m) \in \mathbb{R}$.

Proposition 3.8.2.a $\gamma\colon \mathscr{E} \to \mathbb{R}^4$ *obeys the Lorentz world-force law with respect to* F *iff there is a* $y \in \mathbb{R}^4$ *and* $a, b, \phi \in \mathbb{R}$ *such that* $\forall u \in \mathscr{E}$,

$$\gamma u = y + (a \sin[\omega m u + \phi], a \cos[\omega m u + \phi], bmu, |1 + a^2\omega^2 + b^2|^{1/2} mu).$$

3.8 Particle dynamics

PROOF. Abbreviate $d(u^i \circ \gamma)/du$ by $\dot{\gamma}^i$ and $d^2(u^i \circ \gamma)/du^2$ by $\ddot{\gamma}^i$. Substituting $g = \sum_{\mu=1}^{3} du^\mu \otimes du^\mu - du^4 \otimes du^4$, $F = 2B_0 du^1 \wedge du^2$, and $D_{\partial_i}\partial_j = 0$ into the Lorentz world-force law $D_{\gamma_*}\gamma_* = e\tilde{F}\gamma_*$, we get by algebra:

$$\ddot{\gamma}^3 = \ddot{\gamma}^4 = 0, \qquad \ddot{\gamma}^1 = eB_0\dot{\gamma}^2, \qquad \ddot{\gamma}^2 = -eB_0\dot{\gamma}^1.$$

By elementary integration, using the fact that $m \neq 0$, we obtain: there exist $y \in \mathbb{R}^4$ and $a, b, c, \phi \in \mathbb{R}$ such that

$$\gamma u = y + [a \sin(\omega m u + \phi), a \cos(\omega m u + \phi), bmu, cmu],$$

$\forall u \in \mathscr{E}$. Since $g(\gamma_*, \gamma_*) = -m^2$ is equivalent to

$$(\dot{\gamma}^4)^2 = m^2 + \sum_{\mu=1}^{3} \dot{\gamma}^\mu \dot{\gamma}^\mu,$$

we conclude that $c^2 = 1 + a^2\omega^2 + b^2$. □

To get an intuitive picture of this result, we take $\mathscr{E} = \mathbb{R}$, and tabulate the world-line $\gamma\mathbb{R}$, its projection α into $\{(u^1, u^2, u^3, 0)\}$, and its projection β into $\{(u^1, u^2, 0, 0)\}$.

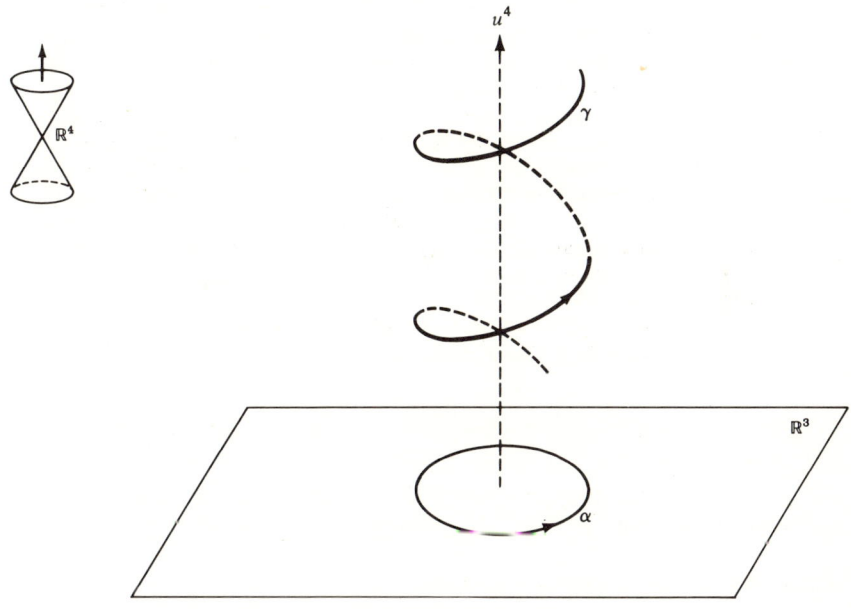

	$a = 0 = b$	$a = 0 \neq b$	$a \neq 0 = b$	$a \neq 0 \neq b$
$\gamma\mathbb{R}$	straight line	straight line	helix	helix
α	point	straight line	circle	helix
β	point	point	circle	circle

3 Electromagnetism and matter

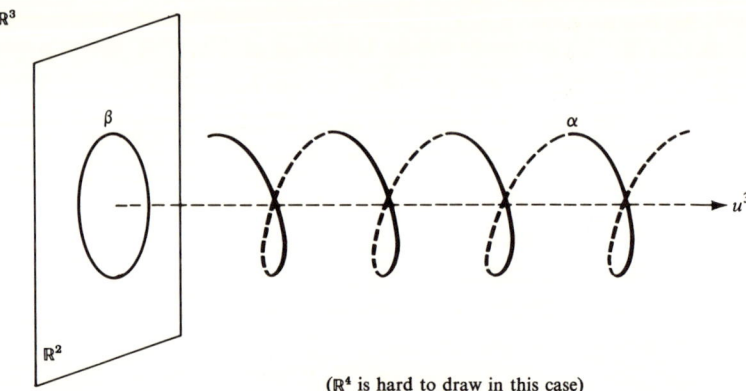

($\mathbb{R}^4$ is hard to draw in this case)

α is the path Newtonian intuition assigns to the particle; in the general case $a \neq 0 \neq b$, one therefore says the particle spirals around the direction u^3 of the magnetic field lines (Example 3.7.2). For example, charged particles in the magnetic field of the earth spiral from pole to pole. Similar motions are also observed in laboratory plasmas and inferred for electrons in a metal subjected to an external magnetic field.

Note that the proper time Δs required to go once around the circle is $\Delta s = m\Delta u = 2\pi/\omega$. $\omega = 2\pi/\Delta s$ is called the (angular) "synchrotron frequency." The term refers to still another application—namely, certain high-energy-particle accelerators that use a magnetic field. Now suppose our radio telescopes receive radio waves from a quasar. It is basically the synchrotron frequency of charge particles moving in a magnetic field near the quasar that determines the (angular) frequency of the received radio waves. However, complicated corrections for the "frequency ratio" (to be defined in Chapter 5) are required to get the actual answers.

The way in which the Lorentz world-force law here interrelates different aspects of nature is satisfying. Perhaps Nature's mottos are "unity in diversity," "less is more." Perhaps her use of Lorentzian metrics and the Lorentz world-force law are prime examples. Perhaps, on the other hand, not. She is every sane human's delight, but no man's whore.

We now give a "present determines the future" result (cf. Proposition 3.2.3). For a particle, the appropriate formulation is to take as given: (M, g, D), a particle type (m, e), an electromagnetic field F on M, an $x \in M$, and a future-pointing $W \in M_x$ such that $g(W, W) = -m^2$.

Theorem 3.8.3. (a) *There is precisely one inextendible curve $\gamma: \mathscr{E} \to M$ such that (1) $\gamma 0 = x$, (2) $\gamma_* 0 = W$, and (3) $D_{\gamma_*}\gamma_* = e\tilde{F}\gamma_*$. (b) $(\gamma, |W|, e)$ is a particle on M. (c) γ depends on W in a C^∞ manner, in the following sense: Let W be a C^∞ vector field defined in a C^∞ imbedded submanifold $\mathscr{N}$ and let γ_y be the inextendible curve of (a) such that $(\gamma_y)_* 0 = Wy$. Then given any $x \in \mathscr{N}$, there exists a neighborhood $\mathscr{U}$ of x in $\mathscr{N}$ and an $\varepsilon > 0$ such that the assignment $(u, y) \to \gamma_y u$ is a well-defined C^∞ mapping: $(-\varepsilon, \varepsilon) \times \mathscr{U} \to M$.*

3.8 Particle dynamics

The neatest proof uses the properties of a second-order differential equation $X: TM \to TTM$ (Bishop–Goldberg p. 246). We give a more elementary version.

Proof. We first prove (a). Let $\{x^i\}$ be coordinate functions around x such that $x^i(x) = 0\ \forall i$, and let $W = W^i(\partial_i x)$, $F = F_{ij} dx^i \otimes dx^j$ ($F_{ij} = -F_{ji}$). If γ is the sought-for curve, let us write: $\gamma^i \equiv x^i \circ \gamma$, $\dot{\gamma}^i \equiv d\gamma^i/du$, $\ddot{\gamma}^i \equiv d^2\gamma^i/du^2$. Then (1)–(3) are, respectively, equivalent to:

(1a) $\gamma^i 0 = 0$,
(2a) $\dot{\gamma}^i 0 = W^i$,
(3a) $\ddot{\gamma}^i + (\Gamma^i{}_{jk} \circ \gamma)\dot{\gamma}^j \dot{\gamma}^k = e(F^i{}_j \circ \gamma)\dot{\gamma}^j$,

where the $\Gamma^i{}_{jk}$ are the connection coefficients defined by

$$D_{\partial_j} \partial_k = \Gamma^i{}_{jk} \partial_i.$$

(3.6.1e). (3a) is a system of ordinary differential equations satisfied by $\{\gamma^i\}$ with prescribed initial conditions (1a) and (2a). Thus both local existence and uniqueness are guaranteed by the basic theorem of such equations. Equivalently, local existence and uniqueness of a γ satisfying (1)–(3) of (a) are now established. The existence of an inextendible such γ follows by standard arguments using Zorn's lemma.

We next prove (b). From (a) we know $|\gamma_* 0| = |W|$. It suffices then to show that

$$\frac{d}{dt} g(\gamma_*, \gamma_*) = 0.$$

The left side equals

$$g(D_{\gamma_*}\gamma_*, \gamma_*) + g(\gamma_*, D_{\gamma_*}\gamma_*) = 2g(D_{\gamma_*}\gamma_*, \gamma_*) = 2g(e\tilde{F}\gamma_*, \gamma_*)$$
$$= 2eF(\gamma_*, \gamma_*) = 0,$$

where we have used (3) of (a) and the skew-symmetry of F. This proves (b).

(c) follows from the proof of (a) together with the observation that solutions of ordinary differential equations are known to depend on their initial conditions in a C^∞ manner. $\square$

In addition to being kicked around by M and F, particles also influence each other, in collisions. Even the emission of light by your lamp can be regarded as due to collisions in which photons are created. We now give the appropriate definitions.

Suppose, in pool, a cue ball hits the eight ball. In the simplest case, each just flies off in a new direction. But one can also have situations where they blow each other to smithereens. The laws we shall give apply equally to both possibilities. In order to state the laws in full generality one takes the following view toward the simple case. At the collision point the incoming balls are both destroyed, a new cue ball is created, and a new eight ball is created. Then the smithereens case is wholly similar except that what is created are the fragments. Our laws will imply, *inter alia*, that at a spacetime point a particle cannot simply appear (or disappear) without any other particles being around. However, they will not preclude particles from sneaking into or out of spacetime "at the edges" (cf. Example 1.4.3).

3 Electromagnetism and matter

Let $\gamma: \mathscr{E} \to M$ be a future-pointing curve and set $b = \text{lub } \mathscr{E}$, with $-\infty < b \leq \infty$. $x \in M$ is a *future endpoint of* γ iff γu approaches x as u approaches b. Since M is Hausdorff, γ has, at most, one future endpoint. Even if b is finite it may have none. *Past endpoints* are defined dually. γ is *endless* iff it has neither a future nor past endpoint.

Let $\mathscr{P}$ be a finite collection $\{(\gamma, m, e)\}$ of particles. To avoid the physically uninteresting possibility of particles that "wiggle themselves to death" assume that each $\gamma \in \mathscr{P}$ that has a future endpoint also contains it—that is, if $\gamma: \mathscr{E} \to M$ has a future endpoint then lub $\mathscr{E} \in \mathscr{E}$, in particular lub $\mathscr{E} < \infty$. Assume the dual condition on past endpoints holds. Then $x \in M$ is defined as a *collision event* for $\mathscr{P}$ iff x is an endpoint, future or past, for some $\gamma \in \mathscr{P}$. Suppose x is a collision event for $\mathscr{P}$. $\gamma \in \mathscr{P}$ is *destroyed* ($\equiv$ *incoming*) *at* x iff x is the future endpoint of γ. In that case $\gamma b = x$ for $b = \max \mathscr{E}$, and $\gamma_* b$ is well-defined. Similarly, $\gamma \in \mathscr{P}$ is *created* ($\equiv$ *outgoing*) *at* x iff x is the past endpoint of γ and then $\gamma_* a$, $a = \text{glb } \mathscr{E}$, is well defined. Given x, we denote the subset of particles destroyed at x by $\text{in}(x) \subset \mathscr{P}$ and define $\text{out}(x)$ dually.

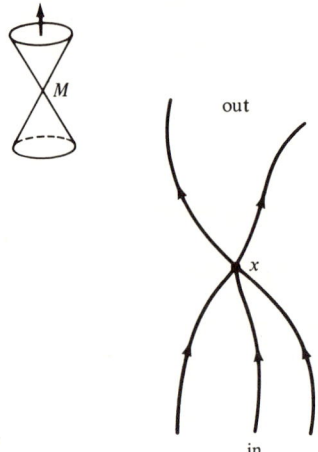

Definition 3.8.4. A finite collection $\mathscr{P} = \{(\gamma, m, e)\}$ of particles on M is *consistent* iff the above restrictions on endpoints hold and for each collision event x for $\mathscr{P}$ the following conditions hold:

(a) *Collision conservation law of energy-momentum*, $\sum_{\text{in}(x)} \gamma_* b = \sum_{\text{out}(x)} \gamma_* a$;
(b) *Collision conservation law of electric charge*, $\sum_{\text{in}(x)} e = \sum_{\text{out}(x)} e$.

The collision conservation law of electric charge is intuitively related to the differential conservation law div $J = 0$ in Section 3.7, but mathematically independent of it.

Newtonian analogue. Definition 3.8.4a is like three Newtonian laws: mass conservation, energy conservation, and momentum conservation. But Definition 3.8.4a is an improvement: it agrees with observations, is observer-independent, and can be applied to rest-mass zero particles. Compare Section 3.1 and Optional exercise 9.2.3.

From a microscopic point of view, a collision is often a complicated process. The conservation laws in Definitions 3.8.4a and b are very important because they lead to predictions even when a collision is imperfectly understood. For example, in 1930–31 collisions were observed that appeared to violate Definition 3.8.4a. Pauli postulated a new kind of particle, the neutrino, which could balance Definition 3.8.4a as an outgoing particle. More than twenty-five years later, neutrinos were first observed "directly,"—that is, as incoming particles.

EXAMPLE 3.8.5. A collision with only one incoming particle is a *decay*. The simplest interesting collision is the decay of a pi-nought meson into two photons. One particle $(\pi, m, 0)$ is incoming, with $m = 273$ electron rest-masses. Two photons α and β are outgoing. In the following, $z \in M$ is the collision event and all vectors lie in M_z. We claim: Definitions 3.8.4a and b hold iff there is a spacelike unit vector $W \in M_z$ such that $g(\pi_*, W) = 0$, $\alpha_* = (\pi_*/2) + (mW/2)$, and $\beta_* = (\pi_*/2) - (mW/2)$.

The "if" assertion is trivial. To show the converse, choose an instantaneous observer (z, Z) with $Z \in \text{span } \pi_*$. Let e_π, e_α, e_β and $\mathbf{p}_\pi, \mathbf{p}_\alpha, \mathbf{p}_\beta$ be the respective energies and 3-momenta observed by (z, Z); note that $Z \in \text{span } \pi_* \Leftrightarrow \mathbf{p}_\pi = 0$. Algebra shows that Definition 3.8.4a and b hold iff $e_\pi = m = e_\alpha + e_\beta$, $e_\alpha = |\mathbf{p}_\alpha| > 0$; $e_\beta = |\mathbf{p}_\beta| > 0$, $\mathbf{p}_\alpha = -\mathbf{p}_\beta$. More algebra gives $e_\alpha = m/2 = e_\beta$, and then the general solution claimed.

> In practice, one usually regards the incoming energy momentum π_* as known and wants to solve for α_* and β_*. Since W is not unique, the collision conservation laws do not give a unique answer. They also do not predict various other features; for example, what is the average proper time $\Delta \bar{u}$ a pi-nought meson lasts before decaying as above? Most of the conservation laws we have introduced and shall later introduce are "essentially universal" but, as here, "give only partial information."
>
> To get more detailed information, one must use experiments and/or quantum theory. For example, both observation and quantum theory indicate that the above decay occurs isotropically in the sense that each possible W is equally probable (cf. Section 2.1). Moreover, observations show $\Delta \bar{u} \simeq 10^{-16}$ seconds; quantum theory can be used to estimate this value, but not to calculate it precisely.

3.8.6 Remarks

(a) In our applications all particles introduced will obey the Lorentz worldforce law. The reason is that in relativity it is not useful to introduce forces *ad hoc*, as the "push of a rigid rod" is introduced in Newtonian theory without worrying too much about the physics involved. Unless one sticks to the actual forces found in nature, the danger of producing nonsense—for example, by implicitly assuming influences that can travel faster than light—is too large. But the only known forces in nature are gravity, electromagnetism, and those that require quantum models (Section 0.1). The Lorentz

3 Electromagnetism and matter

world-force incorporates the first two; the two collision conservation laws above incorporate quantum laws to the extent that quantum laws can be systematically incorporated into macrophysics.

(b) The only particle sets of fundamental interest are the consistent ones. Almost every observed process can be modelled, at least roughly, by a set $(M, F, \mathscr{P})$, where $\mathscr{P}$ is a consistent particle set and each particle in $\mathscr{P}$ obeys the Lorentz world-force law with respect to F. In particular, comparing such models to observations supplies the main motivation for using a Lorentzian metric g, an electromagnetic field F, and the Lorentz world-force law; it also indicates how g and F are in principle measured.

> A sufficient (though not necessary) condition for such a model to be useful is that none of the following plays a dominant role in the process being modelled: the (strong) nuclear interactions (Section 0.1.2); the weak interactions (Section 0.1.2); and a quantum effect known as the Pauli exclusion principle (Messiah [1]).
>
> We give an example of how the use of such models relates to the measurement of g and F. Roughly, the idea is the following. ∀ particle type the Lorentz world-force law determines a second-order differential equation $X: TM \to TTM$ (cf. Bishop–Goldberg, p. 246). Imagine, say, that the electric charge is zero. Then one can appeal to a theorem in geometry (Ambrose, Palais, and Singer [1]; Dombrowski [1]) that X determines at most one symmetric affine connection. If X determines a symmetric affine connection, and if that connection comes from at least one Lorentzian metric g, one gains confidence in the theory and in addition has a good handle on the actual g.

EXERCISE 3.8.7

Verify Theorem 3.8.3 for the special case of Example 3.8.2 by exhibiting γ explicitly in terms of x and W.

EXERCISE 3.8.8

To analyze collisions, one often uses a "center of mass frame." Let $\mathscr{P}$ be a consistent particle set, $z \in M$ be a collision event, (z, Z) be an instantaneous observer, and $\{p\}$ be the set of 3-momenta observed by (z, Z). Z is called a *center of mass frame* at z iff $\sum_{\text{in}(z)} p = 0$. (a) Show that there exists no center of mass frame at z iff all $(\gamma, m, e) \in \text{in}(z) \cup \text{out}(z)$ have $m = 0$ and all their energy-momenta at z are proportional to each other. (b) Show that if a center of mass frame exists, it is unique. (c) Show that the center of mass frame could be defined by the alternate condition $\sum_{\text{out}(z)} p = 0$. (d) Suppose $m \neq 0$ for every incoming particle at z; write the center of mass frame Z as a weighted average of $\{\gamma_* b/|\gamma_* b| \mid \gamma \in \text{in}(z)\}$. [The center of mass frame is sometimes also called the *average world-velocity of* in (z).] (e) For (γ, m, e) in in(z) and Z the center of mass frame, let $V \in Z^\perp$ be the particle's Newtonian velocity observed by (z, Z) (cf. Section 2.1.6). It is called the *random 3-velocity of the particle*. Let e be the energy observed by Z. Show $\sum_{\text{in}(z)} e = \sum_{\text{in}(z)} m/(1 - |V|^2)^{1/2} \geq \sum_{\text{in}(z)} m$. (f) Find the center of mass frame in Example 3.8.5.

EXERCISE 3.8.9

Only in one special case can the concept of particle energy be extended to include what might be called gravitational and electromagnetic potential energy. This special case provides an instructive exercise that will be used in Section 3.11 to analyze difficulties with the concept of total energy in general relativity. Let α be a 1-form, and X be a future-pointing timelike Killing vector field such that $L_X \alpha = 0$. (a) Show that $d\alpha$ is an electromagnetic field F on M which satisfies $L_X F = 0$. (b) Let (γ, m, e) be a particle that obeys the Lorentz world-force law with respect to F. Show that $-g(\gamma_*, X) - (e/2)(\alpha X) \circ \gamma \colon \mathscr{E} \to \mathbb{R}$ is a constant. (c) Let $\mathscr{P}$ be a consistent particle set, each of whose particles obeys the Lorentz world-force law relative to F and let x be a collision event. For each $(\gamma, m, e) \in \mathscr{P}$, let C_γ be the constant in (b). Show $\sum_{\text{in}(x)} C_\gamma = \sum_{\text{out}(x)} C_\gamma$.

> If $|X| \to 1$ at "spatial infinity," C_γ is sometimes called the "total energy" of (γ, m, e), which includes its rest-mass energy, its kinetic energy, its gravitational potential energy, and its electromagnetic potential energy. One then has in mind a (z, Z) such that $z \in \gamma$ and $Z \in \text{span}(Xz)$, so that $-g(\gamma_*, Xz)/|Xz|$ $(= -g(\gamma_*, Z))$ is the particle energy observed by (z, Z), and $-(e/2)(\alpha X)z/|Xz|$ $(= -(e/2)\alpha Z)$ is by definition the electric energy for γ observed by (z, Z). The Newtonian analogue of $-(e/2)\alpha Z$ is the Newtonian electric energy $\tfrac{1}{2} e\varphi$, where φ is the Newtonian electrostatic potential (cf. Section 9.3).

3.9 Matter equations: an example

To model the influence of spacetime, electromagnetism, and matter on matter one needs matter equations. Postponing a general discussion until Section 3.12, we start with an example. Throughout this section $(M, \mathscr{M}, F)$ is a relativistic model, with $\mathscr{M} = \{(m_A, e_A, \mathbf{P}_A, \eta_A) \mid A = 1, \ldots, N\}$ a finite collection of particle flows.

3.9.1 The simple matter equations

In view of the comments in Section 3.8.6, it is often appropriate to assume $\forall A$ that each particle (potentially) in the Ath particle flow obeys the Lorentz world-force law. Since each integral curve of $\mathbf{P}_A$ is such a particle, the formal condition is: (a) $D_{\mathbf{P}_A} \mathbf{P}_A = e_A \tilde{F} \mathbf{P}_A \; \forall A = 1, \ldots, N$.

Suppose in addition there is no creation or destruction of any of the particles (potentially) in any of the flows—no collisions, in particular no decays, no direct influence of matter on matter at all (Section 3.8). This might occur, for example, for a dilute mixture of hydrogen, helium, and photons in intergalactic space. Then, as in Section 3.2, it is appropriate to assume: (b) $\text{div}\,(\eta_A \mathbf{P}_A) = 0 \; \forall A = 1, \ldots, N$. We shall say $(M, \mathscr{M}, F)$ *obeys the simple matter equations* iff (a) and (b) hold.

Let $\hat{T}$ be the stress-energy tensor of $\mathscr{M}$, and $\hat{E}$ be that of F. To motivate the discussion in the next section, we need the following result.

3 Electromagnetism and matter

Proposition 3.9.2. *Suppose* $(M, \mathcal{M}, F)$ *obeys the simple matter equations and Maxwell's equation. Then*
$$\text{div}\,(\hat{T} + \hat{E}) = 0.$$

PROOF. $\text{div}\,\hat{E} = -\tilde{F}J = -\tilde{F}\sum e_A\eta_A P_A$ (Sections 3.7 and 3.5). $\text{div}\,\hat{T} = $ (by Section 3.5) $\text{div}\,\sum \eta_A P_A \otimes P_A = $ (by Exercise 3.6.4d) $\sum \{[\text{div}\,(\eta_A P_A)]P_A + \eta_A D_{P_A} P_A\} = $ (by Section 3.9.1) $0 + \sum \eta_A e_A \tilde{F} P_A$. Thus $\text{div}\,(\hat{T} + \hat{E}) = 0$. □

EXERCISE 3.9.3

Let J be the charge current density of $\mathcal{M} \in (M, \mathcal{M}, F)$. Show: (a) if $(M, \mathcal{M}, F)$ obeys the simple matter equations, $\text{div}\,J = 0$; (b) if $(M, \mathcal{M}, F)$ obeys Maxwell's equations, $\text{div}\,J = 0$.

3.10 Energy-momentum 'conservation'

This section analyzes a point relevant to our later discussion of matter equations. Throughout the section $(M, \mathcal{M}, F)$ is a relativistic model; we abbreviate by writing $\hat{D} = \hat{T} + \hat{E}$, where $\hat{T}$ is the stress-energy tensor of $\mathcal{M}$ and $\hat{E}$ is that of F. It follows that $\hat{D}$ is a stress-energy tensor on M. From the example of the previous section we know that $\text{div}\,\hat{D} = 0$ in at least one relevant situation. But since $\hat{D}$ is a symmetric (2, 0)-tensor, there is in general no natural way to integrate $\hat{D}$ or interpret $\text{div}\,\hat{D} = 0$ via integral conservation laws, in contrast to the situation in Sections 3.2 and 3.7. We begin with the proposition which, for a long time, led physicists to overlook or slur over this distinctly unwelcome difference. Notation is as in Sections 3.0.3 and 3.6.1.

Proposition 3.10.1. *Let X be a Killing vector field and suppose* $\text{div}\,\hat{D} = 0$. *Then* $\text{div}\,(\tilde{D}X) = 0$.

PROOF. Since $\hat{D}$ is symmetric, $D^{ij} = D^{ji}$. We have
$$\text{div}\,(\tilde{D}X) = (D^i{}_j X^j)_{|i} = (D^{ij}X_j)_{|i} = D^{ij}{}_{|i}X_j + D^{ij}X_{j|i}$$
$$= D^{ji}{}_{|i}X_j + \tfrac{1}{2}(D^{ij} + D^{ji})X_{j|i} = 0$$

since $D^{ji}{}_{|i} = 0$ by $\text{div}\,D = 0$, while $X_{i|j} + X_{j|i} = 0$ (Exercise 3.6.6). □

3.10.2 Energy conservation in special relativity

Suppose M is Minkowski space. Then there exists a parallel (covariant constant) reference frame X; for example, $X = \partial_4$. Each such reference frame is Killing (Section 3.6.3). We have $\text{div}\,\hat{D} = 0$ iff $g(X, \text{div}\,\hat{D}) = 0 \,\forall$ such X (cf. Exercise 3.1.6) iff $\text{div}\,(\tilde{D}X) = 0 \,\forall$ such X (cf. Proposition 3.10.1) iff the following integral conservation law holds $\forall$ such X: $\sum_{\mu=1}^{3}\int_{\mathcal{B}_\mu} i(\tilde{D}X)\Omega = 0 \,\forall$ causal box (Section 3.0.2). Now given any space-section $\mathcal{D}$, in particular $\mathcal{B}_2$ above, and any parallel reference frame X, $\int_{\mathcal{D}} i(\tilde{D}X)\Omega$ is defined and interpreted as the *total energy in $\mathscr{D}$ of $\mathscr{M}$ and F for the reference frame X*. The reader can check that if $\mathscr{D}$ happens to be everywhere orthogonal to X this corresponds directly to our Definition 3.3.3 of measured energy density. With this interpretation, there is overwhelming theoretical and empirical evidence that the above integral conservation law, called the *total energy conservation law*, always holds when gravity is negligible. Thus one always assumes div $\hat{D} = 0$ when M is Minkowski space—that is, in special relativity.

> A concrete example of conservation of total energy is that of an antenna, where the energy of the electromagnetic field gets converted to the kinetic energy of the motion of electrons. Although we have formally interpreted the physical significance of div $\tilde{D}X = 0$ only for a timelike X, this equation also has significance for a spacelike X. Indeed, in Minkowski space, there are four Killing vector fields corresponding to translations (∂_4 timelike, $\{\partial_1, \partial_2, \partial_3\}$ spacelike) and six Killing vector fields $u^4\partial_1 + u^1\partial_4$, and so on, corresponding to a natural basis of the Lie algebra of the Lorentz group $SO(3, 1)$. The ten corresponding integral conservation laws are then divided into: one law of total energy conservation, three laws of total momentum conservation (these four correspond to the translations), three laws interpreted as a statement about the center of mass of a system, and three laws of total angular momentum conservation (these six correspond to the Lorentz group). As far as is known, there are no fully analogous conservation laws in general relativity, as we now discuss.

3.10.3 Energy-momentum 'conservation'

In general there are no nonzero Killing vector fields on M. Similarly, suppose M is Einstein–de Sitter spacetime (Example 1.4.3); then there are no timelike ones. To see this let S be the scalar curvature. Then X Killing $\Rightarrow$ (by Section 3.6.3a) $dS(X) = 0 \Rightarrow$ (by the expression of S) $du^4(X) = 0 \Rightarrow$ (by the form of g) X spacelike. In the absence of Killing vector fields the argument in Section 3.10.2 fails miserably.

Nonetheless, in a systematic treatment one always demands div $\hat{D} = 0$ anyway. Among the interrelated motivations are the following. (a) In a region so small curvature is negligible (Section 2.1.2), the laws of special relativity are observed to hold. (b) When spacetime has one of several special properties—for example, has a timelike Killing vector field, an integral formulation can be ressurected (cf. Example 3.8.9). (c) Proposition 3.9.2 and a host of analogous results. (d) The collision conservation law (Definition 3.8.4a). (e) If one postulates Einstein's field equation (Chapter 4), div $\hat{D} = 0$ is a corollary. div $\hat{D} = 0$ is called the *differential energy-momentum 'conservation' law*, where we have added single quotes as a reminder that no honest integral conservation law is in general implied.

> The interrelation between (c), (d), and (e) is very similar to that between Exercise 3.9.3a, Definition 3.8.4b, and Proposition 3.7.5c for the case of a charge-current density.

3 Electromagnetism and matter

As mentioned in Section 3.8, conservation laws have great predictive power. It is a shame to lose the special relativistic total energy conservation law (Section 3.10.2) in general relativity. Many of the attempts to resurrect it are quite interesting; many are simply garbage.

EXERCISE 3.10.4

Suppose: M is Einstein–de Sitter spacetime; $\mathcal{M} = (m, 0, mZ, \eta)$ is a particle flow with $m \neq 0$ and Z the comoving reference frame; and $F = 0$. (a) Show $\hat{T} = \rho Z \otimes Z$, where $\rho = m^2 \eta$ is the energy density a comoving instantaneous observer measures for $\hat{T}$. (b) $(M, \mathcal{M}, F)$ obeys the differential energy-momentum 'conservation' law iff $\rho = (u^4)^{-2} h$ where $h: M \to [0, \infty)$ is a function such that $Zh = 0$ ("the matter energy density decreases as the universe ages").

3.11 Two initial value theorems

We state two theorems similar to Proposition 3.2.3 and Theorem 3.8.3 and prove the harder one. The second, easier, theorem is needed to illustrate one point in the next section. Mathematically, none of the material in the present section will be used at all later. The first theorem concerns Maxwell's equations.

Suppose the following data are given: (a) a spacetime M; (b) a charge-current density J on M such that $\text{div } J = 0$; (c) a coordinate neighborhood $\mathcal{U} \subset M$; (d) an imbedded three-dimensional spacelike submanifold $\phi: \mathcal{W} \to \mathcal{U}$.

In order to state the final item of the given data, we need some notation. Let a 2-form F_0 over ϕ be given. Define a 2-form $\phi^* F_0$ in $\mathcal{W}$ by: $(\phi^* F_0)x = \phi^*(F_0 x)$, $\forall x \in \mathcal{W}$. Let $\hat{F}_0$ be the (2, 0)-tensor field over ϕ physically equivalent to F_0 (Section 2.0.1) and let Ω be the volume element of M; one similarly defines $\phi^*[i(\hat{F}_0)(\Omega \circ \phi)]$ (cf. Section 2.0.1) as a 2-form in $\mathcal{W}$. Suppose that we are also given (e) a 2-form F_0 over ϕ which satisfies: (e_1) $d(\phi^* F_0) = 0$ in $\mathcal{W}$ and (e_2) $d(\phi^*[i(\hat{F}_0)(\Omega \circ \phi)]) = 4\pi \phi^*[i(J)\Omega]$ in $\mathcal{W}$.

Theorem 3.11.1. *Given* (a)–(e). *For $\mathcal{U}$ sufficiently small, there exists a unique 2-form F on $\mathcal{U}$ such that $F \circ \phi = F_0$ and $(\mathcal{U}, F, J)$ obeys Maxwell's equations. Conversely, if $(\mathcal{U}, F, J)$ obeys Maxwell's equations and for $\phi: \mathcal{W} \to \mathcal{U}$ as in* (d), *we define $F_0 = F \circ \phi$, then (e_1) and (e_2) are valid.*

(e_1) and (e_2) are called constraint equations. It is mainly the fact that one must cope with these equations which makes the theorem trickier than most similar ones that arise in physics.

Newtonian analogue. Suppose M is Minkowski spacetime and use the Newtonian notation (Section 9.0) for the electric field $\vec{E}(\vec{x}, t)$, magnetic field $\vec{B}(\vec{x}, t)$, charge density σ, and current density $\vec{j}$. Let $\mathcal{W}$ be a small open set of the hyperplane $u^4 = 0$, regarded as an open set $\mathcal{W}$ in Euclidean 3-space. From Section 9.1 the reader may check the following:

$(e_1) \Leftrightarrow (\vec{\nabla} \cdot \vec{B})(\vec{x}, 0) = 0 \forall \vec{x} \in \mathcal{W}$.
$(e_2) \Leftrightarrow (\vec{\nabla} \cdot \vec{E})(\vec{x}, 0) = 4\pi\sigma(\vec{x}, 0) \forall \vec{x} \in \mathcal{W}$.

The first part of the theorem says that if $\vec{j}$, σ are given for all time t and they obey the equation of continuity, and if $\vec{B}$, $\vec{E}$ are known $\forall \vec{x} \in \mathscr{W}$ at $t = 0$ and they satisfy the equations $\vec{\nabla} \cdot \vec{B} = 0$ and $\vec{\nabla} \cdot \vec{E} = 4\pi\sigma$ at $t = 0$, then there exist unique $\vec{B}(\vec{x}, t)$ and $\vec{E}(\vec{x}, t)$, defined for $|t| < \varepsilon$, such that $\vec{B}(\vec{x}, 0)$ and $\vec{E}(\vec{x}, 0)$ agree with the given values and such that all eight classical Maxwell's equations for $\vec{B}(\vec{x}, t)$ and $\vec{E}(\vec{x}, t)$ are satisfied for $|t| < \varepsilon$.

We remark that the following proof uses some mathematical background which is not in Bishop–Goldberg [1].

Proof of Theorem 3.11.1. Let us first dispose of the second statement. Suppose $(\mathscr{U}, F, J)$ obeys Maxwell's equations. Since $F_0 = F \circ \phi$, $\phi^* F_0 = \phi^* F$, where the right side is just the usual pull-back of forms. Thus $0 = \phi^*(dF) = d(\phi^* F) = d(\phi^* F_0)$, which is ($e_1$). Since div $\hat{F} = 4\pi J$, Section 3.6.2 $\Rightarrow d(i(\hat{F})\Omega) = 4\pi i(J)\Omega$. Applying ϕ^* to both sides and using the fact that $\phi^*(i(F)\Omega) = i(\hat{F}_0)(\Omega \circ \phi)$, we obtain ($e_2$).

To prove the first part of the theorem, we have to introduce special coordinates around $\phi \mathscr{W}$. Because $\phi \mathscr{W}$ is spacelike, the exponential map of the normal bundle of $\phi \mathscr{W}$ in $\mathscr{U}$ is nonsingular at each $x \in \mathscr{W}$. Taking $\phi \mathscr{W}$ to be sufficiently small, we obtain a nonsingular one-one map $\psi: \phi \mathscr{W} \times (-\varepsilon, \varepsilon) \to \mathscr{U}$ such that $\forall x \in \mathscr{W}$, the image of $\{x\} \times (-\varepsilon, \varepsilon)$ is a timelike geodesic with unit tangent vectors orthogonal to $\phi \mathscr{W}$. We may as well let $\mathscr{U}$ be the image of ψ so that ψ becomes a diffeomorphism. Introducing coordinates $\{x^1, x^2, x^3\}$ in $\mathscr{W}$, we obtain coordinates $\{x^1, \ldots, x^4\}$ in $\mathscr{U}$ via ψ such that each x^4-coordinate curve is a timelike geodesic with unit tangent vectors orthogonal to all the level hypersurfaces $x^4 =$ constant, and $\phi \mathscr{W}$ is the hypersurface $x^4 = 0$.

In the remainder of the proof, we adopt two notational conventions. First we shall agree to identify $\mathscr{W}$ with the hypersurface $x^4 = 0$, and simply regard ϕ as the inclusion map. Second, lower-case Greek indices $\alpha, \beta, \gamma, \mu, \nu$ will run from 1 to 3, and any such repeated Greek index (even when both are superscripts or subscripts) will imply summation from 1 to 3.

Now observe that the coordinate vector fields $\{\partial_i\}$ satisfy $g(\partial_4, \partial_4) = -1$ and $g(\partial_\alpha, \partial_4) = 0$. Let $\{X_i\}$ be an orthonormal basis of vector fields such that $X_4 = \partial_4$. Then there exist C^∞ functions $\{a_\beta{}^\alpha\}$ in $\mathscr{U}$ such that

$$X_\alpha = a_\alpha{}^\beta \partial_\beta.$$

Moreover, each $X_\alpha \circ \phi$ is tangent to $\mathscr{W}$. Let us write the unknown 2-form F as

$$F = F_{ij} \omega^i \otimes \omega^j \qquad (F_{ij} = -F_{ji}),$$

where $\{\omega^i\}$ is the basis of 1-forms dual to $\{X_i\}$. Also, let the given charge-current density J be

$$J = J^i X_i.$$

Observe that $F_{\alpha\beta} = F_\alpha{}^\beta = F^{\alpha\beta}$, $F^{\alpha 4} = F_\alpha{}^4 = -F_{\alpha 4}$, $J^\alpha = J_\alpha$, and $J_4 = -J^4$.

Returning to Maxwell's equations, we see that $dF = 0$ and div $\hat{F} = 4\pi J$ are, respectively, equivalent to:

(a$_0$) $\qquad\qquad F_{ij|k} + F_{jk|i} + F_{ki|j} = 0,$

(b$_0$) $\qquad\qquad F^{ij}{}_{|j} = 4\pi J^i.$

3 Electromagnetism and matter

See Section 3.6.1j and Exercise 3.6.4a. Now F_0 in $\mathscr{W}$ is given. If F is to satisfy $F \circ \phi = F_0$, then

$$\phi^* F_0 = \phi^* F = (F_{\alpha\beta}\omega^\alpha \otimes \omega^\beta) \circ \phi,$$
$$\hat{F}_0 = (F^{ij} X_i \otimes X_j) \circ \phi.$$

Thus making use of $F_{\alpha\beta} = F^{\alpha\beta}$, $F_{\alpha 4} = -F_\alpha{}^4$, and so on, we see that (e_1) and (e_2) are, respectively, equivalent to:

(ic$_1$) $\quad\quad\quad\quad F_{\alpha\beta|\gamma} + F_{\beta\gamma|\alpha} + F_{\gamma\alpha|\beta} = 0$ in $\mathscr{W}$,

(ic$_2$) $\quad\quad\quad\quad F_{4\alpha|\alpha} - 4\pi J_4 = 0$ in $\mathscr{W}$.

Next, we note that (a_0) is equivalent to the following two groups of equations:

(a$_1$) $\quad\quad\quad\quad F_{\alpha\beta|\gamma} + F_{\beta\gamma|\alpha} + F_{\gamma\alpha|\beta} = 0$.

(a$_2$) $\quad\quad\quad\quad F_{\alpha\beta|4} + F_{\beta 4|\alpha} + F_{4\alpha|\beta} = 0$.

Similarly, (b_0) is equivalent to the following two groups of equations:

(b$_1$) $\quad\quad\quad\quad F_{4\alpha|\alpha} - 4\pi J_4 = 0$.

(b$_2$) $\quad\quad\quad\quad F_{\alpha 4|4} - F_{\alpha\beta|\beta} + 4\pi J_\alpha = 0$.

Thus (ic$_1$) [respectively (ic$_2$)] is just the restriction of (a$_1$) [respectively (b$_1$)] to $\mathscr{W}$.

Noting that div $\mathbf{J} = 0 \Leftrightarrow g^{ik} J_{i|k} = 0$, we can now state the first part of the theorem in the following form: Given C^∞ functions $\{J_1, \ldots, J_4\}$ in $\mathscr{U}$, which satisfy $g^{ik} J_{i|k} = 0$, and given C^∞ functions $\{F_{12}, \ldots, F_{34}\}$ in $\mathscr{W}$, which satisfy (ic$_1$) and (ic$_2$). Then for $\mathscr{U}$ sufficiently small, there exists a unique set of C^∞ functions $\{F_{12}, \ldots, F_{34}\}$ in $\mathscr{U}$ which assumes the given values in $\mathscr{W}$ and satisfies equations (a_0) and (b_0).

The main step consists of demonstrating the existence and uniqueness of a set of C^∞ functions $\{F_{12}, \ldots, F_{34}\}$ satisfying the initial conditions and only (a$_2$) and (b$_2$) in $\mathscr{U}$. Before making this precise, let us make a further reduction. Let $\{\omega_i{}^j\}$ be the connection forms for $\{\omega^i\}$ so that

$$DX_i = X_j \otimes \omega_i{}^j,$$
$$\omega_i{}^j = \Gamma^j{}_{ki} \omega^k.$$

See Section 3.6.1e. Then again by Section 3.6.1e,

$$F_{ij|k} = X_k F_{ij} - F_{mj} \Gamma^m{}_{ik} - F_{im} \Gamma^m{}_{jk}.$$

Substituting this into (a$_2$) and (b$_2$), and using $X_4 = \partial_4$, $X_\alpha = a_\alpha{}^\beta \partial_\beta$, we see that (a$_2$) is equivalent to:

(a$_3$) $\quad\quad\quad \partial_4 F_{\alpha\beta} = -a_\alpha{}^\gamma \partial_\gamma F_{\beta 4} + a_\beta{}^\gamma \partial_\gamma F_{\alpha 4} + \text{ZOT},$

where ZOT (zeroth order term) is a generic symbol for any expression that is linear in the $\{F_{ij}\}$ and that does not involve any derivatives of $\{F_{ij}\}$. Similarly, (b$_2$) is equivalent to:

(b$_3$) $\quad\quad\quad \partial_4 F_{\alpha 4} = a_\beta{}^\gamma \partial_\gamma F_{\alpha\beta} + \text{ZOT} - 4\pi J_\alpha$.

Now, our precise assertion is: let $\{J_1, \ldots, J_4\}$ be arbitrary C^∞ functions in $\mathscr{U}$, and let $\{F_{12}, \ldots, F_{34}\}$ be arbitrary C^∞ functions in $\mathscr{W}$. If $\mathscr{U}$ is sufficiently small, then there exists a unique set of C^∞ functions $\{F_{12}, \ldots, F_{34}\}$ in $\mathscr{U}$ which assumes the prescribed values in $\mathscr{W}$ and satisfies (a$_3$) and (b$_3$).

For its proof, introduce two column matrices of functions by:

$$\mathfrak{f} = [F_{12} \ \ F_{13} \ \ F_{23} \ \ F_{14} \ \ F_{24} \ \ F_{34}]^t,$$
$$\mathfrak{j} = [0 \ \ 0 \ \ 0 \ \ J_1 \ \ J_2 \ \ J_3]^t,$$

where t denotes transpose. Define three (6×6)-matrices $\mathfrak{a}^1$, $\mathfrak{a}^2$, $\mathfrak{a}^3$ of C^∞ functions by:

$$\mathfrak{a}^\alpha = \begin{bmatrix} 0 & 0 & 0 & a_2^\alpha & -a_1^\alpha & 0 \\ 0 & 0 & 0 & a_3^\alpha & 0 & -a_1^\alpha \\ 0 & 0 & 0 & 0 & a_3^\alpha & -a_2^\alpha \\ a_2^\alpha & a_3^\alpha & 0 & 0 & 0 & 0 \\ -a_1^\alpha & 0 & a_3^\alpha & 0 & 0 & 0 \\ 0 & -a_1^\alpha & -a_2^\alpha & 0 & 0 & 0 \end{bmatrix}$$

for $\alpha = 1, 2, 3$. Note that each $\mathfrak{a}^\alpha$ is symmetric. Then a short calculation shows that (a_3) and (b_3) are equivalent to the following matrix equation:

$$\frac{\partial}{\partial x^4} \mathfrak{f} = \mathfrak{a}^\alpha \left(\frac{\partial}{\partial x^\alpha} \mathfrak{f} \right) + \mathfrak{b} \mathfrak{f} - 4\pi \mathfrak{j},$$

where $\mathfrak{b}$ is a (6×6)-matrix of C^∞ functions coming from ZOTs, and the middle two terms are understood in the sense of matrix multiplication. Since each $\mathfrak{a}^\alpha$ is symmetric, this is a symmetric hyperbolic system of partial differential equations in the sense of Friedrichs [1]. Since $\{\mathfrak{a}^\alpha\}$, $\mathfrak{b}$, and $\mathfrak{j}$ are all C^∞, the fundamental theorem of such systems guarantees that, given $\mathfrak{j}$ and the values of $\mathfrak{f}$ on the hypersurface $\mathscr{W}$, a unique C^∞ solution $\mathfrak{f}$ satisfying the initial values on $\mathscr{W}$ exists in $\mathscr{U}$ provided $\mathscr{U}$ is sufficiently small (Lax [1] and Friedrichs [1]). This proves our assertion.

Thus we have a unique 2-form $F = F_{ij}\omega^i \otimes \omega^j$ ($F_{ij} = -F_{ji}$) whose components $\{F_{ij}\}$ satisfy (a_2) and (b_2) everywhere in $\mathscr{U}$ as well as (ic_1) and (ic_2) in $\mathscr{W}$. It remains to show that, knowing div $J = 0$, we can deduce (a_1) and (b_1). First, define $\eta = dF$. From Section 3.6.1j and (a_2) and (ic_1), we have

$$\eta_{\alpha\beta 4} = 0 \text{ in } \mathscr{U},$$
$$\eta_{\alpha\beta\gamma} = 0 \text{ in } \mathscr{W}.$$

We want to prove $\eta_{\alpha\beta\gamma} = 0$ everywhere in $\mathscr{U}$. Since $d\eta = d(dF) = 0$, we have (again by Section 3.6.1j):

$$\eta_{\alpha\beta\gamma|4} - \eta_{\alpha\beta 4|\gamma} - \eta_{\alpha 4\gamma|\beta} + \eta_{4\beta\gamma|\alpha} = 0.$$

But
$$\eta_{\alpha\beta\gamma|4} = X_4 \eta_{\alpha\beta\gamma} - \Gamma^i{}_{\alpha 4}\eta_{i\beta\gamma} - \Gamma^i{}_{\beta 4}\eta_{\alpha i\gamma} - \Gamma^i{}_{\gamma 4}\eta_{\alpha\beta i}$$
$$= \partial_4 \eta_{\alpha\beta\gamma} - \Gamma^\mu{}_{\alpha 4}\eta_{\mu\beta\gamma} - \Gamma^\mu{}_{\beta 4}\eta_{\alpha\mu\gamma} - \Gamma^\mu{}_{\gamma 4}\eta_{\alpha\beta\mu}.$$

Thus restricting to each x^4-coordinate curve in $\mathscr{U}$, the functions $\{\eta_{\alpha\beta\gamma}\}$ satisfy a homogeneous system of ordinary differential equations (with x^4 as the variable). Since $\eta_{\alpha\beta\gamma} = 0$ in $\mathscr{W}$, the basic uniqueness theorem governing such systems implies that $\{\eta_{\alpha\beta\gamma}\}$ vanish identically on each x^4-coordinate curve. Hence $\eta_{\alpha\beta\gamma} = 0$ everywhere in $\mathscr{U}$, proving (a_1).

The proof of (b_1) is similar. Indeed, define a vector field H by $H = \operatorname{div} \hat{F} - 4\pi J$, where as usual $\hat{F}$ is the $(2, 0)$-tensor field physically equivalent to F. Now div $H = 0$ because div div $\hat{F} = 0$ (Exercise 3.6.4c) and div $J = 0$ (by assumption). Thus $H^i{}_{|i} = 0$. But $(b_2) \Leftrightarrow H^\alpha = 0$ and $(ic_2) \Leftrightarrow H^4 = 0$ on $\mathscr{W}$. Thus $H^4{}_{|4} = 0$ in $\mathscr{U}$ and $H^4 = 0$ in $\mathscr{W}$. Observe that

$$H^4{}_{|4} = 0 \Leftrightarrow X_4 H^4 + \Gamma^i{}_{4i} H^4 = 0$$
$$\Leftrightarrow \partial_4 H^4 + \Gamma^i{}_{4i} H^4 = 0.$$

Hence H^4 satisfies a homogeneous ordinary differential equation on each x^4-coordinate curve, with zero initial condition. Thus $H^4 = 0$, which is equivalent to (b_1). □

3 Electromagnetism and matter

Our second theorem concerns matter equations. It involves neglecting the influence of matter on spacetime and electromagnetism so let F be a given electromagnetic field on M. Let (m, e) be a particle type. Suppose also given an imbedded 3-dimensional spacelike submanifold $\phi: \mathscr{W} \to M$, a C^∞ vector field $\boldsymbol{P}_0$ over ϕ such that $\boldsymbol{P}_0 y$ is future pointing $\forall y \in \mathscr{W}$ and $g(\boldsymbol{P}_0, \boldsymbol{P}_0) = -m^2$, and a positive C^∞ function η_0 over ϕ. Fix $w \in \phi\mathscr{W}$ and let $\mathscr{U}$ be an open neighborhood of M containing w.

Theorem 3.11.2. *For $\mathscr{U}$ sufficiently small, there exists a unique C^∞ vector field $\boldsymbol{P}$ on $\mathscr{U}$ and a unique C^∞ function $\eta: \mathscr{U} \to (0, \infty)$ such that:* (a) $\boldsymbol{P} \circ \phi = \boldsymbol{P}_0$ *and* $\eta \circ \phi = \eta_0$; (b) $\mathscr{M} = (m, e, \boldsymbol{P}, \eta)$ *is a particle flow on* $(\mathscr{U}, g|_{\mathscr{U}})$; *and* (c) $(\mathscr{U}, \mathscr{M}, F|_{\mathscr{U}})$ *obeys the simple matter equations.*

The ideas needed for the proof are in Proposition 3.2.3, Theorem 3.8.3, and Theorem 3.11.1; it is left as an exercise.

3.12 Appropriate matter equations

Let $(M, \mathscr{M}, F)$ be a relativistic model. In general, deciding on appropriate matter equations for the model involves an intricate mixture of empirical and theoretical arguments, as illustrated in Chapter 6 on cosmology. Moreover, in the absence of a universal matter model, the phrase "appropriate matter equations" can be made precise only on a case-by-case basis, as in Section 3.9 and the rest of this chapter. However, two universal criteria are used:

(a) When combined with Maxwell's equations, appropriate matter equations always imply the differential energy-momentum 'conservation' law (Sections 3.9 and 3.10).
(b) Whenever M and F can be regarded as given, one always has a "present determines the future" theorem, as illustrated by Theorem 3.11.2.

EXAMPLE 3.12.1. Suppose $\mathscr{M}$ consists of a pair of particle flows, with $\boldsymbol{P}_1 = 2\boldsymbol{P}_2$, whence $m_1 = 2m_2$, and with $e_1 = 2e_2$. Retaining the Lorentz world-force law equations $D_{\boldsymbol{P}_A}\boldsymbol{P}_A = e_A \tilde{F} \boldsymbol{P}_A$ as matter equations we now assume that the particles of type (m_1, e_1) are decaying into particles of type (m_2, e_2). Then it is sometimes appropriate to replace the assumption in Proposition 3.9.2 that the world densities η_1 and η_2 are conserved by the following matter equations:

(a) $\qquad \mathrm{div}\,(\eta_1 \boldsymbol{P}_1) = -k\eta_1 = -2\,\mathrm{div}\,(\eta_2 \boldsymbol{P}_2), \qquad k \in (0, \infty).$

Roughly: the bigger η_1, the more decays take place. These two equations are perhaps the simplest matter equations that take into account a direct influence of matter on matter.

This is a (highly overidealized) model for the decay of beryllium-8 nuclei into helium-4 nuclei which occurs during "element cooking" in hot stars and probably in the early universe (Weinberg [1]). The proportionality constant k is determined by microphysics.

EXERCISE 3.12.2

In Example 3.12.1, let J be the charge-current density of $\mathcal{M}$, and $\hat{D} = \hat{T} + \hat{E}$ be as in Section 3.10. (a) Show div $J = 0$ ("particles are being created and destroyed but electric charge is not"). (b) Show the universal criterion (a) of this section holds—that is, if $(M, \mathcal{M}, F)$ also obeys Maxwell's equations, div $\hat{D} = 0$. (c) Motivate the factor "2" in Example 3.12.1a directly from the collision conservation laws of Section 3.8. (d) State and prove a theorem that shows that our other universal criterion is also valid, in particular that η_1 cannot "overshoot" and become negative.

PART THREE: OTHER MATTER MODELS

In order to facilitate access to the physics literature we shall later use some other standard matter models. We now define and motivate them. This whole part should be treated as reference material only.

3.13 Examples

Let $(M, \mathcal{M}, F)$ be a relativistic model, $\hat{T}$ be the stress-energy tensor of $\mathcal{M}$, and J be the charge-current density (Section 3.5).

EXAMPLE 3.13.1. DUST. Let $(m, 0, P, \eta)$ be a particle flow on M with $m \mathcal{M} 0$ and η nowhere zero. Then $Z = m^{-1}P$ is a reference frame on M and $\rho = m^{-2}\eta$ is a function: $M \to (0, \infty)$. One can abstract from the examples of Sections 3.5.1 and 3.9.1, throwing away the particle type $(m, 0)$, to get the following definitions. $(M, \mathcal{M}, F)$ is defined as a *dust* iff: $\mathcal{M} = (Z, \rho)$, where Z is a reference frame on M and $\rho: M \to (0, \infty)$ is a C^∞ function; (b) $\hat{T} = \rho Z \otimes Z$ and $J = 0$. Z is then defined as the *comoving reference frame*. $\forall z \in M$, the instantaneous observer (z, Zz) measures energy density $T(Z, Z)z = \rho z$ so ρ is defined as the (comoving) *energy density*. $(M, \mathcal{M}, F)$ obeys the *dust matter equations* iff div $(\rho Z) = 0 = D_Z Z$.

> *Newtonian analogue.* In Section 0.1.10 we introduced Newtonian active-mass per unit euclidean volume, $\rho: \mathbb{R}^3 \to [0, \infty)$, without specifying the mass of the individual particles whose "smoothing out" leads to this ρ.

EXAMPLE 3.13.2. COMPLETE MATTER VACUUM. The preceding example is obtained from our canonical example of a collection of N particle flows by considering the case $N = 1$, $m\eta$ nowhere zero. For $N = 0$ we can "abstract" to get a *complete matter vacuum* model $(M, \mathscr{M}, F)$: $\mathscr{M}$ is the empty set; $\hat{T} = 0 = J$; and the matter "equations" consist of the remark that no matter is created from gravity or electromagnetism. Complete matter vacuum is interpreted as absence of all matter, including "test matter" (cf. Section 3.5). If also $F = 0$, one has *complete vacuum*.

EXAMPLE 3.13.3. SUPERPOSITIONS. If $(M, \mathscr{M}_1, F)$ and $(M, \mathscr{M}_2, F)$ are relativistic models, then it makes sense to define their *superposition* $(M, \mathscr{M}, F)$: $\mathscr{M}$ is the pair $\{\mathscr{M}_1, \mathscr{M}_2\}$; and, in the obvious notation, $\hat{T} = \hat{T}_1 + \hat{T}_2$, $J = J_1 + J_2$. Then $\mathscr{M}_1$ and $\mathscr{M}_2$ are the *components of* $\mathscr{M}$. If we are given some matter equations for $(M, \mathscr{M}_A, F)$, $A = 1, 2$, it makes sense formally to require that $(M, \mathscr{M}, F)$ obey the same matter equations. The superposition is then defined as *collision-free* and this corresponds to assuming no direct interaction between $\mathscr{M}_1$ and $\mathscr{M}_2$. The generalization to the *finite superposition* of $N \geq 1$ relativistic models with M and F fixed is the obvious one; the canonical examples are given by Sections 3.5.1 and 3.9.1 as before.

> The addition of stress-energy tensors and charge-current densities is ultimately motivated by the collision conservation laws of energy-momentum and of charge (Definition 3.8.4a and b). In Chapter 5, we shall show how to superimpose an infinite number of relativistic models.

EXAMPLE 3.13.4. QUASI-GAS. For brevity, we refer to $\mathscr{M}$ as a *quasi-gas* on M iff $\mathscr{M}$ is a finite superposition of particle flows on M (Section 3.5) obeying the following two "generic" conditions: (a) $\mathscr{M}$ is a nonempty set; (b) ∀ particle-flow in $\mathscr{M}$, the world density is nowhere zero. Here "quasi" refers to the fact that a true gas is, roughly speaking, described by an infinite collection of particle flows (cf. Section 5.7).

EXERCISE 3.13.5.

Let $(M, \mathscr{M}, F)$ be a dust with co-moving reference frame Z and stress-energy tensor $\hat{T}$, (z, X) be an instantaneous observer. Show $X = Zz$ iff T is spatially isotropic for (z, X) as defined in Section 2.1, iff the energy density (z, X) measures for $\hat{T}$ is smaller than the energy density any other instantaneous observer at z measures for $\hat{T}$.

3.14 Normal stress-energy tensors

When analyzing complicated situations it is often useful, and correct, to assume that a stress-energy tensor singles out a reference frame in a natural way. For example, we do not know a fully detailed, realistic matter model for the earth. But for each point in the history of the earth, the concept of being at rest with respect to the matter at that point pretty clearly has at least one precise formulation (cf. Exercise 3.13.5). We now give some corresponding formal results.

3.14 Normal stress-energy tensors

Let $\hat{T}$ be a stress-energy tensor on M and suppose $x \in M$. $\hat{T}$ is defined to be *normal at x* iff $\tilde{T}X$ is timelike $\forall$ causal $X \in M_x$. For example, the matter stress-energy tensor of a dust (Example 3.13.1) is normal at each $x \in M$. A nonzero $X \in M_x$ is called an *eigenvector of $\hat{T}$ at x* iff $\tilde{T}X = aX$ for some $a \in \mathbb{R}$.

Proposition 3.14.1. *If a stress-energy $\hat{T}$ is normal at $x \in M$, then $\hat{T}$ has a timelike eigenvector which is unique up to nonzero multiples.*

PROOF. Let $\mathscr{S}$ be the unit sphere on M_x relative to an arbitrary positive definite inner product on M_x. Let $\mathscr{A}$ be the set of nonspacelike one-dimensional subspaces of M_x. Then $\mathscr{A} \cap \mathscr{S}$ has two components, each diffeomorphic to the closed unit ball in $\mathbb{R}^3$ (Exercise 1.1.9). Take one of these components and call it $\mathscr{C}$. By assumption, the endomorphism $\tilde{T}: M_x \to M_x$ carries $\mathscr{A}$ into itself and hence $\hat{T}$ induces a continuous map f of $\mathscr{C}$ into itself. By Brouwer's fixed point theorem, $f: \mathscr{C} \to \mathscr{C}$ has a fixed point, say X. Since $\hat{T}$ is normal at x, the image f consists only of timelike vectors. Hence X is timelike, and it follows from the definition of f that $\tilde{T}X = aX$ for some $a \in \mathbb{R}$.

It remains to show that X is unique up to a nonzero multiple. Suppose not, let Z be another timelike vector in M_x such that $\tilde{T}Z = bZ$ and $\{X, Z\}$ are linearly independent. Let T be the $(0, 2)$-tensor field physically equivalent to $\hat{T}$ as usual. Then using the symmetry of $\hat{T}$, we have

$$ag(X, Z) = g(\tilde{T}X, Z) = T(X, Z) = T(Z, X) = g(\tilde{T}Z, X)$$
$$= bg(X, Z).$$

Since $g(X, Z) < 0$ (Exercise 1.1.9c), $a = b$. Thus $W = \text{span}\{X, Z\}$ is a two-dimensional timelike subspace on which $\tilde{T}$ acts as a pure magnification by a factor of a. In particular, if Y is a lightlike vector in W, then $\tilde{T}Y = aY$, contradicting the normality of $\hat{T}$. □

> Needless to say, the preceding proposition could be proved directly by linear algebra without recourse to Brouwer's fixed point theorem. The result also follows from Exercises 3.14.4 and 3.14.5, or else can be deduced immediately from the canonical forms of a stress-energy tensor (Optional exercise 8.1.5).
>
> Normality is one of several standard conditions to impose on a stress-energy tensor. Exercises 3.14.4, 3.14.5, Section 4.3, and Optional exercises 8.1 give some of the other conditions and their interrelations.

A stress-energy tensor $\hat{T}$ on M is called *normal* iff it is normal at every $x \in M$. A nowhere zero vector field X on M is called an *eigenvector field of $\hat{T}$* iff there is a function f on M such that $\tilde{T}X = fX$; f is then called the *eigenfunction of $\hat{T}$ corresponding to X*.

Corollary 3.14.2. *A normal stress-energy tensor T on M possesses a unique future-pointing unit timelike eigenvector field.*

For a dust (Example 3.13.1) the eigenvector field is just Z.

Proof. First note in Proposition 3.14.1 that the solution X of $(T_j{}^i - \delta_j{}^i a)X^j = 0$ is unique (up to scalar multiples). For if we assume a second linearly independent solution, say Z, considering span $\{X, Z\}$ again gives a contradiction.

Now define a (not necessarily C^∞) vector field X on M and a (not necessarily C^∞) function f on M by: Xx is the unique future pointing unit timelike eigenvector of T at x, and fx is the corresponding eigenvalue. It suffices to prove that X and f are C^∞.

Fix $x \in M$ and let $\mathcal{U}$ be a coordinate neighborhood around x with coordinate functions $\{x^1, \ldots, x^4\}$. In $\mathcal{U}$, let $\hat{T} = T^{ij} \partial_i \otimes \partial_j$, where the $\{T^{ij}\}$ are C^∞ functions. Since $(\tilde{T} - f)X = 0$, $\det\{T_j{}^i - \delta_j{}^i f\} = 0$ identically in $\mathcal{U}$. Since Xz is the unique solution of the homogeneous system of linear equations $(T_j{}^i z - \delta_j{}^i fz)(X^j z) = 0$, where $z \in \mathcal{U}$ and $i = 1, \ldots, 4$, $\{T_j{}^i z - \delta_j{}^i fz\}$ is a matrix of rank 3 and hence fz is a simple root of the characteristic equation in λ: $\det\{T_j{}^i z - \delta_j{}^i \lambda\} = 0$. Since the coefficients of this polynomial equation are C^∞ functions of z (because the $\{T_j{}^i\}$ are), a standard theorem about polynomials says that a simple root must also be a C^∞ function of z. Thus f is a C^∞ function of z in $\mathcal{U}$. Since being C^∞ is a local property, f is C^∞ on M.

Standard arguments using the fact that the above matrix has rank 3 now show X is also C^∞. □

EXERCISE 3.14.3

Let (m, e, P, η) be a particle flow on M, and let $\hat{T}$ be its stress-energy tensor. Suppose $\omega \in M_z{}^*$, $\eta z \neq 0$ ω is causal and $\hat{T}(\omega, \omega) = 0$. Show that Pz is then physically equivalent to $a\omega$ for some $a \in \mathbb{R}$ and $m = 0$.

EXERCISE 3.14.4

Let $\hat{T}$ be a stress-energy tensor on M. Thus for $x \in M$ $\tilde{T}x: M_x \to M_x$ is self-adjoint with respect to gx (Exercise 1.0.6). Show $\hat{T}$ is normal at x iff there is an orthonormal basis (X_1, X_2, X_3, Z) of M_x such that $\hat{T}x = \sum_{\mu=1}^{3} p_\mu X_\mu \otimes X_\mu + \rho Z \otimes Z$ with $p_\mu \in \mathbb{R}$ and $\rho > |p_\mu|$ $\forall \mu \in (1, 2, 3)$.

EXERCISE 3.14.5

Let $\hat{T}$ be a symmetric (2, 0)-tensor on a Lorentzian vector space V and let $\tilde{T}: V \to V$ have the usual meaning. Recall that a subspace W of V is an *invariant subspace of $\tilde{T}$* iff $\tilde{T}W \subset W$. Show: (a) If W is an invariant subspace of $\tilde{T}$, so is $W^\perp$. (b) If W is a spacelike invariant subspace of $\tilde{T}$, then $\tilde{T}|_W$ is diagonalizable. (c) If W is a lightlike invariant subspace of $\tilde{T}$, then $\tilde{T}|_W$ possesses a lightlike eigenvector. (d) $\tilde{T}$ possesses an invariant subspace W of dimension 1 or 2. [Hint: If $a + \sqrt{-1}b$ is an eigenvalue of $\tilde{T}$, let $(v + \sqrt{-1}w)$ be a corresponding complex eigenvector, where $v, w \in V$. Then $W = \text{span}\{v, w\}$ would do.] (e) $\hat{T}(\omega, \omega) > 0$ $\forall$ causal $\omega \in V^*$ iff $\tilde{T}$ has a unique timelike invariant subspace and $\hat{T}(\omega, \omega) \geq 0$ $\forall$ such ω.

3.15 Perfect fluids

A perfect fluid matter model can be obtained by abstracting from a quasi-gas (Example 3.13.4) of a certain kind. We use the notation of Section 3.0.3 and define a stress-energy tensor $\hat{T}$ on a spacetime M to be *spatially isotropic at* $z \in M$ iff there is one instantaneous observer (z, Z) for whom T is spatially isotropic (Section 2.1.7).

Proposition 3.15.1. *Let $\mathcal{M}$ be a quasi-gas on M whose stress-energy tensor $\hat{T}$ is spatially isotropic at each $z \in M$. Then:*

(a) *$\hat{T}$ is normal.*
(b) *$\hat{T} = \rho Z \otimes Z + p(\hat{g} + Z \otimes Z)$, where Z is the reference frame which is an eigenvector field of T, $\hat{g}$ is the $(2, 0)$-tensor field physically equivalent to g, and ρ, p are functions on M such that $\rho > 0$ and $\rho \geq 3p \geq 0$. T is spatially isotropic for (z, Z) iff $Z = Zz$.*
(c) *The following are equivalent:* (1) $\rho z = 3pz$ *for one* $z \in M$; (2) $\rho = 3p$; *and* (3) *each component of $\mathcal{M}$ has zero rest-mass.*

Proof. Fix a $z \in M$ and let (z, Z) be an instantaneous observer for whom T is spatially isotropic. By Exercise 2.1.12,

(i) $\hat{T}z = aZ \otimes Z + b\hat{h}$, where $a, b \in \mathbb{R}$ and $\hat{h}$ is physically equivalent to the projection tensor (Section 2.1.5). On the other hand, Section 3.5.1 gives:
(ii) $\hat{T} = \sum \eta_A P_A \otimes P_A$.

Comparing (i) and (ii) gives the following results.

(iii) $a = T(Z, Z) > 0$ (Section 3.3.1, Proposition 3.5.2).
(iv) $-a + 3b = $ trace $Tz = \sum (\eta_A z)(-m_A^2)$;

since $\eta_A z > 0 \, \forall A$ this implies $a \geq 3b$, where equality holds iff $m_A = 0 \, \forall A$.

(v) $3b = h^{ij}T_{ij} = \sum (\eta_A z)h(P_A, P_A)$ (Section 3.6.1, Exercise 3.1.5);

since $Z^\perp$ is spacelike this gives $b \geq 0$.

We can now dispose of the proposition. Suppose $X \in M_z$. (i) implies

$$\tilde{T}X = (a + b)g(X, Z)Z + bX.$$

Suppose X is causal. Then

$$g(\tilde{T}X, \tilde{T}X) = [-(a + b)^2 + 2b(a + b)][g(X, Z)]^2 + b^2 g(X, X)$$
$$= (-a^2 + b^2)[g(X, Z)]^2 - b^2|X|^2 < 0$$

since $a > b$ by (iii) and (iv), while $g(X, Z)$ is nonzero as usual. Thus $\tilde{T}X$ is timelike $\forall$ nonspacelike X. This implies $\hat{T}$ is normal at $z \, \forall z \in M$, thereby proving (a). By Corollary 3.14.2, $\hat{T}$ has a unique eigenvector field Z which is a reference frame. Since the above expression of $\tilde{T}X$ implies $\tilde{T}Z = -aZ$, Z is a unit timelike future-pointing eigenvector of

$\hat{T}$ at z. Therefore $Z = Zz$ by uniqueness. Since this holds $\forall z \in M$, the above expression of $\hat{T}z$ implies
$$\hat{T} = \rho Z \otimes Z + p(\hat{g} + Z \otimes Z),$$
where ρ and p are the (not necessarily C^∞) functions on M such that, in the preceding notation, $\rho z = a$ and $pz = b$ at each such $z \in M$. Since $\hat{T}Z = -\rho Z$, $-\rho$ is the eigenfunction of $\hat{T}$ corresponding to Z and is therefore C^∞. Since trace $T = -\rho + 3p$ by (iv), $(-\rho + 3p)$ is also C^∞, and hence so is p. The previous inequalities for a and b now give: $\rho > 0$, $\rho \geq 3p \geq 0$, and $\rho = 3p$ iff $m_A = 0 \,\forall A = 1, \ldots, N$. This proves (b). This also proves that in (c), (1) $\Leftrightarrow$ (3). (2) $\Rightarrow$ (1) trivially. If (1) holds, then in our notation, $a = 3b$, $\Rightarrow m_A = 0 \,\forall A \Rightarrow$ (3). $\square$

For $\hat{T}$ as in the preceding proposition, let T be the $(0, 2)$-tensor field physically equivalent to $\hat{T}$. Then $\forall z \in M$, $\rho z = T(Zz, Zz)$ = the energy density of $\hat{T}$ observed by (z, Zz) (see discussion after Definition 3.3.4). Accordingly, we define ρ to be the *energy density of the quasi-gas* $\mathcal{M}$. p is called the *pressure* of the quasigas $\mathcal{M}$. We refer the reader to Exercise 3.15.6 for a physical interpretation of the pressure p in terms of "random velocities."

Retaining our preceding assumptions on $\mathcal{M}$, suppose the relativistic model $(M, \mathcal{M}, F)$ obeys Maxwell's equations and the differential energy-momentum 'conservation' law, while the charge-current density J of $\mathcal{M}$ is zero. Then, by Proposition 3.7.4c, div $\hat{T} = 0$. We now abstract, much as in the case of dust, to get the following.

Definition 3.15.2. A relativistic model $(M, \mathcal{M}, F)$ is a *perfect fluid* model iff the following hold. (a) $\mathcal{M} = (Z, \rho, p)$, where Z is a reference frame on M, and ρ, p are C^∞ functions on M with $\rho > 0$ and $\rho \geq 3p \geq 0$. (b) The charge-current density J of $\mathcal{M}$ is zero and the stress-energy tensor of $\mathcal{M}$ is $\hat{T} = \rho Z \otimes Z + p(\hat{g} + Z \otimes Z)$. Then by definition: $\mathcal{M}$ is a *perfect fluid on M*; Z is *the comoving reference frame*; ρ is *the energy density*, and p is *the pressure* of $\mathcal{M}$. By definition: $(M, \mathcal{M}, F)$ *obeys the perfect fluid matter equations* iff div $\hat{T} = 0$; and $\mathcal{M}$ is *rest-mass zero* iff $\rho = 3p$.

The motivation for the terminology follows Proposition 3.15.1 and our discussion line by line.

> The intuitive picture of a perfect fluid is that an enormous number of particle flows are present, with lots of random collisions. Now random collisions tend to establish isotropy; the following heuristic argument will serve as an illustration.
>
> In Newtonian physics, imagine two billiard balls which collide; suppose the initial velocities are $\vec{v}$ and $-\vec{v}$. If the collision is head on, the balls simply recoil, with final velocities $-\vec{v}$ and $\vec{v}$. But if the collision is slightly off center, the balls fly off in directions different from span $\vec{v}$. In general, random collisions take us from a highly anisotropic situation (all velocities in span $\vec{v}$) to a more nearly isotropic one (given any direction there is some chance a final velocity lies along that direction).
>
> Thus, the stress-energy tensor $\hat{T}$ of a perfect fluid comes out spatially isotropic at each point of the spacetime. Roughly, $\hat{T}$ is obtained by averaging the energy momenta of all components at each point.

3.15 Perfect fluids

We now analyze the matter equation div $\hat{T} = 0$ above. Let Z be a reference frame on M, and let ρ and p be functions on M satisfying $\rho + p > 0$. Define a (2, 0)-tensor by

$$\hat{T} = \rho Z \otimes Z + p(\hat{g} + Z \otimes Z).$$

Let **grad** p denote the vector field physically equivalent to dp.

Proposition 3.15.3. div $\hat{T} = 0$ iff: (a) div $(\rho Z) = -p$ div Z; and (b) $D_Z Z = -(\rho + p)^{-1}\{(Zp)Z + \mathbf{grad}\, p\}$.

PROOF. div $\hat{T} = 0$

$\Leftrightarrow (\rho Z^i Z^j + p g^{ij} + p Z^i Z^j)_{|j} = 0,$
$\Leftrightarrow Z^i_{|j}(\rho Z^j) + Z^i(\rho Z^j)_{|j} + p_{|j} g^{ij} + p_{|j} Z^i Z^j + p Z^i_{|j} Z^j + p Z^i Z^j_{|j} = 0,$
$\Leftrightarrow \rho D_Z Z + (\text{div}\, \rho Z)Z + \mathbf{grad}\, p + (Zp)Z + p D_Z Z + p(\text{div}\, Z)Z = 0,$
$\Leftrightarrow \{\text{div}\, (\rho Z) + p\, \text{div}\, Z\}Z + \{(\rho + p) D_Z Z + \mathbf{grad}\, p + (Zp)Z\} = 0.$

Note that the sum inside the second set of braces is orthogonal to Z, for the following reasons: $g(D_Z Z, Z) = \frac{1}{2} Z g(Z, Z) = \frac{1}{2} Z 1 = 0$. Moreover $g(\mathbf{grad}\, p + (Zp)Z, Z) = g(\mathbf{grad}\, p, Z) + (Zp)g(Z, Z) = Zp - Zp = 0$. Thus the above equation holds iff the sum inside each set of braces vanishes separately. In other words, div $\hat{T} = 0 \Leftrightarrow$ (a) and (b) hold. □

Very roughly, Equation 3.15.3a says that energy densities are increased or decreased in a way determined by the work pressure does; very roughly, Equation 3.15.3b says that mass density times acceleration is a pressure gradient.

Unfortunately the matter Equations 3.15.3 do not lead to an initial value theorem except in special cases such as that discussed in Exercise 3.15.7 below. They must be supplemented by "equations of state" whose structure depends on the details of the physics; a huge number of special cases arise. We will not consider or need them.

EXERCISE 3.15.4

Let $\hat{T}$ be a normal stress-energy tensor on M. Show: (a) $\hat{T}$ is spatially isotropic at every $x \in M$ iff there exists a reference frame Z on M and functions ρ and p on M satisfying $\rho > |p|$ and $\rho + p > 0$, such that $\hat{T} = \rho Z \otimes Z + p(g + Z \otimes Z)$. (b) If $\hat{T}$ is spatially isotropic, then $\hat{T}(\omega, \omega) > 0 \,\forall$ nonspacelike 1-form ω.

EXERCISE 3.15.5

Show that there exists a spacetime M and a quasi-gas $\mathcal{M}$ on M whose stress-energy tensor is spatially isotropic at every $x \in M$ and whose pressure is nowhere zero. (Hint: The simplest case is a quasi-gas on Minkowski space with six components.)

EXERCISE 3.15.6

Let $\mathcal{M}$ be a quasi-gas on M whose components all have the same particle-type (m, e), and whose stress-energy tensor $\hat{T}$ is spatially isotropic at every $x \in M$. Let $\gamma_A: \mathscr{E}_A \to M$ be a particle potentially in the Ath component and let Z be the

3 Electromagnetism and matter

reference frame guaranteed by Proposition 3.14.1. The *random 3-velocity* V_A *of the Ath particle flow* at $z \in M$ is by definition the Newtonian velocity of γ_A observed by (z, Zz); *the energy* e_A *of Ath particle flow* is the energy of γ_A observed by (z, Zz) (Sections 2.1 and 3.2). Note that $e_A > 0$ because of the wrong-way Schwarz inequality (Exercise 1.1.10). Denote the world density of the Ath component by η_A, and the pressure and energy-density of the quasi-gas by p and ρ, respectively. Show:

$$(3pz/\rho z) = \frac{\sum_{A=1}^{N} \eta_A(z) e_A^2 |V_A|^2}{\sum_{A=1}^{N} \eta_A(z) e_A^2} \leq 1.$$

Roughly speaking, this exercise says:

$$3p = \rho \cdot |\text{average random 3-velocity}|^2,$$

which corresponds to the intuitive notion that random velocity creates pressure. In cosmology, the random 3-velocity of the galaxies can be observed and typically $|V_A| < 10^{-2}$. Thus $p \ll \rho$. This exercise should be compared with Exercise 3.8.8.

EXERCISE 3.15.7

For a perfect fluid with rest-mass zero, formulate and prove an initial value theorem analogous to Theorem 3.11.2.

EXERCISE 3.15.8

(a) Show that a perfect fluid with zero pressure is a dust. (b) Let $\mathcal{M}$ be a quasigas on M and W be a reference frame on M. Suppose that the energy-momentum $\boldsymbol{P}_A$ of each component of $\mathcal{M}$ satisfies $\boldsymbol{P}_A = f_A W$, where $f_A > 0$. Show that $\mathcal{M}' = (W, \sum f_A^2 \eta_A)$ is a dust on M.

EXERCISE 3.15.9

Proposition 3.15.3 can be generalized to include electromagnetic interactions. Suppose $\mathcal{M}$ is as in Definition 3.15.2 except that $\boldsymbol{J} = \sigma \boldsymbol{Z}$, where $\sigma \colon M \to \mathbb{R}$ is a C^∞ function. Suppose $(M, F, \boldsymbol{J})$ obeys Maxwell's equations and we have the equation $\operatorname{div}(\hat{\boldsymbol{T}} + \hat{\boldsymbol{E}}) = 0$, where $\hat{\boldsymbol{E}}$ is the stress-energy tensor of F. Find, and interpret roughly, the way in which Proposition 3.15.3 is modified.

The Einstein field equation

4

We now consider the influence of matter and electromagnetism on spacetime.

4.0 Review and notation

Throughout this chapter $(M, \mathscr{M}, F)$ is a relativistic model (Section 3.5); $G = \text{Ric} - \frac{1}{2}g\text{S}$ is the Einstein tensor of the spacetime M (Section 1.0.2); $\hat{G}$ is the (2, 0)-tensor field physically equivalent to G (Section 1.0.1); $\hat{T}$ is the stress-energy tensor of the matter model $\mathscr{M}$ (Section 3.5); and $\hat{E}$ is the stress-energy tensor for the electromagnetic field F (Section 3.7). T and E are the (0, 2) tensor fields physically equivalent to $\hat{T}$ and $\hat{E}$, respectively.

4.1 The Einstein field equation

Einstein argued that the stress-energy of matter and electromagnetism influences spacetime. His specific postulate is summarized in the following definition.

Definition 4.1.1. $(M, \mathscr{M}, F)$ obeys the *Einstein field equation* iff $G = T + E$.

Although many modifications of Einstein's field equation have been suggested, it is generally accepted as a basic postulate of current macrophysics. Many motivations for it have also been suggested. None of the motivations is wholly convincing or simpler than the postulate itself.

> Detailed discussions of the motivations are given by Misner–Thorne–Wheeler [1] and by Weinberg [1]. We briefly outline some of the main ones.

4 The Einstein field equation

(a) In an appropriate limit the Einstein field equation becomes the Poisson equation $\nabla^2 \phi = \rho/2$, which governs gravity in Newtonian physics (Optional exercise 9.3.2).

(b) Assuming the equation leads to results consistent with observation (Section 4.2, Chapters 6 and 7).

(c) The equation has the following consistency property: div $\hat{G} = 0$ (Exercise 3.6.4b) and, for appropriate matter equations, div $(\hat{T} + \hat{E}) = 0$ (cf. Sections 3.9 and 3.10).

(d) If T and E are regarded as given, $G = T + E$ is a system of second-order quasilinear partial differential equations (for the components of the metric tensor g; see Exercise 4.1.5). This is a standard situation in physics.

(e) The equation $G = 0$, corresponding to a vacuum model (Example 3.13.2), comes from a variational principle and leads to a "present determines the future" theorem (Misner–Thorne–Wheeler [1], Weinberg [1]). Both of these properties are also typical in physics. Generalizations to the nonvacuum case exist (cf. Lichnerowicz [1]).

We shall here emphasize the physical consequences and geometric interpretation of Einstein's field equation, rather than historical or philosophical arguments. Section 4.2 indicates how local measurements of relative accelerations ("tidal forces") can be used to check the equation empirically. Section 4.3 discusses a basic effect the equation predicts: gravity tends to pull causal geodesics together.

Exercise 4.1.2

Suppose $(M, \mathcal{M}, F)$ obeys Einstein's equation. Using Proposition 3.7.4, show trace $\hat{T} = -S$.

Exercise 4.1.3

In special relativity, one often uses models of the following kind. $M \in (M, \mathcal{M}, F)$ is Minkowski space, $F \neq 0$, and $T = 0$. Show that then the model does not obey the Einstein field equation.

The idea is that the influence of F on M is negligibly small. Generally speaking, one does not have a fully systematic model unless the model obeys: (a) Maxwell's equations (Definition 3.7.1); (b) Einstein's field equation; and (c) appropriate matter equations (Section 3.12). Only then is one taking into account all the mutual influences. But it is often convenient to be less systematic, just as one often neglects Jupiter's gravity when analyzing the motion of the earth around the sun.

Exercise 4.1.4

Suppose $(M, \mathcal{M}, F)$ obeys Einstein's equation, $F = 0$, and $\mathcal{M}$ is a particle flow (Definition 3.2.1) with a nowhere zero world density. Show that the model obeys the simple matter equations in Section 3.9.1.

Generalizing from this and similar cases, it is sometimes asserted that Einstein's and Maxwell's equations "imply the matter equations." This is very misleading. In more complicated cases, some of the matter equations are independent of Einstein's and Maxwell's equations. This occurs, for example, in the models of Section 3.9.1 when N exceeds one.

EXERCISE 4.1.5

Show that in local coordinates the components of G can be expressed as sums and products of the components of g and $\dot{g}$, first derivatives of the components of g, and the second derivatives of the components of g. Show that the second derivatives of the components of g appear linearly.

4.2 Ricci flat spacetimes

We now analyze how Einstein's field equation is in principle checked empirically. For brevity we take the special case where the equation becomes the geometric condition that spacetime be Ricci flat. Using the notation of Section 4.0, we have the following more physical characterization of this case.

Proposition 4.2.1. *Suppose* $(M, \mathcal{M}, F)$ *obeys Einstein's equation. M is Ricci flat iff* $T = 0 = F$.

PROOF. Suppose $T = 0 = F$. Then $E = 0$, so $G = 0$; by Exercise 1.4.7, **Ric** $= 0$. Conversely, suppose $0 = G = T + E$. Suppose $x \in M$, and let $\omega \in M_x{}^*$ be causal. Then $\hat{T}(\omega, \omega) + \hat{E}(\omega, \omega) = 0$. By Definition 3.3.1, $\hat{T}(\omega, \omega) \geq 0$, and by Proposition 3.7.4b, $\hat{E}(\omega, \omega) \geq 0$. Thus $\hat{T}(\omega, \omega) = 0 = \hat{E}(\omega, \omega)$ $\forall$ causal $\omega \in M_x{}^*$. By Exercise 3.3.4, $T = 0 = E$ at each $x \in M$. It then follows that $T = 0 = F$ (Exercise 3.7.9). □

Thus **Ric** $= 0$ in a region corresponds to vanishing stress-energy and negligible influence of matter and electromagnetism there on spacetime. Vacuum models (Example 3.13.2) which obey the Einstein field equation are the main examples. We thus define a spacetime as *vacuum* iff its Ricci tensor vanishes identically. We now show how freely falling observers can measure whether or not **Ric** $= 0$.

> The class of Ricci flat spacetimes is quite extensive. Minkowski space is flat and hence Ricci flat. The Schwarzschild spacetimes of Example 1.4.2 are Ricci flat but not flat (Chapter 7).
> Assuming Einstein's field equation, the nonvanishing of the Ricci tensor implies $T + E = G \neq 0$ (Exercise 1.4.7) and hence matter or an electromagnetic field would be present. For this reason, the analysis of a non-Ricci flat spacetime often requires detailed discussions of matter theories. It is therefore sometimes convenient to focus attention on a Ricci flat open submanifold (of a larger spacetime which may be not Ricci flat).
> As an example, the normal Schwarzschild spacetime is simple and

4 The Einstein field equation

can be used to get a good model for the history of the exterior of the earth. However, a complete model for the region including the interior and exterior of the earth would require a detailed knowledge of terrestrial matter composition, temperature variations in the earth's core, ferromagnetism, and so on. Things then become complicated.

Let Q be a geodesic reference frame defined on an open subset $\mathscr{U} \subset M$. For $z \in \mathscr{U}$, set $Z = Qz$ and let $M_z = R \oplus T$ be the associated orthogonal decomposition (Section 2.1.3). Thus R is the local rest space for the instantaneous observer (z, Z). Let $\psi_z : R \to R$ be the self-adjoint linear transformation defined above Proposition 2.3.3. Thus if $\gamma : \mathscr{E} \to M$ is an observer in Q with $\gamma 0 = z$ and W is any neighbor of γ in Q, then $\psi_z W(0)$ is the 3-acceleration of W relative to γ at proper time zero. We define the *mean relative-acceleration of Q* to be the function $\alpha : \mathscr{U} \to \mathbb{R}$ with rule $\alpha z = (1/3)$ trace ψ_z $\forall\, (z, Z)$ as above. The next proposition gives a geometric interpretation of α.

There is a conceptually more satisfactory definition of α explicitly showing it as a mean over the unit sphere $\mathscr{S}^2$ of R. Let ζ denote the standard volume element of $\mathscr{S}^2$ (Section 0.0.9) and let σ be the volume of $\mathscr{S}^2$—that is, $\sigma = \int_{\mathscr{S}^2} \zeta = 4\pi$. Then one may define:

$$\alpha z = \frac{1}{\sigma} \int_{X \in \mathscr{S}^2} g(\psi_z X, X)\zeta.$$

(See Optional exercise 8.1.11.)

Proposition 4.2.2. *The mean relative-acceleration of a geodesic reference frame Q equals $-(1/3)$ Ric (Q, Q).*

PROOF. For (z, Z) as above, let $\{X_i\}$ be an orthonormal basis of M_z such that $X_4 = Z$; let $\{\omega^i\}$ be the dual basis and R be the curvature tensor. Then for the rest space we have $R = \text{span} \{X_1, X_2, X_3\}$; we also have $R(\omega^4, Z, Z, X_4) = R(\omega^4, Z, Z, Z) = 0$. Using the definitions of ψ_z and Ric we now get: $\alpha z = (1/3)$ trace $\psi_z = (1/3) \sum_{\mu=1}^{3} \omega^\mu(\psi_z X_\mu) = (1/3) \sum_{\mu=1}^{3} R(\omega^\mu, Z, Z, X_\mu) = (1/3) \sum_{i=1}^{4} R(\omega^i, Z, Z, X_i) = -(1/3)$ Ric (Z, Z). Since this holds for each $z \in \mathscr{U}$, $\alpha = -(1/3)$ Ric (Q, Q). □

Put intuitively, the situation is this. Imagine you are at the center of a big, freely falling elevator. There is a whole bushel of freely falling apples floating around the elevator. The Newtonian velocity of an apple does not matter; what counts is the apple's observed relative 3-acceleration toward you or away from you. Now look at all those apples that are at unit distance; average the observed acceleration over apples. Out pops the Ricci tensor—specifically, $-(1/3)$ Ric (Z, Z), where Z is your own world velocity.

Proposition 4.2.3. Ric $= 0$ *iff the mean relative-acceleration of every locally defined geodesic reference frame vanishes.*

Proof. The "only if" half follows directly from Proposition 4.2.2. Conversely, suppose $\alpha = 0\,\forall Q$ as above. By Proposition 3.3.4, it suffices to show Ric $(Z, Z) = 0\,\forall$ unit timelike vector $Z \in M_z$, $\forall z \in M$.

4.2 Ricci flat spacetimes

Let $\mathcal{N}$ be a 3-dimensional spacelike submanifold through z such that the given Z is orthogonal to $\mathcal{N}$ at z. Let Z be a unit vector field defined on $\mathcal{N}$ such that $Zz = Z$ and Zy is orthogonal to $\mathcal{N}$ at y $\forall y \in \mathcal{N}$. Define a map $\phi: \mathcal{N} \times (-\varepsilon, \varepsilon) \to M$ by $\phi(y, t) = \gamma t$, where γ is the geodesic issuing from y with $\gamma 0 = y$ and $\gamma_* 0 = Zy$. ϕ is a C^∞ map by virtue of the C^∞ dependence of solutions of ordinary differential equations on their initial conditions (cf. the proof of Theorem 3.8.3). At $(z, 0)$, ϕ_* is the identity on $\mathcal{N}_z$ and carries (d/dt) onto Z. Thus ϕ_* is nonsingular at $(z, 0)$. By the inverse function theorem, ϕ is a diffeomorphism in a neighborhood of $(z, 0)$. We may assume $\mathcal{N}$ and ε to be so small that ϕ is a diffeomorphism on $\mathcal{N} \times (-\varepsilon, \varepsilon)$ itself. Let $Q = \phi_*(d/dt)$. Then Q is a geodesic reference frame in image ϕ, and $Q|_\mathcal{N} = Z$. Let α be the mean relative-acceleration of Q. By Proposition 4.2.2 and by hypothesis,

$$\text{Ric}(Z, Z) = -3(\alpha z) = 0,$$

as desired. □

Thus in vacuum we should have zero mean relative-acceleration.

Experimentally this result has been checked, directly and indirectly. In some cases, such as the gravitational field of the sun, accuracies as high as one part in ten million are attained. However, in these same cases the Newtonian limit, discussed below, agrees with the relativistic results to almost as high an accuracy. Thus the Einstein field equation is known to be an extremely accurate approximation; but the deviations from Newtonian theory it predicts have at best been observed to an accuracy of 1%.

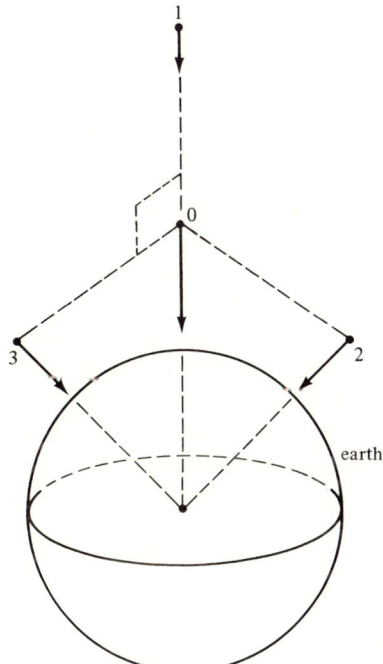

Figure 4.2.4a

4 The Einstein field equation

On an intuitive level, the above physical interpretation of **Ric** = 0 can also be given in Newtonian physics. Imagine four Newtonian point particles placed at Newtonian rest in the Newtonian gravitational field of the earth (Figure 4.2.4a). We assume that 0 and 1 are collinear with the center of the earth, while the plane determined by 0, 2, 3 is perpendicular to the line joining 0 and 1. We further assume that the lines joining 0, 3 and 0, 2 are perpendicular. Now allow all four particles to fall freely. Because 0 is closer to the earth than 1, 0 (Newtonian) accelerates more than 1 and outruns 1. On the other hand, since 0, 2, and 3 follow converging lines toward the earth center, both 2 and 3 accelerate toward 0. Suppose 0 regards herself as at rest, then she sees the pattern in Figure 4.2.4b of the accelerations of her neighbors relative to herself.

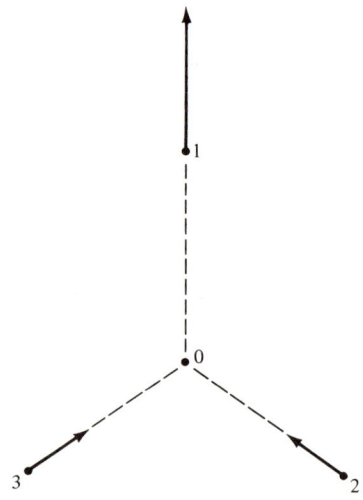

Figure 4.2.4b

Suppose we let $\vec{\alpha}_\mu$ ($\mu = 1, 2, 3$) be the Newtonian acceleration of μ relative to 0 and let $\vec{e}_\mu$ be the unit vector from 0 pointing toward μ. Then we see from the diagram that the dot products obey: $\vec{\alpha}_1 \cdot \vec{e}_1 > 0$, $\vec{\alpha}_2 \cdot \vec{e}_2 < 0$, and $\vec{\alpha}_3 \cdot \vec{e}_3 < 0$. It is then not unreasonable to expect that, if 1, 2, and 3 are equidistant from 0, the sum of these dot products should be zero. In fact, a computation using Newton's inverse square law shows that then, $-\vec{\alpha}_1 \cdot \vec{e}_1 = \frac{1}{2}\vec{\alpha}_2 \cdot \vec{e}_2 = \frac{1}{2}\vec{\alpha}_3 \cdot \vec{e}_3 > 0$, which implies $\sum_\mu \vec{\alpha}_\mu \cdot \vec{e}_\mu = 0$. This is the intuitive Newtonian analogue of Proposition 4.2.3.

We now generalize to show that within Newtonian theory one has the following: active mass density negligible iff $\nabla^2 \phi = 0$ (corresponding to stress-energy negligible iff **Ric** = 0, Proposition 4.2.1); and $\nabla^2 \phi = 0$ iff Newtonian mean relative-accelerations zero (corresponding to the relativistic result, Proposition 4.2.3).

In fact, Poisson's equation $\nabla^2 \phi = \frac{1}{2}nm$ gives: nm negligible iff $\nabla^2 \phi = 0$ (Section 0.1.9; we will use Sections 0.1.5 to 0.1.10 throughout the following; see also Section 2.1.2).

For the second Newtonian result, suppose $\vec{x}, \vec{y}: \mathbb{R} \to \mathbb{R}^3$ are freely falling Newtonian point particles. Thus (Section 0.1.7):
$$\ddot{\vec{x}} = -\vec{\nabla}\phi \circ \vec{x}, \qquad \ddot{\vec{y}} = -\vec{\nabla}\phi \circ \vec{y}.$$
Let $\vec{n} = \vec{y} - \vec{x}$ and assume that the particles are nearby—that is, that $|\vec{n}|$ is so small the Taylor series following are applicable. For the relative acceleration $\ddot{\vec{n}} = \ddot{\vec{y}} - \ddot{\vec{x}}$ we have, in components:
$$n^\alpha = -\left(\frac{\partial\phi}{\partial u^\alpha}\right)(x^1 + n^1, x^2 + n^2, x^3 + n^3)$$
$$-\left(\frac{\partial\phi}{\partial u^\alpha}\right)(x^1, x^2, x^3)$$
$$= -\sum_{\beta=1}^{3}\left(\frac{\partial^2\phi}{\partial u^\alpha \partial u^\beta}\right)(x^1, x^2, x^3)n^\beta + \text{SMALL}.$$
where SMALL is the finite Taylor series correction term. Thus for each $t \in \mathbb{R}$, to first order:
$$\ddot{\vec{n}}(t) = -\Psi_t \vec{n}(t),$$
where Ψ_t is that linear transformation whose matrix elements are
$$(\partial^2\phi/\partial u^\alpha \partial u^\beta)(x^1(t), x^2(t), x^3(t)).$$

Thus $(1/3)$ trace Ψ_t is the Newtonian mean relative-acceleration for freely falling particles near $\vec{x}$ at time t. Now trace $\Psi_t = (\nabla^2\phi)[\vec{x}(t)]$. Thus we have: $\nabla^2\phi = 0$ iff trace $\Psi_t = 0$ for all $\vec{x}$ and all t as above. Thus Proposition 4.2.3 has a full analogue within Newtonian theory. In this sense empirical checks of Newtonian theory also check Einstein's field equation.

EXERCISE 4.2.5

Show that the curvature tensor of a spacetime can be completely determined by relative-acceleration measurements. Specifically, suppose $x \in M$ and gx are given. Let $R \in T_3{}^1(M_x)$ and $R' \in T_3{}^1(M_x)$ both have the algebraic properties of a curvature tensor (Section 1.0.2). Show that if $R_{ZX}Z = R'_{ZX}Z \;\forall$ unit timelike $Z \in M_x$ and $\forall\; X \in Z^\perp \subset M_x$, then $R = R'$.

EXERCISE 4.2.6

Show that for each instantaneous observer $(z, Z) \in TM$ there exists a geodesic, irrotational reference frame Z, defined on some sufficiently small open neighborhood of z, such that $Zz = Z$. (Hint: See the proof of Proposition 4.2.3).

4.3 Gravitational attraction and the phenomenon of collapse

Newtonian intuition says that gravity attracts, rather than repels. The Einstein field equation makes two corresponding predictions. First, for most of the matter models used in physics, the mean relative-acceleration (Section 4.2) of any geodesic reference frame is negative ("inward"). This section will

4 The Einstein field equation

start by giving a specific example. Second, under certain circumstances, the attraction leads to a catastrophic collapse. To formulate precisely, let alone prove, the main theorems on collapse requires rather sophisticated global techniques, due mainly to Penrose, Hawking, and Geroch (Hawking–Ellis [1], Chapters 8 to 10). In this section we give instead an earlier collapse theorem, due to Raychaudhuri (*ibid.*), which exhibits some of the main ideas in their simplest form.

> Optional exercises 8.3 are intended as an introduction to the global techniques used in the general theorems.

4.3.1 Assumptions

Throughout this section, $(M, \mathcal{M}, F)$, **G**, **Ric**, S, **T**, **E**, $\hat{T}$, $\hat{E}$, and $\hat{G}$ are as in Section 4.0. To see explicitly in what sense gravity attracts, we now assume that $(M, \mathcal{M}, F)$ is a perfect fluid (Definition 3.15.2) which obeys the Einstein field equation, leaving the similar results for other cases to Exercise 4.3.7. Thus the matter variables are the energy density ρ, the pressure p, and the comoving reference frame **Z**. We have $\rho > 0$, $\rho \geq 3p \geq 0$, and $\hat{G} = (\rho + p)\mathbf{Z} \otimes \mathbf{Z} + p\hat{g} + \hat{E}$. Let **Q** be any geodesic reference frame defined on an open subset $\mathcal{U} \subset M$; let $\alpha: \mathcal{U} \to \mathbb{R}$ be the mean relative-acceleration of **Q**. Gravitational attraction shows up by the fact that α is negative.

Proposition 4.3.2. $\alpha \leq -(1/6)(\rho + 3p) - (1/3)E(\mathbf{Q}, \mathbf{Q}) < 0$; *moreover, at* $z \in \mathcal{U}$, *the first inequality becomes an equality iff* $\mathbf{Q}z = \mathbf{Z}z$.

PROOF. Since trace $\hat{E} = 0$ (Proposition 3.7.4), trace $\hat{G} = -\rho - p + 4p = 3p - \rho$. Section 1.0.2 and Exercise 1.4.7 give **Ric** $= (\rho + p)\omega \otimes \omega + \frac{1}{2}(\rho - p)g + E$, where ω is physically equivalent to **Z**. Thus **Ric**$(\mathbf{Q}, \mathbf{Q}) = \frac{1}{2}\rho\{[\omega(\mathbf{Q})]^2 + [\omega(\mathbf{Q})]^2 - 1\} + p\{[\omega(\mathbf{Q})]^2 + \frac{1}{2}\} + E(\mathbf{Q}, \mathbf{Q})$. By the wrong-way Schwarz inequality (Exercise 1.1.10), $[\omega(\mathbf{Q})]^2 \geq 1$, where equality holds iff $\mathbf{Z}|_\mathcal{U} = \mathbf{Q}$; since **E** is a stress-energy tensor, $\hat{E}(\mathbf{Q}, \mathbf{Q}) \geq 0$, By Proposition 4.2.3, $\alpha = -\frac{1}{3}$ **Ric**$(\mathbf{Q}, \mathbf{Q})$. Both halves of the proposition now follow algebraically. □

> Note that if $\mathbf{Q}z \neq \mathbf{Z}z$, the mean relative acceleration at z is even more negative than if $\mathbf{Q}z = \mathbf{Z}z$. Roughly, the reason is the following. If $\mathbf{Q}z \neq \mathbf{Z}z$, the particles in the fluid have, on the average, some extra Newtonian velocity in the rest frame $(\mathbf{Q}z)^\perp$ (cf. Exercise 3.8.8). From the point of view of $(z, \mathbf{Q}z)$, this corresponds, very roughly, to extra kinetic energy, extra active-mass, and thus extra gravity.
>
> Note that the pressure contributes a negative term in Proposition 4.3.2. This contradicts our Newtonian intuition that pressures tend to push things apart. The following gives a rough explanation.
>
> It is indeed often true that a pressure gradient ***grad*** *p* (Section 3.15) tends to push things apart. This can best be made plausible by ana-

4.3 Gravitational attraction and the phenomenon of collapse

lyzing Proposition 3.15.3b in some special cases, but the following shows the idea. Suppose we have a gas in a tube with a moveable piston and have air outside. Then it is the pressure difference between air and gas—corresponding to a pressure gradient—rather than the air pressure alone or gas pressure alone which determines how the piston moves.

But general relativity, unlike Newtonian theory, predicts that a pressure *per se* has an additional, quite different role. Roughly, such a pressure corresponds to some extra kinetic energy (cf. Exercise 3.8.8) which generates an attractive gravitational field. Now in typical laboratory situations, p is negligible compared to ρ but **grad** p dominates. Then our Newtonian intuition gives qualitatively correct results. But in extreme collapse situations it can (presumably) happen that the general relativistic, attracting, gravitational effect of p indicated by Proposition 4.3.2 is more important than the standard effects of **grad** p. Then a higher pressure merely speeds up the collapse.

To analyze collapse in a specific situation, we now consider the special case of Proposition 4.3.2 where the pressure p is zero (hence $\mathscr{M}$ is a dust; see Exercise 3.15.8) and the comoving reference frame Z is irrotational (Section 2.3). Roughly the idea is that then there is neither a pressure gradient **grad** p nor a "centrifugal" effect of rotation to counterbalance the attraction effects (Proposition 4.3.2) of gravity (see the interpretation of irrotationality after Proposition 2.3.4). This gives a drastic simplification, which obviates the need for global hypothesis in the elementary collapse theorem (Theorem 4.3.4, below) we want to prove. Formally then we have the following simplifications (assuming zero electromagnetic field henceforth):

$D_Z Z = 0$ and div $(\rho Z) = 0$ (Proposition 3.15.3; in particular, all the observers in Z are geodesics); Z is irrotational (Section 2.3); and $S = \rho$ (Exercise 4.1.2).

Lemma 4.3.3

(a) $Z(\text{div } Z) \leq -\rho/2 - (\text{div } Z)^2/3$.
(b) $Z\rho = -\rho(\text{div } Z)$.

In Optional exercise 8.1.10, we indicate an alternate, slick proof of the lemma.

PROOF. To prove (a) let $\{X_1, X_2, X_3\}$ be vector fields defined on an open set $\mathscr{U}$ of M such that $\{X_1, X_2, X_3, Z\}$ is orthonormal in $\mathscr{U}$. It suffices to verify the inequality in $\mathscr{U}$. In the following, Greek letters α, β, μ, ν will run from 1 to 3, and any such repeated index will imply summation from 1 to 3. Also, $\perp$ will stand for "is orthogonal to."

We first note that $D_{X_\alpha} Z \perp Z$, and $[Z, X_\alpha] \perp Z$. Indeed, Z being a geodesic reference frame implies:

$$g(D_Z X_\alpha, Z) = Zg(X_\alpha, Z) - g(X_\alpha, D_Z Z) = 0,$$
$$g(D_{X_\alpha} Z, Z) = \tfrac{1}{2} X_\alpha g(Z, Z) = 0,$$

4 The Einstein field equation

and
$$g([Z, X_\alpha], Z) = g(D_Z X_\alpha - D_{X_\alpha} Z, Z) = 0.$$
In particular, we may write
$$D_{X_\alpha} Z = f_\alpha^\beta X_\beta$$
for some functions $\{f_\alpha^\beta\}$ in $\mathcal{U}$. Let $\{\omega^1, \omega^2, \omega^3, \omega\}$ be the dual basis of $\{X_1, X_2, X_3, Z\}$. Then each ω^α is physically equivalent to X_α and ω is physically equivalent to $-Z$. This implies:
$$\text{div } Z = \omega^\alpha(D_{X_\alpha} Z) + \omega(D_Z Z)$$
$$= g(D_{X_\alpha} Z, X_\alpha),$$
or
$$\text{div } Z = f_\alpha^\alpha.$$

Finally, if X is a vector field on $\mathcal{U}$, denote the vector field $-D_X Z$ by $A_Z X$. Thus $A_Z X_\alpha = -f_\alpha^\beta X_\beta$. Note that this notation is consistent with that of Proposition 2.3.4. By that proposition, $X \perp Z \Rightarrow A_Z X \perp Z$. Moreover, since Z is irrotational, $g(A_Z X, W) = g(X, A_Z W) \; \forall \; X, W \perp Z$. In particular, this implies $f_\beta^\alpha = f_\alpha^\beta$. With these preparations, we have:

$$Z(\text{div } Z) = Zg(D_{X_\alpha} Z, X_\alpha)$$
$$= g(D_Z D_{X_\alpha} Z, X_\alpha) + g(D_{X_\alpha} Z, D_Z X_\alpha)$$
$$= g(R_{ZX_\alpha} Z + D_{[Z, X_\alpha]} Z, X_\alpha) + g(D_{X_\alpha} Z, D_Z X_\alpha)$$
$$= -\text{Ric}(Z, Z) + g(D_{[Z, X_\alpha]} Z, X_\alpha) + g(D_{X_\alpha} Z, D_Z X_\alpha)$$
$$= -\text{Ric}(Z, Z) - g(A_Z[Z, X_\alpha], X_\alpha) + f_\alpha^\beta g(X_\beta, D_Z X_\alpha)$$
$$= -\text{Ric}(Z, Z) - g([Z, X_\alpha], A_Z X_\alpha) + f_\alpha^\beta g(D_Z X_\alpha, X_\beta)$$
$$= -\text{Ric}(Z, Z) + f_\alpha^\beta g([Z, X_\alpha], X_\beta) + f_\alpha^\beta g(D_Z X_\alpha, X_\beta)$$
$$= -\text{Ric}(Z, Z) + f_\alpha^\beta g(D_Z X_\alpha + [Z, X_\alpha], X_\beta)$$
$$= -\text{Ric}(Z, Z) - f_\alpha^\beta g(D_{X_\alpha} Z, X_\beta) + 2 f_\alpha^\beta g(D_Z X_\alpha, X_\beta),$$

the last equality is because $[Z, X_\alpha] = D_Z X_\alpha - D_{X_\alpha} Z$. Since $\{X_\alpha\}$ is orthonormal, $g(D_Z X_\alpha, X_\beta) = Zg(X_\alpha, X_\beta) - g(X_\alpha, D_Z X_\beta) = -g(D_Z X_\beta, X_\alpha)$, so that $g(D_Z X_\alpha, X_\beta)$ is skew-symmetric in α and β. But we have already observed that f_α^β is symmetric in α and β. Thus the sum $f_\alpha^\beta g(D_Z X_\alpha, X_\beta)$ must vanish. Consequently,

$$Z(\text{div } Z) = -\text{Ric}(Z, Z) - f_\alpha^\beta f_\alpha^\beta$$
$$= -\text{Ric}(Z, Z) - \{(f_1^1)^2 + (f_2^2)^2 + (f_3^3)^2\} - \sum_{\alpha \neq \beta} (f_\alpha^\beta)^2$$
$$\leq -\text{Ric}(Z, Z) - \{(f_1^1)^2 + (f_2^2)^2 + (f_3^3)^2\}$$
$$\leq -\text{Ric}(Z, Z) - \tfrac{1}{3}(f_\alpha^\alpha)^2 \quad \text{(Schwarz inequality)}$$
$$= -\text{Ric}(Z, Z) - \tfrac{1}{3}(\text{div } Z)^2$$
$$\leq -\tfrac{1}{2}\rho - \tfrac{1}{3}(\text{div } Z)^2$$

where the last inequality follows from Propositions 4.2.2 and 4.3.2.
For (b) we note that $\text{div}(\rho Z) = Z\rho + \rho \text{ div } Z$ (Section 3.0.2). □

To interpret the lemma, we need an interpretation of div Z. Let A_Z be the linear transformation of Proposition 2.3.4 which assigns negative 3-

velocities to neighbors. The preceding proof (or a computation using the definition of divergence in Section 3.0.3) shows that div $Z = -$ trace A_Z. In this sense div Z measures how much the dust is expanding (compare the definition of mean relative-acceleration in Section 4.2 and Optional exercise 8.1.11). More specifically, div $Z > 0$ in $\mathcal{U} \subset M$ indicates that the observers in Z are (on the average) spreading apart in $\mathcal{U}$; similarly div $Z < 0$ indicates coming together. The two halves of the lemma can now be interpreted as follows: (a) the gravity generated by the energy density ρ of the dust drives the expansion-measuring function (div Z) toward less positive or more negative values; (b) as the dust expands or contracts, the energy density decreases or increases correspondingly.

Now, retaining all our above assumptions, let $\gamma: \mathscr{E} \to M$ be an observer in Z. Then by Lemma 4.3.3a, $(d/du)[(\text{div } Z)\gamma u] < 0$. Thus there is at most one $u \in \mathscr{E}$ for which (div $Z)\gamma u = 0$. Suppose (div $Z)\gamma u > 0$ for all $u \in \mathscr{E}$. By introducing the spacetime with the opposite time orientation one can reduce this, mathematically, to the case (div $Z)\gamma u < 0$. However, the preceding interpretation of (div Z) means that physically there is a big difference. Indeed, the positive divergence case is of interest primarily in cosmology, where one is considering expansion from indefinitely large energy densities near the big bang (cf. Corollary 1.4.6, Example 2.3.6, and Proposition 2.3.7); the case where the divergence is negative for at least one proper time u is relevant when analyzing matter collapsing toward a black hole. But formally, we can and shall assume $0 \in \mathscr{E}$ and (div $Z)\gamma 0 < 0$ without essential loss of generality.

Then a rather grim future awaits the freely falling observer γ. Even if M is maximal, he must leave M in a given finite proper time. Moreover, his last moments may be plagued by unbounded energy density and scalar curvature. All these interpretations follow from the next theorem.

Theorem 4.3.4. *Let* $\gamma: [0, a) \to M$ *be an observer in* Z *with* (div $Z)\gamma 0 = b \in (-\infty, 0)$. *Then* $a \leq 3/|b|$; *moreover, if* $a = 3/|b|$,

$$\lim_{u \to a} \rho\gamma u = \infty = \lim_{u \to a} S\gamma u.$$

Proof. Let $f = $ div $Z \circ \gamma$ and let $h = \rho \circ \gamma$. Then f and h are functions defined on $[0, a)$, with $h > 0$. By Lemma 4.3.3:

$$f' < -\tfrac{1}{3}f^2, \qquad (\ln h)' = -f$$

where prime denotes differentiation. Since $f0 < 0$ and $f' < 0$, $f < 0$ on $[0, a)$. Thus $(1/f)' = -f'/f^2 > \tfrac{1}{3}$. For $u \in [0, a)$ integrating this inequality from 0 to u gives $-(1/b) + [1/f(u)] > u/3$. Thus $f(u) < 3b/(3 + bu)$. $\lim_{u \to 3/|b|} 3b/(3 + bu) = -\infty$; since f is defined on $[0, a)$, this gives $a \leq 3/|b|$.

Suppose now $a = 3/|b|$. We have $(\ln h)'(u) = -f(u) > -3b/(3 + bu) = -3[\ln (3 + bu)]'$ for all $u \in [0, a)$. Thus $\ln [h(u)/h(0)] > 3 \ln [3/(3 + bu)]$ and thus $h(u) > h(0)[3/(3 + bu)]^3$. Thus $\lim_{u \to a} \rho\gamma u = \lim_{u \to 3/b} h(u) = \infty$. Since $S = \rho$, the scalar curvature also approaches infinity. □

4 The Einstein field equation

The conceptual content of this theorem and its proof from the point of view of Riemannian geometry is the following. Notation as above, suppose γ is infinitely extendible in both directions in terms of its parameter. Since Z is irrotational, it is locally synchronizable (Proposition 2.3.5). Hence there exists a function h defined in a neighborhood of $x = \gamma 0$ such that each of its level sets $\{h = \text{constant}\}$ is an imbedded spacelike hypersurface everywhere orthogonal to Z. Let $\mathcal{N}$ be such a hypersurface passing through x. The integral curves of Z are just geodesics orthogonal to $\mathcal{N}$. Since Z is a nowhere zero vector field, its integral curves "do not crowd together": the exponential map exp of the normal bundle $N(\mathcal{N})$ of $\mathcal{N}$ is everywhere nonsingular on the fibre (span Zx) of $N(\mathcal{N})$ over x. In particular, if $\mathbf{Y}$ is the (4, 0)-tensor field in $N(\mathcal{N})$ dual to the canonical volume element of $N(\mathcal{N})$ and if $\mathbf{Y}_0$ denotes $\exp_* \mathbf{Y}$, then $\mathbf{Y}_0$ is nowhere zero on γ. A computation reveals that $L_Z \mathbf{Y}_0 = (\text{div } Z) \mathbf{Y}_0$. Thus the function $f = (\text{div } Z) \circ \gamma$ is defined over $\mathbb{R}$. However, the basic inequality in the above proof leads to $f' < -\frac{1}{3}f^2$ on $\mathbb{R}$, which implies $(1/f)' > \frac{1}{3}$ on $\mathbb{R}$. By an integration, this leads to the fact that f must be infinite or $-\infty$ at a finite value, contradiction.

One should also note that div Z at $y \in \mathcal{N}$ is just the trace of the second fundamental form of $\mathcal{N}$ at y with respect to Zy. Using this fact and the positivity of $\text{Ric}(Z, Z)$, the reader familiar with the Morse theory of focal points (cf. Bishop–Crittendon [1]) can also write out a direct proof showing f cannot be nowhere zero. Indeed, if, say, $(\text{div } Z)x < 0$ and $\gamma 0 = x$, then the index form on $\gamma[0, b]$ for a sufficiently large b must be negative on a suitably chosen vector field along γ on $[0, b]$; compare the usual proof of Myers Theorem. Then $\mathcal{N}$ has a focal point c along γ before b. This implies $f(c) = \infty$, by definition of the exponential map of $N(\mathcal{N})$. Contradiction.

EXERCISE 4.3.5

Let M be a spacetime and let Z be a geodesic irrotational reference frame on M such that $\text{Ric}(Z, Z) \geq 0$. Show: (a) If $\text{Ric}(Z, Z)x > 0$ for some $x \in M$, then an observer in Z through x cannot have all of $\mathbb{R}$ as his proper time domain. (b) If Z is complete, then $\text{Ric}(Z, Z) = 0 = \text{div } Z$.

EXERCISE 4.3.6

Let Z be an irrotational geodesic reference frame on a spacetime M. At each $x \in M$, if R denotes the local rest space of (x, Zx), then we have the self-adjoint linear transformation $A_Z : R \to R$ of Proposition 2.3.4 defined by $A_Z X = -D_X Z$. Now define a (not necessarily C^∞) function $h: M \to [0, \infty)$ by $hx = \sum_{\alpha=1}^{3} \lambda_\alpha^2$, where $\{\lambda_\alpha\}$ are the eigenvalues of A_Z. (a) Show that h is C^∞. (b) Elaborate on the proof of Lemma 4.3.3 to prove the following special case of the *Raychaudhuri equation*:

$$Z(\text{div } Z) = -\text{Ric}(Z\ Z) - h.$$

(c) Use (b) and Exercise 4.3.5b to show that if Z is complete and $\text{Ric}(Z, Z) \geq 0$, then Z is a parallel vector field.

Since the deRham decomposition theorem is known to hold for semi-Riemannian manifolds (see, for example, Wu [1]), (c) above leads to

4.3 Gravitational attraction and the phenomenon of collapse

the following geometric theorem: Suppose M is a spacetime which admits a geodesic irrotational reference frame Z which is complete and has the property that **Ric** $(Z, Z) \geq 0,$ then locally M is isometric to a direct product $M' \times \mathbb{R}$, where M' is a 3-dimensional Riemannian manifold. If M is simply connected and complete, then M is globally isometric to a direct product as above.

EXERCISE 4.3.7

Let $\hat{T}$ be a stress-energy tensor on a spacetime M. Define $\hat{W} = \hat{T} - \frac{1}{2}(\text{trace } \hat{T})\hat{g}$, where $\hat{g}$ is the (2, 0)-tensor field physically equivalent to g; $\hat{W}$ is then a symmetric (2, 0)-tensor field. We say $\hat{T}$ obeys the *timelike convergence condition* iff $\hat{W}(\omega, \omega) > 0$ $\forall$ timelike 1-form ω. Show: (a) The stress-energy tensor of a quasigas (Example 3.13.4) obeys the timelike convergence condition. (b) Let $(M, \mathcal{M}, F)$ be a relativistic model which obeys the Einstein field equation, and let $\hat{T}$ obey the timelike convergence condition. Then **Ric** $(Z, Z) > 0$ $\forall$ timelike vector field Z. (c) Assumption as in (b), let Z be a geodesic irrotational reference frame on M. Then no observer in Z can have all of $\mathbb{R}$ as domain.

(c) accounts for the nomenclature of "timelike convergence." Indeed, from a geometric standpoint (as expounded in the fine-print section after the proof of Theorem 4.3.4), (c) holds because the neighboring geodesics of each timelike geodesic γ tangent to Z must converge on γ "sooner or later."

5 Photons

Most of our information about the solar system, stars, galaxies, and the universe as a whole comes from observing photons. This chapter analyzes the lightlike objects—vectors, curves, submanifolds, and so on—used in interpreting photon observations. Throughout the chapter, (M, g) is a spacetime. We shall say a curve $\gamma: \mathcal{E} \to M$ goes *from* $x \in M$ *to* $y \in M$ iff $\mathcal{E} = [a, b]$, $x = \gamma a$ and $y = \gamma b$.

5.0 Mathematical preliminaries

5.0.1 Causality

Suppose $x, z \in M$. By definition, x *chronologically precedes* z iff there exists a future-pointing timelike curve γ from x to z. Here "chronologically" refers to the fact that the arclength of γ models comoving clock time (Section 2.1). We thus get a binary *chronology relation* $\ll$ on M, with $x \ll z$ iff x chronologically precedes z. Similarly, define a *causality relation* by $x \leq z$ iff $x = z$ or there exists a future pointing (and thus causal) curve from x to z. Intuitively, $x \leq z$ iff z can get some information about x. Given $z \in M$, the *chronological* (respectively, *causal*) *past of* z is the set of points in M that chronologically (respectively, causally) precede z. In discussions of chronology and causality we will, as a mnemonic, generally use alphabetical order—for example, $x \leq y \ll z$.

Formally, $\ll$ is a subset of $\leq$ which in turn is a subset of $M \times M$. Locally, both $\ll$ and $\leq$ have many of the properties suggested by the notation; globally, both can be extremely subtle, with $\ll$ always beautiful and $\leq$ often messy (Section 8.3). Computationally both are often calculated *via* the following trick. Suppose (M, g') is conformal to (M, g); then $g' = fg$, $f > 0$, and a given curve γ is future pointing for (M, g) iff it is future pointing

5.0 Mathematical preliminaries

for (M, g') (Exercise 3.7.8). Thus $\leq \, = \, \leq'$; similarly $\ll \, = \, \ll'$. One says both relations are *conformally invariant*.

In discussions of chronology and causality, one generally has dual definitions—for example, *chronologically follows*, corresponding to reversing the time orientation of (M, g), and dual results. These will generally be taken for granted.

5.0.2 Geodesics

Let $\gamma: \mathscr{E} \to M$ be a geodesic, $\gamma^\sim$ be a positive affine reparametrization of γ (Section 0.0.6); then $\gamma^\sim$ is a geodesic. Let $[\gamma]$ denote the corresponding equivalence class of geodesics. For example suppose γ is a $1-1$ future-pointing geodesic and $\gamma^\sim$ is a curve; $\gamma^\sim \in [\gamma]$ iff $\gamma^\sim$ is a future-pointing geodesic whose image in M coincides with that of γ. If we fail to single out a representative in an equivalence class $[\gamma]$ of causal geodesics the connotation will be that some kind of genuine equivalence is involved. For example, in Chapter 6 the cosmological red shift is the same whether one is observing radio, microwave, visible, or x-ray photons, and this corresponds to the fact that cosmological red shift is a property of a whole equivalence class of lightlike future-pointing geodesics (cf. Sections 5.4 and 6.0.7).

Let $\gamma: \mathscr{E} \to M$ be an inextendible causal geodesic. Then γ_* is nowhere zero and the following hold:

a. $g(\gamma_*, \gamma_*) = $ constant.
b. $g(\gamma_*, K) = $ constant $\forall$ Killing vector field K (Section 3.6.3).
c. $\phi \circ \gamma$ is an inextendible geodesic $\forall$ isometry $\phi: M \to M$.
d. Given $u \in \mathscr{E}$, the initial data $(\gamma u, \gamma_* u)$ determines γ uniquely.
e. Suppose an isometry ϕ leaves the initial data invariant—that is, $\phi \gamma u = \gamma u$, $\phi_* \gamma_* u = \gamma_* u$. Then $\phi \circ \gamma = \gamma$ by (c) and (d). We shall manage to avoid computing connection coefficients in finding the explicit geodesics needed here by systematic use of (a)–(e).

> In spacetimes less highly idealized than our examples, the isometry group $\mathscr{G}M$ is trivial (Section 8.4), and (b), (c), and (e) become useless. Then one must normally fall back on local basis computations as in the proof of Theorem 3.8.3.

5.0.3 The Gauss Lemma

Suppose $b > a > 0$, $\varepsilon \in (0, \infty)$. Let $\mathscr{D} \subset \mathbb{R}^2$ be the subset $[a, b] \times [-\varepsilon, \varepsilon]$, $\sigma: \mathscr{D} \to M$ be a C^∞ map—that is, there is a C^∞ extension with open domain. Then $T_1 = \sigma_* \partial_1$ and $T_2 = \sigma_* \partial_2$ are vector fields over σ (Section 2.0). Assume: (a) $|T_1| = $ constant; (b) $\forall v \in [-\varepsilon, \varepsilon]$, $\sigma_v: [a, b] \to M$, defined by $\sigma_v u = \sigma(u, v)$, is a geodesic. The *Gauss Lemma* asserts that then $g(T_1, T_2)$ is constant along each geodesic σ_v (compare Section 2.0.3e).

> Since many of the standard texts give the proof only for the case that σ is an imbedding, we give the general proof. Since $[a, b]$ is connected

125

5 Photons

it suffices to show that $\partial_1 g[T_1, T_2] = 0$. Let D^* be the induced connection $\sigma^* D$. Then $\partial_1 g(T_1, T_2) = g(D^*_{\partial_1} T_1, T_2) + g(T_1, D^*_{\partial_1} T_2) =$ (by b)$g(T_1, D^*_{\partial_1} T_2)$. Now since D is symmetric and $[\partial_1, \partial_2] = 0$, $D^*_{\partial_1} T_2 = D^*_{\partial_2} T_1$ (cf. Bishop–Goldberg p. 231). Thus $\partial_1 g(T_1, T_2) = g(T_1, D_{\partial_2} T_1) = (1/2)\partial_2 g(T_1, T_1) =$ (by a) 0. □

5.0.4 Simply convex neighborhoods

We review without proof certain properties of the exponential map (Section 0.0.8). A good reference is Helgason [1], pp. 32–36.

Suppose $x \in M$. An open neighborhood $\mathcal{U}_0$ of the origin in M_x is defined as *normal* iff: (a) $\exp_x|_{\mathcal{U}_0}$ is defined and is a diffeomorphism onto its image; and (b) if $0 \leq t \leq 1$ then $tX \in \mathcal{U}_0 \ \forall X \in \mathcal{U}_0$. A *simply convex neighborhood* in M is a nonempty open set $\mathcal{U} \subset M$ such that, $\forall x \in \mathcal{U}$, $\mathcal{U} = \exp_x \mathcal{U}_0$ for some normal neighborhood $\mathcal{U}_0 \subset M_x$. The *Whitehead Lemma* asserts that given $x \in M$ and a neighborhood $\mathcal{W}$ of x there is a simply convex neighborhood $\mathcal{U}$ of x such that $\mathcal{U} \subset \mathcal{W}$. For example, the simply convex neighborhoods in Minkowski spacetime are just the nonempty open subsets that are convex in the $\mathbb{R}^4$ sense. More generally, suppose $\mathcal{U} \subset M$ is a simply convex neighborhood, $x, z \in \mathcal{U}$, and $x \neq z$. Then our definitions above imply there is a unique geodesic of the form $\gamma: [0, 1] \to \mathcal{U}$ from x to z.

5.0.5 The geometric energy function

We give some results needed in discussing causality and photons. Throughout this subsection, $(\mathcal{U}, g)$ is a spacetime such that $\mathcal{U}$ itself is simply convex; $\forall x \in \mathcal{U}$, ϕ_x denotes the diffeomorphism $\phi_x = (\exp_x|_{\mathcal{U}_0})^{-1}: \mathcal{U} \to \mathcal{U}_0$, $\mathcal{U}_0$ as in Section 5.0.4. Treating $\mathcal{U}$ as a spacetime in its own right (temporarily) relieves us of the worry that a C^∞ curve might leave $\mathcal{U}$ and then sneak back in. For example, $\forall$ pair of distinct points $x, z \in \mathcal{U}$ there is now a unique equivalence class (Section 5.0.2) of geodesics from x to z. Recall that the function $K: T\mathcal{U} \to \mathbb{R}$, defined by $K(x, X) = g(X, X)$ as in Proposition 1.2.1, is C^∞. Moreover $\mathcal{U}^\dagger = \{(x, \phi_x z) \mid x, z \in \mathcal{U}\}$ is an open subset of $T\mathcal{U}$ and $\exp: \mathcal{U}^\dagger \to \mathcal{U} \times \mathcal{U}$, defined by $\exp(x, X) = (x, \exp_x X)$, is a diffeomorphism (cf. Bishop–Crittenden [1], p. 109).

Thus the function $\Phi = K \circ (\exp)^{-1}: \mathcal{U} \times \mathcal{U} \to \mathbb{R}$ is C^∞. $\Phi(x, x) = 0 \ \forall x \in \mathcal{U}$. Let $x, z \in \mathcal{U}$ be distinct, and let $\gamma: [0, 1] \to \mathcal{U}$ be the unique geodesic from x to z. Directly from the definitions we have:

(a) $\Phi(x, z) = g(\phi_x z, \phi_x z) = \int_0^1 g(\gamma_* u, \gamma_* u)\, du = \Phi(z, x)$.
(b) $\Phi(x, z)$ is negative, zero, or positive according as γ is timelike, lightlike, or spacelike, respectively.

We define Φ as the *geometric energy function*.

> The term "energy," appropriate in geometric versions of prerelativistic mechanics, is misleading in relativity. Specifically, the spacelike case $\Phi > 0$ is of no interest as usual, so assume γ future pointing. Then γ is a particle (Definition 3.1.1) and (a) gives $\Phi(x, z) = -m^2$, where m

is the rest-mass of γ. Even a rest-mass can be interpreted as total energy only in one very special case (Section 3.1.2), and m^2 is not in any sense an energy. (b) indicates "causality" function would be the right term in relativity, but we stick to the standard term with "geometric" added as a reminder.

Now fix $x \in \mathcal{U}$ and let $\Phi_x: \mathcal{U} \to \mathbb{R}$ be the C^∞ function $K \circ \phi_x$—that is, $\Phi_x z = \Phi(x, z)$. We have $d\Phi_x = (\phi_x)^* \, dK$, which implies, via the algebraic results in Exercise 0.0.10 and the fact that ϕ_x is a diffeomorphism, that $d\Phi_x$ is different from zero everywhere except at x. Denote by R the vector field physically equivalent to $d\Phi_x$. Roughly speaking, R is "radial." For example, suppose x is the origin of Minkowski space. The reader may check that then $\Phi_x = (u^1)^2 + (u^2)^2 + (u^3)^2 - (u^4)^2$ whence $d\Phi_x = 2(u^1 \, du^1 + \cdots - u^4 \, du^4)$ and $R = 2 \sum_{i=1}^4 u^i \, \partial_i$. More generally, suppose x, z, and γ are as in (a); we get:

(c) $Rz = 2\gamma_* 1$.

Proof. $d\Phi_x z \neq 0$ so there is, locally, a C^∞ imbedded 3-submanifold $\mathcal{B} \subset \mathcal{U}$ which is a level surface of Φ_x and contains z. Let $W \in \mathcal{U}_z$ be tangent to $\mathcal{B}$. Then $g(R, W) = d\Phi_x(W) = W\Phi_x = 0$; thus Rz is orthogonal to $\mathcal{B}$.

$\gamma_* 1 \in \mathcal{U}_z$ is nonzero; we claim $\gamma_* 1$ is likewise orthogonal to $\mathcal{B}$. Indeed, given W as above, let $\eta: [-\varepsilon, \varepsilon] \to \mathcal{B}$ be a C^∞ curve, where $\varepsilon > 0$, such that $\eta 0 = z$ and $\eta_* 0 = W$. In the notation of Section 5.0.3, define $\mathcal{D} = [0, 1] \times [-\varepsilon, \varepsilon]$. Define a map $\sigma: \mathcal{D} \to \mathcal{U}$ by $\sigma(u, v) = \exp_x [u\phi_x(\eta v)]$. Thus $\sigma(0, v) = x$ and $\sigma(1, v) = \eta v$. The reader may check (shrinking ε for convenience if he likes) that all the hypotheses of the Gauss lemma (Section 5.0.3) are satisfied, with $T_2(0, 0) = 0$, $T_2(1, 0) = W$, $g(T_1, T_1) = \Phi_x z =$ constant, and $T_1(1, 0) = \gamma_* 1$. Thus $g(W, \gamma_* 1) = g(0, \gamma_* 0) = 0$. This holds $\forall W$ so $\gamma_* 1$ is orthogonal to $\mathcal{B}$. Since $\mathcal{B}$ is 3-dimensional, we get $Rz = a\gamma_* 1$, $a \neq 0$. It remains to show $a = 2$.

Suppose γ is not lightlike. Then $\Phi_x z \neq 0$, say $\Phi_x z = b$. On unravelling the definitions we get $(\Phi_x \circ \gamma)(u) = bu^2 \, \forall u \in [0, 1]$ and $g(\gamma_*, \gamma_*) = b$. Thus $ab = g(Rz, \gamma_* 1) = d\Phi_x(\gamma_* 1) = (\Phi_x \circ \gamma)'(1) = 2b$. Since $b \neq 0$, $a = 2$. If γ is lightlike, $a = 2$ by continuity. $\square$

5.0.6 No escape

The study of "who can communicate with whom," especially the study of the chronology relation (Section 5.0.1), is central in many of the deeper current applications (cf. the references in Section 8.3). One relevant lemma says, roughly speaking, that there is no way to escape from the chronological future of a point. We now state and prove a simple local version of this lemma. As is typical, one starts with an algebraic result from Sections 1.1 and 1.2. Let $\mathcal{T}_x^+ \subset M_x$ be the set of future-pointing timelike vectors at $x \in M$, with $\mathcal{T}_x^-$ defined dually. Then $\mathcal{T}_x^+$ has the following properties: (a) it is open; (b) its boundary consists of the zero vector and the set $\mathcal{L}_x^+$ of future-pointing lightlike vectors; (c) its closure intersects the closure of

5 Photons

$\mathcal{T}_x^-$ only at the zero vector and $\mathcal{T}_x^+ \cap \mathcal{T}_x^-$ is empty; (d) if $X \in \mathcal{T}_x^+$ and $Y \in \mathcal{L}_x^+ \cup \mathcal{T}_x^+$, $X + Y \in \mathcal{T}_x^+$.

Now let $(\mathcal{U}, g)$ be a geodesically convex spacetime and $\forall x \in \mathcal{U}$ define $I_x^\pm = \exp_x \{\mathcal{T}_x^\pm \cap (\phi_x \mathcal{U})\}$, notation as in Section 5.0.5. Then the topological properties (a) and (c) go through verbatim with $\mathcal{T}_x^\pm$ replaced by $I_x^\pm$ and the zero vector replaced by x since $\exp_x$ is a homeomorphism on its domain $\phi_x \mathcal{U}$. An analogue of (b) and (d) is the following.

Lemma. *Let $\beta: (a, c) \to \mathcal{U}$ be a future pointing timelike curve whose image does not contain x, and suppose $y = \beta b$, $b \in (a, c)$. Then: (b') y is on the boundary of I_x^+ iff the unique geodesic $\alpha: [0, 1] \to \mathcal{U}$ from x to y is future pointing lightlike. (d') Suppose such a y exists, then $\forall u \in (b, c)$ $z = \beta u$ obeys $z \in I_x^+$—that is, the unique geodesic $\gamma: [0, 1] \to \mathcal{U}$ from x to z is future pointing timelike. (e) Thus y and b are unique.*

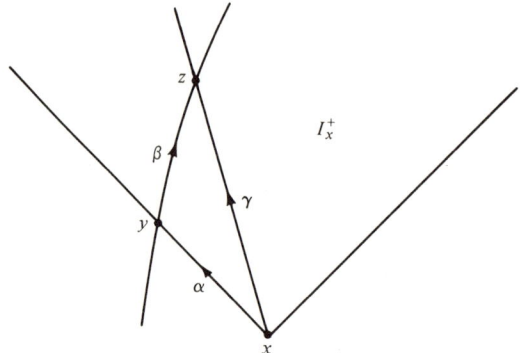

The generalization to the case where x lies on the image of β will be left as part of Exercise 5.0.10.

In (d') we have $x \leq y \ll z$. The generalization of (d') to arbitrary spacetime reads: $x \ll z$ iff there exists a y such that $x \leq y \ll z$ (Section 8.3).

PROOF. (b') follows directly from the algebraic result (b) and the fact that ϕ_x, $\exp_x$ are homeomorphisms. Suppose now such a y exists. Then $\Phi(x, y) = 0$ by Section 5.0.5b. Define the function $f = \Phi_x \circ \beta: (a, c) \to \mathbb{R}$. $fb = 0$ and $f'b = (\Phi_x \circ \beta)'(b) = d\Phi_x(\beta_* b) = g(Ry, \beta_* b) < 0$ since $\beta_* b$ is timelike and Ry is lightlike, with both future pointing (Sections 1.1 and 5.0.5). Thus there is a $d \in (b, c)$ such that $fu < 0 \forall u \in (b, d)$; indeed, choosing d so that $d - b$ is sufficiently small we have $\beta u \in I_x^+ \forall u \in (b, d)$, since $fu < 0$ implies $\beta u \in I_x^+ \cup I_x^-$ by Section 5.0.5b and we have $\beta u \notin I_x^-$ for u sufficiently close to b by (c) above and the fact that $\exp_x$ is a homeomorphism. Suppose there were a lub $\bar{d}$ of those $u \in (b, c)$ for which $\beta u \in I_x^+$. Then we would have $\beta \bar{d}$ on the boundary of I_x^+ and thus $f\bar{d} = 0$. But then $f'u = g(R\beta u, \beta_* u) < 0 \forall u \in (b, \bar{d})$ so $fd < 0$. Contradiction. Thus $\beta u \in I_x^+ \forall u \in (b, c)$. Thus (d') holds: (e) is now trivial. □

EXERCISE 5.0.7

Suppose $(\mathbb{R}^4, g)$ is Minkowski space and $x, z \in \mathbb{R}^4$. Show $x \ll z$ iff

$$\left[\sum_{\rho=1}^{3}(u^\rho z - u^\rho x)^2\right]^{1/2} < u^4 z - u^4 x,$$

and $x \leq z$ iff the same condition with "$<$" replaced by "$\leq$" holds.

In view of Exercise 3.7.8 and our remarks on conformal invariance in Section 5.0.1, this result is applicable in a less trivial case (Section 6.0.13).

EXERCISE 5.0.8. GEODESICS AND REFLECTIONS

Let (M, g) be Einstein–de Sitter spacetime and suppose $z = (0, 0, 0, a) \in M$, $a \in (0, \infty)$. (a) Show that the "spatial reflection" $(u^1, u^2, u^3, u^4) \to (-u^1, u^2, u^3, u^4)$ determines an isometry $\phi: M \to M$ with $(\phi z, \phi_* \partial_4 z) = (z, \partial_4 z)$. (b) Show from (a) and Section 5.0.2 that if $\gamma: (0, \infty) \to M$ is a geodesic and $(\gamma b, \gamma_* b) = (z, \partial_4 z)$ for some $b \in (0, \infty)$ then $u^1 \circ \gamma = 0$. (c) Generalizing, show in (b) that $u^2 \circ \gamma = 0 = u^3 \circ \gamma$ as well. (d) Now show, without computing connection coefficients but using Section 5.0.2a instead, that each integral curve of ∂_4 in Einstein–de Sitter spacetime is a geodesic. (e) Let $\lambda: \mathscr{E} \to M$ be a geodesic such that for one $u \in \mathscr{E}$, $du^2(\lambda_* u) = 0 = du^3(\lambda_* u)$; show that $du^2(\lambda_*) = 0 = du^3(\lambda_*)$ by generalizing the "reflection" argument just given. Thus if λ starts out tangent to a (u^1, u^4) plane it remains in this plane. (f) By similar arguments show there exist "purely radial" geodesics on a normal Schwarzschild spacetime—for example, $\mathbf{P} \circ \gamma = $ North Pole in $\mathscr{S}^2$.

EXERCISE 5.0.9. GEODESICS AND CONSTANTS OF THE MOTION

(a) Let $\gamma: \mathbb{R} \to \mathbb{R}^2$ be a smooth curve. Suppose $du^1(\gamma_*)$ and $du^2(\gamma_*)$ are known functions on $\mathbb{R}$. Show that then γ is uniquely determined by its initial values $\gamma 0$. (b) Let γ, γ' be curves $(0, \infty) \to M$, where M is Einstein–de Sitter spacetime. Suppose $(\gamma 1, \gamma_* 1) = (\gamma' 1, \gamma'_* 1)$ and $u^2 \circ \gamma = 0 = u^2 \circ \gamma'$, $u^3 \circ \gamma = 0 = u^3 \circ \gamma'$. Suppose γ is a freely falling particle (cf. Exercise 5.0.8e). (i) Show from Section 5.0.2 that $g(\gamma_*, \gamma_*) = -m^2$ and since ∂_1 is Killing $g(\gamma_*, \partial_1) = $ constant $= a$ (say). (ii) Suppose $g(\gamma'_*, \gamma'_*) = -m^2$ and $g(\gamma'_*, \partial_1) = a$; show $\gamma' = \gamma$. Thus, in this case, Section 5.0.2a–e suffices to determine the geodesics. (c) In Exercise 5.0.8f state and prove a similar uniqueness theorem, using the Killing vector field $\partial/\partial t$.

EXERCISE 5.0.10

In Sections 5.0.5 and 5.0.6 show: $x \ll z$ iff $z \in I_x^+$ iff $x \in I_z^-$ iff $\mathbf{R}z$ is timelike future pointing.

5.1 Photons

Recall that a photon is a particle with zero rest-mass (Definition 3.1.1). Thus suppose $\gamma: \mathscr{E} \to M$ is a particle. γ is a photon iff γ is lightlike iff one instantaneous observer on $\gamma \mathscr{E}$ observes Newtonian speed $c = 1$ iff all instantaneous observers on $\gamma \mathscr{E}$ observe unit Newtonian speed ("even an observer who runs

5 Photons

away from the photon as fast as he can still observes the overhauling speed as the speed of light").

5.1.1 Free fall

The electric charge of a photon is zero (Example 3.1.4). Thus the only photons of interest are geodesic—that is, freely falling (Section 3.8.6; "electromagnetic fields do not interact directly with photons; other matter can influence a photon only by destroying it; thus each photon of interest is subject only to gravity").

5.1.2 Emission and absorption

Let $\lambda: \mathscr{E} \to M$ be a photon. Suppose λ is created within spacetime rather than having been always around—that is, the curve λ has a past endpoint $x \in M$ (Section 3.8). Then one sometimes says λ is *emitted* or *sent out* at x; these redundant synonyms for "created" in Section 3.8 are just part of the colorful language developed for dealing with these colorful particles. Similarly, one sometimes says λ is *emitted by*, for example, an instantaneous observer or star at x. If λ has a future endpoint $z \in M$ λ is *absorbed* and *received* at z as well as being destroyed there.

> Additional terms—for example, "scattering"—will not be needed except in informal comments. Within our framework "scattering" just refers to a collision in which one photon is destroyed and one is created.

5.1.3 Pointwise concepts

(a) Let $\lambda: \mathscr{E} \to M$ be a $1-1$ photon and $Y = \lambda_* u$ be its energy-momentum at $x = \lambda u$, $u \in \mathscr{E}$. Suppose an instantaneous observer (z, Z) is actually present and observing; then the existence of a distinguished local time axis, span Z, gives rise as usual to a number of auxiliary concepts. We have $Y = e(Z - U)$, where $e = -g(Y, Z) > 0$ is the energy (Section 3.1.2) that (z, Z) measures and $U \in Z^\perp$ is the *spatial direction* (z, Z) *measures* for Y and for λ. Intuitively, U is the direction from which (z, Z) sees λ coming. The collection of all such U is (z, Z)'s private *celestial sphere* $\mathscr{S}_z = \{U \in Z^\perp \mid g(U, U) = 1\}$. $\mathscr{S}_z$ is isometric, in the natural way, to the unit 2-sphere (Section 0.0.9). To get an intuitive picture of your celestial sphere, go out at night and look at "the bowl of the sky." Exercise 5.1.4 concerns the interrelation between different celestial spheres at the same z.

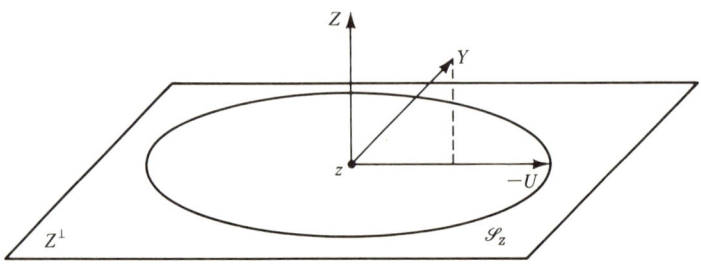

(b) Let $h \cong (6.3 \times 10^{-43} \text{ seconds})^2$ be the universal constant called *Planck's constant*. We define the *frequency (z, Z) measures* for Y as $f = e/h$. The *wavelength (z, Z) measures* is $\lambda = f^{-1}$.

The terms "frequency" and "wavelength" are motivated by alternate, wave models for light (cf. the comment below Exercise 3.7.7).

Following a suggestion of Planck, Einstein postulated $e = hf$ in 1905. Here we have reversed the historical order, defining f in terms of e. $f\lambda = 1 = $ speed of light merely corresponds to the usual relation between frequency, wavelength, and speed of any wave motion.

The above number $\sqrt{h} = 6.43 \times 10^{-43}$ seconds is regarded as the natural time scale for quantum gravity. For example, in current cosmology one usually assumes quantum gravitational effects can be neglected at times later than about 10^{-42} seconds after a big bang (cf. Section 6.6.5).

(c) If the frequency f measured by (z, Z) is less than about 10^9 (seconds)$^{-1}$, this corresponds to *radio waves* for (z, Z); the range 10^9 (second)$^{-1} \leq f \leq 3 \times 10^{14}$ (seconds)$^{-1}$ corresponds to *microwaves, millimeter radiation*, and *infrared radiation*; 3×10^{14} (seconds)$^{-1} \leq f \leq 10^{15}$ (seconds)$^{-1}$ corresponds to *visible light* ($=$ *optical radiation*) for (z, Z), with light that is red for (z, Z) having a smaller f, and thus a larger λ, than light that is blue or violet for (z, Z); and at still larger frequencies one has *ultraviolet radiation*, then *x-rays* and finally γ *rays*. We shall use *radiation* or *electromagnetic radiation* or sometimes simply *light* to refer generically to all ranges.

(d) Let λ, Y, and (z, Z) be as in (a). To say, for example, that λ is "blue" would be sheer nonsense, even in a pointwise argument, since there will always exist some instantaneous observer at z for whom the measured frequency corresponds to red. But suppose Y' is future pointing with span $Y' = $ span Y. Then by the above definitions the ratio f/f' of measured frequencies is the same for all instantaneous observer at z so "Y is bluer than Y'," meaning $f > f'$ $\forall$ instantaneous observer at z, can make sense.

To extend these pointwise concepts to the world line of one observer we need only take into account the behavior of gyroscopes (Section 2.2). For example, suppose you see a shooting star sending out orange photons; in what sense do you see it move? Regard yourself as an observer $\gamma: \mathscr{E} \to M$. Thus you have a collection of celestial spheres $\{\mathscr{S}_{\gamma_* u} \subset (\gamma_* u)^\perp \mid u \in \mathscr{E}\}$. By using gyroscopes—that is, Fermi–Walker parallelism—you can identify all these celestial spheres to get a single copy, say $\mathscr{S}_\gamma$. In principle an actual gyroscope, a Foucault pendulum, or your ear fluids are used; in practice, reference to the "fixed earth" gives a good approximation. By correlating proper times of arrival of orange photons with their directions (Section 5.1.3a) you get a curve on $\mathscr{S}_\gamma$ and thus perceive motion.

EXERCISE 5.1.4

Let $(\mathscr{S}^2, h)$ be the unit 2-sphere (Section 0.0.9). A diffeomorphism $\phi: \mathscr{S}^2 \to \mathscr{S}^2$ is *conformal* iff $\phi^* h = fh$ where $f: \mathscr{S}^2 \to (0, \infty)$; compare Exercise 3.7.8. (a) Let (z, Z), (z, Z') be instantaneous observers, and let $\mathscr{S}_Z, \mathscr{S}_{Z'} \subset M_z$ be their

5 Photons

celestial spheres. Show there exists a conformal diffeomorphism $\phi: \mathscr{S}_z \to \mathscr{S}_{z'}$ which, for each photon energy-momentum $Y \in M_z$, carries the spatial direction (z, Z) measures for Y to that which (z, Z') measures for Y. ("Aberration," if Z' is "rushing forward" he sees the pattern "bunched up" ahead and "thinned out" behind, in a systematic way.) (b) (Optional.) Show the collection of conformal diffeomorphisms $\mathscr{S}^2 \to \mathscr{S}^2$ is a six-dimensional Lie group.

> The connected component of the identity is isomorphic to that of the Lorentz group (Section 8.4.2b).

Exercise 5.1.5

Let M be Minkowski space, γ be an observer, and λ be a freely falling photon. Show: (a) Both curves are $1 - 1$; they intersect at most once. (b) If γ is an inextendible geodesic and x is not on the world line of γ then there is exactly one $\lambda: [0, 1] \to M$ from x to a point $\lambda 1$ on the world line of γ. Then u^4 is smaller at emission than at reception. (c) Dually, there is exactly one such λ from γ's world line to x. (d) In (b) suppose γ is not a geodesic but is endless (Section 3.8). Show no λ from γ's world line to x need exist. Hint: Look at the picture here.

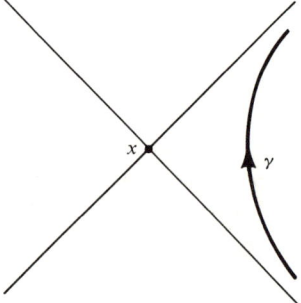

We now indicate that the situation need not be so simple. Actually, in general, almost anything can happen (Section 8.3).

Exercise 5.1.6

(a) Suppose (M, g') is conformal to (M, g) (Exercise 3.7.8) and λ is a lightlike geodesic for (M, g). Show there is a reparametrization of λ which is a lightlike geodesic of (M, g'). (b) Use (a) and Exercises 5.0.7 and 3.7.8 to show the following. In Exercise 5.1.5 let M be Einstein–de Sitter spacetime, γ be an inextendible comoving observer. Then Exercise 5.1.5a and b go through verbatim but in (c) "exactly one" must be replaced by "at most one." Thus γ can get information from any point if he waits long enough but there are points blithely unaware of his existence.

Exercise 5.1.7

Let $[\lambda]$ be a geodesic equivalence class (Section 5.0.2) of freely falling photons with each $\lambda \in [\lambda]$ a $1 - 1$ curve. Suppose $z \in M$ is on the image. Give examples of relevant physical quantities which are defined: (a) given an instantaneous observer (z, Z) but no distinguished representative in $[\lambda]$; (b) given $\lambda \in [\lambda]$ but no distinguished Z; (c) given both. Answers: (a) The spatial direction (z, Z) measures for $[\lambda]$; often this is the most important measurement. (b) Only the energy-momentum λ_*. (c) Energy, 3-momentum, frequency, wavelength, color, spatial direction.

5.2 Light signals

Define a *light signal* as an equivalence class (Section 5.0.2) of freely falling photons; the motivations are indicated in Section 5.1.1 and Exercise 5.1.7b.

EXAMPLE 5.2.1. Let $(\mathbb{R}^4, g)$ be Minkowski space. Each inextendible light signal has exactly one representative $\lambda: \mathbb{R} \to \mathbb{R}^4$ of the form $\lambda u = (v, 0) + u(w, 1)$ where $v, w \in \mathbb{R}^3$ and $w \cdot w = 1$ ($\mathbb{R}^3$ inner product). Conversely, $\forall$ pair $(v, w) \in \mathbb{R}^3 \times \mathbb{R}^3$ with $w \cdot w = 1$ there is a unique light signal, say $[\lambda]_{(v,w)}$. For example one here has a new equivalence relation among the light signals themselves: $[\lambda]_{(v,w)} \sim [\lambda]_{(v',w')}$ iff $w = w'$ and $(v - v') \cdot w = 0$.

> Two light signals are said to meet at infinity iff they are equivalent in this sense. This concept can be extended to certain kinds of nonflat spacetimes and is then very useful (Hawking–Ellis [1]).

EXAMPLE 5.2.2. Let (M, g) be Einstein–de Sitter spacetime (Section 1.4). Thus $M = \mathbb{R}^3 \times (0, \infty)$, $g = (u^4)^{4/3} \sum_{\mu=1}^{3} du^\mu \otimes du^\mu - du^4 \otimes du^4$. There exist light signals lying entirely within the $(0, 0, u^3, u^4)$ plane (Exercise 5.0.8). As inspection of g suggests, analyzing one of these suffices to give complete information on all freely falling photons in (M, g) (Exercise 5.2.5 following).

Define $\lambda: (0, \infty) \to M$ by $\lambda u = (0, 0, 3u^{1/5}, u^{3/5})$. We claim λ is an inextendible, freely falling photon.

PROOF. λ is C^∞ and future pointing. Moreover, $g(\lambda_*, \lambda_*) = (u^4 \circ \lambda)^{4/3} \times [du^3(\lambda_*)]^2 - [du^4(\lambda_*)]^2$. Now $u^4 \lambda u = u^{3/5}$, $[du^3(\lambda_*)](u) = (3/5)u^{-4/5}$ and $[du^4(\lambda_*)](u) = (3/5)u^{-2/5}$. Thus $g(\lambda_*, \lambda_*) = 0$; thus λ is a photon. To see that λ is geodesic, note that ∂_3 is a Killing vector field (Section 3.6.3) and $g(\lambda_*, \partial_3) = (u^4 \circ \lambda)^{4/3}[du^3(\lambda_*)] = 3/5$, a constant. By Exercise 5.0.9, this suffices. Finally, λ is inextendible since the scalar curvature obeys $S\lambda u \to \infty$ as $u \to 0$. □

We define λ as *the standard photon* on (M, g), its equivalence class (Section 5.0.2) as *the standard light signal*.

Let $\beta: \mathscr{E} \to M$ be an observer on a spacetime M. From which points can β receive light signals and to which ones can he send them? For Minkowski space the situation is simple (Exercise 5.1.5); for Einstein–de Sitter spacetime it is a little more complicated (Exercise 5.1.6); in a black hole situation there are points that can receive light signals from "outside" observers, but cannot send any (Section 7.5); and so on. Locally, the situation remains simple, as the following proposition shows.

Proposition 5.2.3. *Suppose $u_1 \in \mathscr{E}$ is given. There exists an open interval $\mathscr{F} \subset \mathscr{E}$ containing u_1 and an open neighborhood $\mathscr{W}$ of βu_1 such that, $\forall x \in \mathscr{W} - \beta\mathscr{F}$ the following holds. There exist $u_0, u_2 \in \mathscr{F}$, a light signal $[\lambda]$ from x to βu_2 and a light signal $[\lambda']$ from βu_0 to x; $u_0, u_2, [\lambda]$ and $[\lambda']$ are unique.*

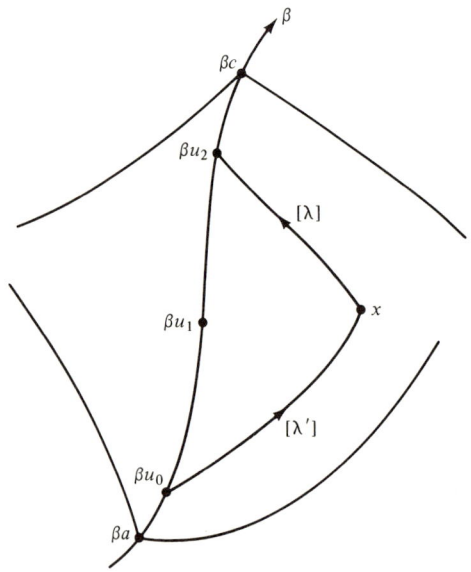

PROOF. Let $\mathscr{U}$ be a simply convex open neighborhood of βu_1, $[a, c]$ be an interval in $\mathscr{E}$ containing u_1 as an interior point such that $\beta[a, c] \subset \mathscr{U}$.

Regard $(\mathscr{U}, g)$ as a spacetime as in Sections 5.0.5 and 5.0.6, and adopt the notation of those subsections.

Let $\mathscr{W} = (I_{\beta c}^-) \cap (I_{\beta a}^+)$. Thus $\mathscr{W}$ is an open neighborhood of $\beta(a, c)$, in particular of βu_1 (Exercise 5.0.10). Define $\mathscr{F} = (a, c)$. We claim $\mathscr{W}$ and $\mathscr{F}$ have the required property.

In fact, suppose $x \in \mathscr{W} - \beta\mathscr{F}$. Then $\beta c \in I_x^+$ since $x \in I_{\beta c}^-$. Similarly $\beta a \in I_x^-$, in particular $\beta a \notin$ closure I_x^+. Since β is continuous, the lemma in Section 5.0.6 shows there exists a unique $u_2 \in (a, c)$ such that the geodesic $\lambda: [0, 1] \to \mathscr{U}$ from x to βu_2 is future pointing lightlike. Thus $u_2, [\lambda]$ exist and are unique. The dual argument proves $u_0, [\lambda']$ exist and are unique. □

EXAMPLE 5.2.4. RADAR. Let (M, g) be two-dimensional Minkowski space (Section 0.2), $\gamma: \mathbb{R} \to M$ be the observer defined by $\gamma u = (0, u)$. For $l, \theta \in (0, \infty)$, $\bar{\gamma}: \mathbb{R} \to M$, defined by $\bar{\gamma} u = (l - u \sinh \theta, u \cosh \theta)$, is another observer. γ can use radar to observe $\bar{\gamma}$ as follows. At proper time u_0, γ emits a light signal $[\lambda']$ which travels to $x \in \bar{\gamma}\mathbb{R}$, is "reflected" there, and returns to γ at proper time u_2 as a light signal $[\lambda]$. Working algebraically, the reader can check that if [λ'] is emitted prior to the collision event z, $u_2 = u_0 e^{-2\theta} + 2l e^{-\theta} \cosh \theta$. Thus if γ knows *a priori* that $\bar{\gamma}$ is freely falling, two radar measurements (u_0, u_2), (u_0', u_2') suffice to fix l and θ and thus to determine the world line of $\bar{\gamma}$.

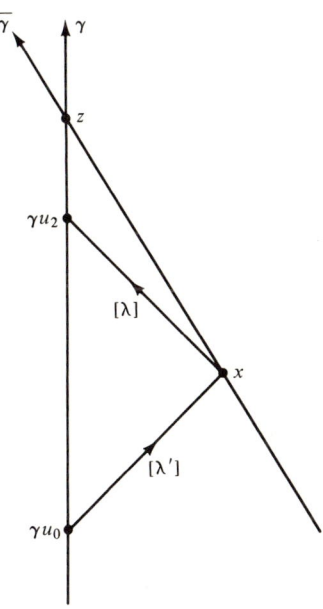

In more general cases γ may need a whole function $u_2(u_0)$ and must also observe directions (Section 5.1.3). Since meter sticks do not really make sense in general relativity and are not used in actual astronomical measurements, one regards a radar set as the basic distance measuring device (cf. Exercise 2.3.12d and the comment below it).

EXERCISE 5.2.5

Let (M, g) be Einstein–de Sitter spacetime; thus $M = \mathbb{R}^3 \times (0, \infty)$. Show: (a) If $\tilde{\psi}: \mathbb{R}^3 \to \mathbb{R}^3$ is a Euclidean isometry, $\psi: M \to M$, defined by $\psi(v, t) = (\tilde{\psi}v, t)$, is an isometry. (b) If $\gamma: \mathscr{E} \to M$ is a geodesic there is an isometry $\psi: M \to M$ such that $\gamma \circ \psi$ lies in the plane $u^1 = 0 = u^2$. (c) If γ in (b) is an inextendible, freely falling photon there exists a ψ such that $[\gamma \circ \psi] = [\lambda]$, λ as in Example 5.2.2. (Hint: See Section 6.0.5 if you get stuck.)

EXERCISE 5.2.6 ORTHOGONALITY AND RADAR

Let $\gamma: \mathscr{E} \to M$ be an observer, $u_1 \in \mathscr{E}$, $\mathscr{W}$ be a neighborhood of γu_1 with the properties in Proposition 5.2.3, $\alpha: [0, a) \to \mathscr{W}$ be a curve which intersects γ's world line at the parameter value 0 with $\alpha 0 = \gamma u_1$. Define a function $f: (0, a) \to \mathbb{R}$ by $f = u_1 - \frac{1}{2}(u_2 + u_0)$, where u_2 and u_0 are as in Proposition 5.2.3 with $x = \alpha s$

5 Photons

$\forall s \in (0, a)$. (a) Suppose M is Minkowski spacetime and γ, α are geodesics. Show $f = 0$ iff γ and α are orthogonal. (b) Dropping the geodesic restriction on both show that $\lim_{s \to 0} f(s) = 0$ and show that $\lim_{s \to 0} f'(s) = 0$ iff $g(\gamma_* 1, \alpha_* 0) = 0$. (c) Show that (b) remains valid when M is an arbitrary spacetime.

EXERCISE 5.2.7

(a) Given $x \in M$, show there exists an open neighborhood $\mathcal{U}$ of x with the following property. $\forall z \in \mathcal{U}$, there exists $y \in \mathcal{U}$ such that there exists a lightlike geodesic from x to y and there exists a lightlike geodesic from y to z. (b) Suppose $N \subset M$ is an open proper submanifold. Show there exists a point x on the boundary of N. (c) By combining (a) and (b), prove Proposition 1.3.2. (d) Use Proposition 1.3.2, Example 5.2.2, and Exercise 5.2.5 to show Einstein–de Sitter spacetime is maximal.

5.3 Synchronizable reference frames

In Section 2.3 we defined synchronizable reference frames without making explicit the empirical significance of synchronizability. Now that we have light signals at our disposal, we can fill in this gap. Throughout the section Z is a reference frame on spacetime M and ζ is the physically equivalent 1-form.

Recall the following results from Section 2.3: Z is locally proper time synchronizable iff Z is geodesic and irrotational, proper time synchronizable iff there is a function $t: M \to \mathbb{R}$ such that $\zeta = -dt$. Moreover, suppose Z is proper time synchronizable and t is as above. (a) Each level surface of t is a 3-manifold everywhere orthogonal to Z. (b) The function t is unique up to a single additive constant. (c) Let $\gamma: \mathcal{E} \to M$ be an inextendible integral curve of Z; thus γ is an observer in Z. Then there is a constant $a_\gamma \in \mathbb{R}$ such that $t\gamma u = u - a_\gamma$—that is, $t \circ \gamma$ agrees with proper time u up to a constant. Hence the term proper time function for t in Section 2.3.

If Z is proper time synchronizable, a set $\{\gamma\}$ of observers in Z can empirically synchronize their clocks, by radar, as follows. Suppose first that by sheer luck the observers already have $a_\gamma = 0$ $\forall \gamma$ and for one t. Consider any two sufficiently nearby observers $\gamma, \gamma' \in \{\gamma\}$. Here "sufficiently nearby" means Proposition 5.2.3 is applicable. It also means so nearby that the approximation $u_1' = \frac{1}{2}(u_0 + u_2)$ below holds to within other empirical inaccuracies.

> Since we are discussing actual measurements the standard mathematics vs. physics ambiguity (Section 2.1.2) on infinitesimals here comes into play and Einstein's comment applies. To make the discussion mathematically rigorous, we would need to assume an infinite number of observers present. Rather than adopt this wildly unrealistic approach we couch the discussion in physicists' language.

Suppose γ and γ' communicate by radar. Say γ emits his light signal at his proper time $u_0 = t\gamma u_0$, the signal strikes γ' at his proper time $u_1' = t\gamma' u_1'$ and returns to γ at $u_2 = t\gamma u_2$ (cf. the figure in Proposition 5.2.3). By the orthogonality property (a) above and Exercise 5.2.6, u_1' is the average:

$u_1' = \frac{1}{2}(u_0 + u_2)$ within the limit of empirical accuracy. By subsequent communication γ and γ' learn that this *consistency condition* held. Now imagine each observer in Z continually performing measurements of this kind with each nearby neighbor. By noting that the consistency condition always holds in this large set of interlocking measurements they can conclude that orthogonal surfaces $t = $ constant must exist and that t is indeed a proper time function.

In the more general case that $a_{\gamma'} \neq 0$ for some of the observers in Z, all that is required is that one observer, say γ, take the lead. γ autocratically decides that $t\gamma u = u$. He transmits his fiat to his neighbors—for example, by demanding $t\gamma' u_1'$ above be given by $t\gamma' u_1' = \frac{1}{2}(u_0 + u_2)$. They in turn inform their neighbors, and so on. This determines a function t on (part of) M. Somewhat angry at first, the other observers find that at worst they need only change the origin of their own proper time to get agreement with t and that now all further radar measurements as above give consistency.

EXERCISE 5.3.1

Generalize the above discussion to the case that Z is synchronizable but not proper time synchronizable by assuming one autocrat γ and other observers who regard the consistency condition, $t_1 = \frac{1}{2}(t_0 + t_2)$ ∀ nearby radar measurement, as more important than insisting on their own proper time.

> In practice, γ may have some justification—for example, as the "observer" in the center of a star or as an "observer at infinity." Note that unless the chronological future $\{z \in M \mid x \ll z$ for some x on γ's world line$\}$ is all of M, t is not radar determined on all of M by signals that travel no faster than light.

EXERCISE 5.3.2

Take a rotating reference frame on Minkowski spacetime and show explicitly how the radar measurements can lead to an inconsistency with the assumption that a time function exists (cf. Exercise 2.3.15).

5.4 Frequency ratio

Suppose we observe a photon from a star. By measuring its energy we can assign it a frequency (Section 5.1). Suppose we see first one photon, then another, then another, and so on. By counting how many photons arrive during one second of our own proper time we get a conceptionally different "frequency": number per unit proper second. We now show these two concepts are consistent. In practice, both kinds of measurement are key tools in analyzing, roughly speaking, "how fast the star is moving" or "how much gravity is between us and it."

Formally, the situation of interest is the following. We have a rectangle $\mathscr{D} = [a, b] \times [-\varepsilon, \varepsilon]$ and a map $\sigma: \mathscr{D} \to M$ as in Section 5.0.3. $\forall v \in [-\varepsilon, \varepsilon]$ the curve $\sigma_v: [a, b] \to M$ is a freely falling photon ("from the star to us").

Moreover the curves $\gamma_A: [-\varepsilon, \varepsilon] \to M$, defined by $\gamma_A v = \sigma(A, v)$ for $A = a$ or b, are future-pointing timelike. Intuitively, γ_a describes the history of the star and γ_b describes our history; we are not insisting either be parametrized via proper time (arclength). In particular, we then have that σ_0 is a freely falling photon from $\sigma_0 a = \gamma_a 0 = \sigma(a, 0)$ to $\sigma_0 b = \gamma_b 0 = \sigma(b, 0)$. Moreover, $Z_A = \gamma_A \cdot 0 / |\gamma_A \cdot 0|$ is an instantaneous observer at $\sigma_0 A$ $\forall A = a$ or b.

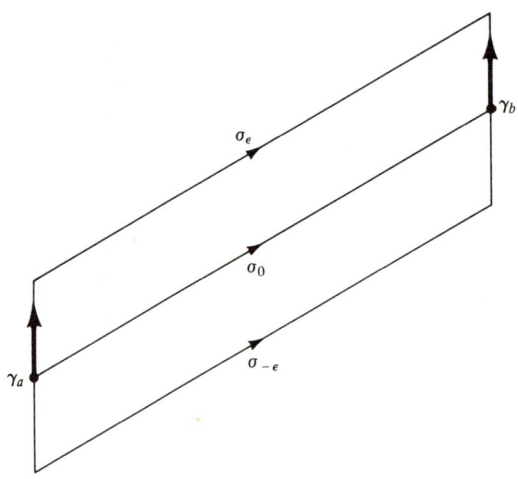

Now given any freely falling photon $\lambda: [a, b] \to M$ and any pair of instantaneous observers $(\lambda a, Z_a)$, $(\lambda b, Z_b)$, the respective frequencies measured are $f_A = -g(\lambda_* A, Z_A)/h$ (Section 5.1.3.b). Moreover, the reader can check that the ratio, f_a/f_b, of frequencies at emission and reception comes out the same if λ is replaced by any other photon in the light signal $[\lambda]$. Thus $\forall$ triple λ, Z_a, Z_b as above we define the *frequency ratio for* $([\lambda], Z_a, Z_b)$ as $\imath = g(\lambda_* a, Z_a)/g(\lambda_* b, Z_b)$ ($= f_a/f_b$; sometimes the more explicit phrase *ratio of emitted to observed frequency* is used). What we will show is that when there also exists a rectangle as above with $\lambda = \sigma_0$, the frequency ratio $\imath$ can also be interpreted via a more geometric counting experiment.

In fact, suppose σ is as above and some finite number of the curves σ_v are occupied by actual photons, say $\sigma_0, \sigma_d, \sigma_{2d}, \ldots$ with $d = 10^{-6}\varepsilon$. Regarding the image of γ_b as our own history, we then count 10^6 photons in the parameter interval $[0, \varepsilon)$, corresponding to our proper time $\int_0^\varepsilon |\gamma_b \cdot v|\, dv$. Similarly an observer on the star counts 10^6 photons in his proper time $\int_0^\varepsilon |\gamma_a \cdot v|\, dv$. Thus the proper time ratio for equal photon numbers is $\int_0^\varepsilon |\gamma_a \cdot v|\, dv / \int_0^\varepsilon |\gamma_b \cdot v|\, dv$, independent of which photon curves are actually occupied. Applying the mathematics versus physics comments in Section 2.1.2, we may take the limit $\varepsilon \to 0$ to get the *proper time ratio for σ at σ_0* as $|\gamma_a \cdot 0|/|\gamma_b \cdot 0| \in (0, \infty)$.

5.4 Frequency ratio

Proposition 5.4.1. *The frequency ratio $\imath$ for $[(\sigma_0)], Z_a, Z_b]$ is the inverse of the proper time ratio for σ at σ_0.*

PROOF. $\imath = [g(\sigma_0 \cdot a, \gamma_a \cdot 0)/g(\sigma_0 \cdot b, \gamma_b \cdot 0)][|\gamma_b \cdot 0|/|\gamma_a \cdot 0|]$. The first square bracket is one, by Gauss' lemma (Section 5.0.3). □

In general, one simply has a measured frequency ratio interpreted as above. In some (very special) cases one can also interpret $\imath$ via "motion" or "cosmological red shift" or a "photon losing energy as it struggles upward in a gravitational field," and so on. We give two examples, emphasizing the extra structure one needs before such additional interpretations become legitimate.

EXAMPLE 5.4.2. DOPPLER EFFECT IN 2-DIMENSIONAL MINKOWSKI SPACE. When (M, g) is Minkowski space, distant parallelism makes sense—that is, there exists a global basis consisting of covariant constant vector fields (Exercise 2.3.12). In this case, and really only in this case, can one talk of two instantaneous observers, (z_a, Z_a) and (z_b, Z_b) with $z_a \neq z_b$, being at rest or moving with respect to each other. Then the frequency ratio being different from 1 indicates motion, as we now illustrate, in the simplest case.

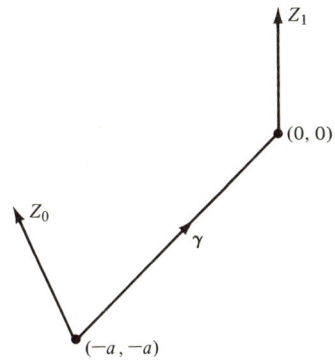

Let $(\mathbb{R}^2, g)$ be 2-dimensional Minkowski space, $\lambda: [0, 1] \to \mathbb{R}^2$ be a freely falling photon, $(\lambda 0, Z_0)$ and $(\lambda 1, Z_1)$ be instantaneous observers. Without loss of generality, we may suppose $\lambda 1 = (0, 0) \in \mathbb{R}^2$, $Z_1 = \partial_2 \lambda 1$, and $\lambda_* = (a, a) \in \mathbb{R}^2_{\lambda u}$, $a \in (0, \infty)$ (cf. Section 0.2). Then $\lambda 0 = (-a, -a) \in \mathbb{R}^2$ and, for some $\theta \in \mathbb{R}$, $Z_0 = (\sinh \theta, \cosh \theta) \in \mathbb{R}^2_{\lambda 0}$. Then, identifying $\mathbb{R}^2_{\lambda 0}$ with $\mathbb{R}^2_{\lambda 1}$, which is kosher when distant parallelism is defined, the Newtonian velocity Z_1 observes for Z_0 is $\tanh \theta = v$. Note that here $\theta < 0$ iff $v < 0$ iff, intuitively speaking, Z_0 is receding from Z_1 in the sense of the picture. On the other hand $f_1 = -g(\lambda_*, Z_1)/h = a/h$ and $f_0 = -g(\lambda_*, Z_0)/h = a(\cosh \theta - \sinh \theta)/h$. Thus the frequency ratio for $([\lambda], Z_0, Z_1)$ is $\imath = \cosh \theta - \sinh \theta = [(1 - v)/(1 + v)]^{1/2}$. In particular, the ratio of emitted to observed frequency is bigger than 1 iff Z_0 is "*receding*" from Z_1.

5 Photons

EXAMPLE 5.4.3. Let $\lambda: [a, b] \to M$ be a restriction of the standard photon on Einstein–de Sitter spacetime. Thus $(\lambda a, \partial_4 \lambda a)$ and $(\lambda b, \partial_4 \lambda b)$ are instantaneous comoving observers at emission and reception. We compute the frequency ratio $\imath$ for $([\lambda], \partial_4 \lambda a, \partial_4 \lambda b)$ in terms of the emission time $t_a = u^4 \lambda a$ and reception time $t_b = u^4 \lambda b$; note $t_b > t_a > 0$ by Exercise 5.1.6.

We have $\lambda u = (0, 0, 3u^{1/5}, u^{3/5})$ $\forall u \in [a, b]$ with $b > a > 0$. Thus $\lambda_* u = (3/5)[u^{-4/5} \partial_3(\lambda u) + u^{-2/5} \partial_4(\lambda u)]$ and $u^4 \lambda u = u^{3/5}$. Thus $g(\lambda_* u, \partial_4) = -(3/5)u^{-2/5} = -(3/5)(u^4 \lambda u)^{-2/3}$. By the definitions $\imath = f_a/f_b = (t_b/t_a)^{2/3} > 1$.

Intuitively speaking, $\imath > 1$ is consistent with the "expansion of the universe" (Proposition 2.3.7c), since in Example 5.4.2, $\imath > 1$ corresponds to recession. However, this intuitive argument cannot be made precise, except in a neighborhood so small the curvature is negligible (Section 2.1.2), since there are no covariant constant timelike vector fields on Einstein–de Sitter spacetime (Exercise 2.3.12) and thus no wholly natural way to define "relative velocity" for distant instantaneous observers. This last comment applies also to interpreting the "gravitational red shift" of Chapter 7 as "due to gravity rather than to relative motion."

EXERCISE 5.4.4

Let $\lambda: [a, b] \to M$ be a freely falling photon, $(\lambda a, Z_a)$ and $(\lambda b, Z_b)$ be instantaneous observers. Show that a map $\sigma: \mathscr{D} \to M$ as in this section exists in the following three cases. (a) $\lambda[a, b]$ is contained in a simply convex neighborhood. (b) M is Einstein–de Sitter spacetime and Z_a, Z_b are comoving as in Example 5.4.3. (c) M is a normal Schwarzschild spacetime, λ is "purely radial" (i.e., $P \circ \lambda =$ constant) and Z_a, Z_b are stationary as defined in Section 2.1. (d) (Hard.) Find a case where no such σ exists.

EXERCISE 5.4.5

Let $\gamma_0, \gamma_1: (0, \infty) \to M$ be comoving observers on Einstein–de Sitter spacetime with $\gamma_0 u = (0, 0, c, u)$, $c \in (0, \infty)$, and $\gamma_1 u = (0, 0, c + \delta, u)$, $\delta \in (0, \infty)$. Let $\lambda: [a, b] \to M$ be a freely falling photon, from some point on γ_0's world line, which is received by γ_1 ("us-now") at proper time $u_1 \in (0, \infty)$. Show: (a) $t\lambda a = (u_1^{1/3} - \delta/3)^3 > 0$; (b) $\imath = [3u_1^{1/3}/(3u_1^{1/3} - \delta)]^2 > 1$; (c) $u_1 > \delta^3/27$. (Hint: cf. Example 5.2.2 and Exercise 5.2.5.)

(c) Says that γ_1 cannot receive any light signal from γ_0 until u_1 gets big enough. δ should not be regarded as in any sense a distance nor even as a directly observable quantity.

5.5 Photon distribution functions

Astronomers observe many photons from each known star, many stars, many photons from extended sources such as gas clouds in our galaxy, many photons which have no identifiable source and seem to be just wandering around the universe, and so on. The total number of photons involved is

5.5 Photon distribution functions

so large that the models of Chapter 3—for example, a finite collection of photons or superposition of photon beams—are sometimes inconvenient. We devote the rest of Chapter 5 to a powerful alternate model. In the present, very informal section we specify how a single instantaneous observer analyzes nearby photons using the model. The generalization to an observer-independent description of photons on all of M will require considerable formal machinery, given in the next two sections.

Thus through this section let (z, Z) be an instantaneous observer, $\mathscr{L}_z^+$ be the future lightcone in M_z, $Z^\perp$ be (z, Z)'s rest space, and $\mathscr{S}_z$ be (z, Z)'s celestial sphere. To each photon energy-momentum $Y \in \mathscr{L}_z^+$, (z, Z) can assign a measured energy $e = -g(Y, Z) \in (0, \infty)$ and measured spatial direction $U \in \mathscr{S}_z$ (Section 5.1). We therefore introduce (z, Z)'s *direction-energy space* $P_z = \mathscr{S}_z \times (0, \infty)$ and denote the projection onto the second factor by $e: P_z \to (0, \infty)$.

We will presently use integrals on P_z to represent—for example, the energy density (Section 3.3.3) due to many photons—so we next make some remarks on integration.

5.5.1 Integration

(a) Let N be a C^∞ manifold and let $\varLambda$ be a volume element on N. Thus N orientable. We take N, and open submanifolds of N, as oriented via $\varLambda$. Define an *integration region* in N to be a compact subset $\mathscr{K} \subset N$ such that the *interior* $\mathscr{K}^0$ is nonempty and $\mathscr{K}$, $\partial\mathscr{K}$ are piecewise C^∞ (Section (3.0.1). This is sufficient to insure that $\int_\mathscr{K} f\varLambda$ is a well-defined real number whenever $f: N \to \mathbb{R}$ is continuous (Bishop–Goldberg, Section 4.8). The integral is nonnegative if f is. (b) If $\mathscr{K} \subset \mathscr{S}_z$ is an integration region, the area Ω of $\mathscr{K}$ is defined as the *solid angle* of $\mathscr{K}$. Thus $\Omega = \int_\mathscr{K} \zeta$, where ζ is the natural volume element of $\mathscr{S}_z$, obtained by regarding $(Z_\perp, g|_{Z^\perp})$ as Euclidean 3-space (cf. Sections 0.0.9 and 2.1.2). (c) On (z, Z)'s direction-energy space P_z the natural volume element is $\pi_z = \zeta \wedge e^2\,de$; the extra factor e^2 merely corresponds to a transition from rectangular to spherical coordinates (Exercise 5.5.6c). We take P_z oriented by π_z.

Now to analyze nearby photons, (z, Z) can choose in $Z^\perp$ an ordinary solid ball B, centered at the origin of $Z^\perp$ and so small that B can also be regarded as a subset of M (cf. our comments in Section 2.1.2 ff. on actual measurements, tangent spaces, and negligible curvature). If a photon's world line intersects B, the tangent Y can be regarded as a vector in M_z (Exercise 0.0.10). Then $Y \in \mathscr{L}_z^+$ so (z, Z) measures a point $(U, e) \in P_z$ for Y as in Section 5.1.3. If $\mathscr{K} \times [a, b] \subset P_z$ is given where $\mathscr{K}$ is an integration region in $\mathscr{S}_z$, then (z, Z) can count the number of incoming photons with measured energies e in the range $[a, b]$ and measured spatial directions U in $\mathscr{K}$. Knowing this for all $\mathscr{K}$ and for all $[a, b]$, (z, Z) completely determines the distribution of incoming photons relative to his own direction-energy space.

5 Photons

If the number of photons intersecting the solid ball B is enormous, then the preceding clumsy counting procedure must be replaced by a "smoothed-out" idealization in much the same way the world density of a particle flow is used to describe the distribution of molecules in a cold, streaming gas (see Section 3.2). Thus we introduce a *photon distribution function for* (z, Z), which is formally defined as a function $F_Z: P_Z \to [0, \infty)$. Physically, $e^2 F_Z$ models the number of photons per unit spatial volume per unit solid angle per unit energy interval. The rest of this section will be devoted to a discussion of this function, to be denoted exclusively by F_Z.

> Imagine you are in your backyard on a beautiful sunny day. All around you light of many colors is whizzing about in all directions. Even the tiny fraction that enters your eye presents a rich picture. F_Z is essentially that picture. Indeed, suppose z is the center of your eye at one moment, that Z is tangent to your history, that B models your whole eye, while U points toward a rose and e corresponds to the color red (Section 5.1). Then $e^3 F_Z(U, e)$ is essentially the brightness of the rose as perceived by you (compare Exercises 5.5.6 and 5.5.7 following).
>
> Suppose we replace a large collection of Newtonian point particles by a Newtonian number density (cf. Chapter 3). Then we have in mind averages over regions large enough to contain many particles but so small that conditions do not change drastically within one region. In addition to the restriction of negligible curvature already mentioned above, the solid ball B should be neither too large nor too small in roughly the sense just indicated. If no such B exists, (z, Z) should use a different model.
>
> We are using F_Z to model photons irrespective of "polarization." As a result, some of our equations below will differ from those found in some texts by a factor of 2.

5.5.2 Physical interpretation

The following definition indicates somewhat more specifically how F_Z is measured. Let $\mathcal{K} \subset \mathcal{S}_z$ be an integration region. Then with $0 < a < b < \infty$ $\mathcal{H} = \mathcal{K} \times [a, b] \subset P_Z$ is also an integration region so $n = \int_{\mathcal{H}} F_Z \pi_Z$ is a finite, nonnegative number. It is defined as *the number of photons* (z, Z) *measures per unit spatial volume with measured energies in the range* $a \leq e \leq b$ *and measured spatial directions in the region* $\mathcal{K}$ of his celestial sphere. For example, suppose F_Z vanishes outside $\mathcal{H}$, then $n = \int_{P_Z} F_Z \pi_Z =$ the total number of photons (z, Z) measures per unit spatial volume, irrespective of energies or directions. In this case, if the ordinary 3-volume of the B of Section 5.5.1 is 7, then n would be (approximately) $\frac{1}{7}$ of the total number of photons whose world lines intersect B. The basic interpretation just given leads to an enormous number of derived concepts and definitions. The rest of the section is denoted to a few of the most important, and to some examples. However, we emphasize that, given (z, Z), there is really only one basic quantity involved, namely F_Z.

> In astronomy one can for most purposes identify all the astronomers who ever lived, regarding them as a single instantaneous observer (z, Z).

5.5 Photon distribution functions

On this view literally billions of man-hours have been spent measuring F_Z, and no other function of 3 variables describes as much of nature. Hence the plethora of auxiliary concepts that follow.

Our definition (Section 5.5.2) in effect makes precise the earlier phrase "per unit energy interval per unit solid angle." For a more precise version of "per unit spatial volume" see Exercise 5.7.7b.

$Z^\perp \times P_Z$ is often called (z, Z)'s "phase space"; compare Exercise 5.5.6.

Much as in Section 3.2, we do not require that nV, where V is the ordinary 3-volume of B, be an integer.

5.5.3 Energy density and spectra

(a) According to our interpretation (Section 5.5.2), eF_Z is the energy per unit volume of $Z^\perp \times P_Z$. More specifically, for each interval $[a, b] \subset (0, \infty)$, define a nonnegative number $u[a, b]$ by

$$u[a, b] = \int_{\mathscr{S}_Z \times [a, b]} eF_Z \pi_Z = \int_{\mathscr{S}_Z} \zeta \int_a^b e^3 F_Z \, de,$$

where the second equality comes from Section 5.5.1c and Fubini's theorem. Then $u[a, b]$ is interpreted as the measured energy density (Section 3.3.3) of those photons described by F_Z that have measured energies in the range $[a, b]$. Since F_Z is nonnegative, the double limit $\lim_{a \to 0, b \to \infty} u[a, b]$ exists and is either ∞ or a nonnegative number; call this limit u. u is then interpreted as the *total energy density (z, Z) measures for the photons modelled by F_Z*.

In practice, the assumption that u is finite fails only for certain models suggested by quantum theory but usually regarded as pathological. Thus in the more formal treatment of Section 5.7, we shall build this assumption into the definition.

(b) Primarily for historical reasons, the integrand $e^3 F_Z: P_Z \to [0, \infty)$ in (a) is also given a special name, being defined as the *specific intensity (z, Z) measures* for the photons. A slight variation of this function is, however, very useful. Recall from Section 5.1.3 that $e = hf$, where f is the frequency (z, Z) measures for the photon and h is Planck's constant. Thus $e^3 F_Z(U, e) = h^3 f^3 F_Z(U, hf)$. This leads to the definition: for a fixed $U \in \mathscr{S}_Z$, $S: (0, \infty) \to [0, \infty)$ is defined to be

$$S(f) = h^4 f^3 F_Z(U, hf).$$

In other words, S/h is just the specific intensity restricted to $\{U\} \times (0, \infty)$. The function S is defined as the *energy spectrum (z, Z) measures in the U direction*. The term comes from the fact that if S is independent of U, $4\pi \int_0^\infty S \, df$ is, by the equations in (a), the energy density u.

(c) An enormous amount is known about the energy spectra produced by various kinds of physical processes—for example, emission or absorption of photons by a gas or by electrons moving in a magnetic field. Conversely, suppose (z, Z) measures an energy spectrum in the U direction and

5 Photons

finds many sharp local maxima and minima ("spectral lines") as in the figure. Often he can determine in detail what the physical conditions at the emitter and in the intervening spacetime must be in order to produce the observed energy spectrum. Most of our information about the physics of stars is obtained in this way.

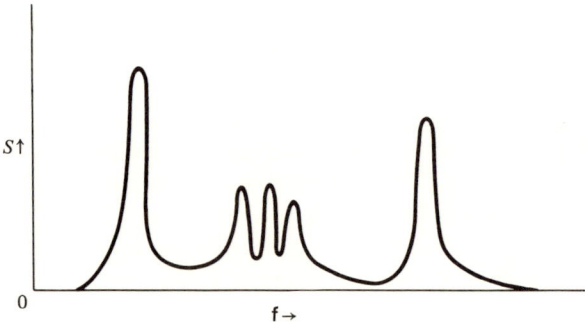

(d) Often (z, Z) can assign an empirical frequency ratio $\imath$ to a distant-early galaxy, roughly as follows. He measures the spectrum S for U in the direction toward the galaxy center. From the general shape of S he infers what physical process produced a particular local maximum. He can then work out what frequency f_0 a hypothetical observer at the galaxy, at rest with respect to the galaxy, would have measured for the photons at that local maximum. Comparing with the actually observed frequency f gives $\imath$ as f_0/f. Now a frequency ratio is the same for all the photons in a given light signal (Section 5.4). Thus $\imath$ typically comes out independent of which particular local maximum (z, Z) uses. This serves as a check on the argument, in particular on the implicit assumption that local physics at the galaxy is the same as local physics near here-now.

EXAMPLE 5.5.4. PLANCK PHOTON DISTRIBUTION FUNCTIONS. The most important examples of photon distribution functions are those that correspond, intuitively speaking, to thermal equilibrium. We define them, compute their energy density and add some heuristic comments on the concept of thermal equilibrium. (a) Define the *Planck function* $P: (0, \infty) \to (0, \infty)$ by $Pu = 2h^{-3}[(\exp u) - 1]^{-1}$, where h is Planck's constant. Let k be the universal constant called *Boltzmann's constant*; in our units $k \simeq 10^{-74}$ seconds per Kelvin. F_Z is defined as *Planck with temperature* T iff $F_Z(U, e) = P(e/kT)$ $\forall (U, e) \in P_Z$ and for some $T \in (0, \infty)$.

> The numerical value 10^{-74} is due to our conventional choice of temperature units and time units. Sometimes one uses temperature units such that $k = 1$. "Kelvins" (formerly called degrees Kelvin) refers to the standard temperature scale based on absolute zero.

5.5 Photon distribution functions

A Planck distribution function with temperature T can be produced—for example, by a gas that is held at temperature T while it emits light. The main properties of Planck distribution functions were known by 1900 from experiments and preliminary theoretical arguments, but it was Planck who first guessed the above explicit expression of the functions themselves. In 1901, he gave a deeper theoretical analysis. He applied statistical arguments to an intentionally oversimplified model of atoms emitting and absorbing light and found that the empirical formula can be derived by introducing his constant h and introducing the concept of a photon. His analysis assumed that collisions in which photons are created or destroyed can occur freely.

Nowadays, various equivalent terms are used for Planck distribution functions: "thermal [equilibrium] spectrum," "blackbody radiation," and so on. The term "blackbody" refers, roughly speaking, to the fact that if a photon of any frequency can be destroyed in a collision at the surface of a body, the body may absorb all incident light and appear black.

(b) Let F_z be Planck with temperature T. The graph shows its specific intensity $I = e^3 P(e/kT)$ as a function of e. 4π times the enclosed area is the energy density (Section 5.5.3a)—that is,

$$u = \int_{P_z} e F_z \pi_z = 4\pi \int_0^\infty I\, de$$

$$= 8\pi h^{-3} \int_0^\infty \frac{e^3\, de}{[\exp(e/kT)] - 1}$$

$$= 8\pi h^{-3}(kT)^4 \int_0^\infty \frac{u^3\, du}{(\exp u) - 1} = a_0 T^4,$$

where a_0 is the *blackbody constant*, $a_0 = 8\pi^5 k^4/(15h^3) \cong 1.4 \times 10^{-41}$ (seconds)$^{-2}$ (Kelvin)$^{-4}$ in our units. If the x and y axes are, respectively, stretched and contracted by a (multiplicative) factor h, the graph becomes that of the energy spectrum (Section 5.5.3b). In observations, the characteristic shape of such a Planck spectrum stands out, literally, like a sore thumb (cf. Section 5.5.3c).

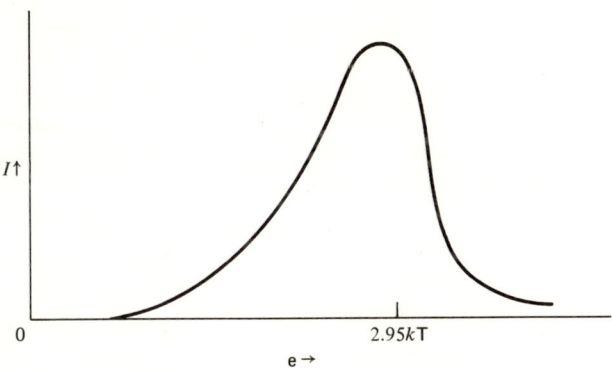

(c) Speaking very generally, and rather vaguely, some equilibrium pattern of particle motions tends to form whenever many particles undergo frequent collisions; one can then define the concept of the *temperature* of the system of particles. The collisions not only tend to establish spatial isotropy, exemplified in our case by the fact that a Planck F_Z is independent of direction U. They also tend to share out the available energy according to a simple, standard, "thermal equilibrium" pattern which depends only on the type of particles involved and a few parameters—for example, temperature, "chemical potential" for electric charge, and so on (cf. Ehlers [1]). For example, the entire graph in (b) for photons in thermal equilibrium is uniquely determined given a single parameter, the temperature T.

> An intuitive reason why collisions tend to establish a standard energy pattern can be roughly seen from the billiard ball example we used below Definition 3.15.2 to indicate why collisions tend to establish spatial isotropy. Indeed, whenever two billiard balls collide, the one with the larger (Newtonian kinetic) energy tends, on the average, to give the other a bit of its own energy. Thus imagine a billiard table with negligible friction and a number of balls on it. Suppose only one ball is moving: a thoroughly nonthermal-equilibrium pattern. It is intuitively rather clear that after many collisions most of the balls will be moving with energies close to the average energy, with minor deviations from the average energy not too infrequent. Then the situation approximates the thermal equilibrium energy pattern for billiard balls.

Once such a thermal equilibrium pattern is set up it tends to maintain itself whether or not collisions occur until some variable—for example, the density of particles—is changed. Then, if collisions occur often enough, a thermal equilibrium pattern at a new temperature quickly emerges.

5.5.5. *Astrophysical quantities*

For later reference we describe a few basic concepts in astronomy. Suppose (z, Z) is observing photons from a specific object, say the moon, a star or a distant-early galaxy.

(a) Physically, the *absolute luminosity* $L > 0$ of the object is defined as the total energy it sends out in photons during one second of its own proper time. A precise mathematical model, in a special case, is given in Section 6.7 following.

(b) Let $\mathcal{M}$ be a matter model which represents the photons from the object and let $\hat{T}$ be the stress-energy tensor of $\mathcal{M}$. The *apparent luminosity* ℓ (z, Z) *measures* for the object is the energy density (z, Z) measures for $\hat{T}$ (Section 3.3.3)—that is, $\ell = T(Z, Z)$. In very simple situations the absolute luminosity is typically $4\pi\ell$, where ℓ is the apparent luminosity measured by an instantaneous observer "at rest" with respect to the object and at unit distance from it (cf. Exercise 6.7.7).

(c) To your eye, the moon appears as an extended patch on the sky, but a star looks like a point. Assume, as is normally an excellent idealization,

5.5 Photon distribution functions

that there exists a smallest integration region $\mathscr{K} \subset \mathscr{S}_z$ with the property that an observed photon comes from the particular object only if the corresponding spatial direction U is in $\mathscr{K}$. Then the solid angle subtended by $\mathscr{K}$ is defined as the *apparent size* of the object for (z, Z) [and also as the "solid angle it subtends" for (z, Z)]. Thus when you observe a star, $\mathscr{K}$ is so small that your eye cannot distinguish $\mathscr{K}$ from a point.

(d) (z, Z) has many different photon detection devices available—his eyes, an optical telescope combined with a photographic plate, a radio telescope, an x-ray counter, and so on. Each such device responds to photons in a somewhat different way. To be more specific, fix $\mathscr{K}$ and suppose $\mathscr{M}$ in (b) can be regarded, as far as (z, Z) is concerned, as a photon distribution function F_Z, with $F_Z = 0$ outside of $\mathscr{K} \times (0, \infty)$. Then if (z, Z) simply looks at the object corresponding to this $\mathscr{K}$, his eye typically responds rather directly to ℓ_V, the *apparent luminosity in the visible*, defined by $\ell_V = \int_{\mathscr{K} \times [a,b]} eF_Z \pi_Z$, where $[a, b]$ corresponds to the visible range—that is, $(a/h) \cong 3 \times 10^{14}$ (seconds)$^{-1}$, $(b/h) \cong 10^{15}$ (seconds)$^{-1}$ (Section 5.1.3c). Very roughly, the eye simply "adds up" all the photon energies in an appropriate energy range, corresponding to the integral just given. For comparison, note here that the apparent luminosity for (z, Z) is given by Sections 5.5.3a and 5.5.5b as $\ell = \int_{\mathscr{K} \times (0, \infty)} eF_Z \pi_Z = u \geq \ell_V$.

On the other hand some detection devices respond rather directly to the specific intensity (Section 5.5.3b). And a radio telescope looking at a quasar typically responds to another quantity: $\forall e \in (0, \infty)$ the *specific flux* is $F_e = e^3 \int_{\mathscr{K}} F_Z(U, e) \zeta$. The radio telescope is usually "tuned" to a specific frequency f and responds to F_{hf}. And so on for other detectors, *ad nauseam* (cf. Emming [1]).

Exercise 5.5.6

Let $P_z = \mathscr{S}_z \times (0, \infty)$ be (z, Z)'s direction-energy space. (a) Define a map $\pi_Z: P_Z \to M_z$ by $\pi_Z(U, e) = e(Z - U) \forall (U, e) \in \mathscr{S}_z \times (0, \infty)$ (cf. Section 5.1.3). Show π_Z is a diffeomorphism onto the future lightcone $\mathscr{L}_z^+$. (b) Define a map $\rho: P_Z \to Z^\perp$ by $\rho = v_Z \circ \pi_Z$, where $v_Z: M_z \to Z^\perp$ is the orthogonal projection along Z—that is, $v_Z W = W + g(Z, W)Z \forall W \in M_z$. Show $\rho(U, e) = -eU$, and show ρ is a diffeomorphism onto $Z^\perp - \{0\}$. (c) In (b) show that, regarding $Z^\perp$ as $\mathbb{R}^3$, we have $\rho^*(du^1 \wedge du^2 \wedge du^3) = -\zeta \wedge e^2 \, de = -\pi_Z$.

We mention, without attempting to explain in detail a few of the further phrases used to describe F_Z in the literature. It is photon number "per mode" or "per unit volume in phase space" or "per unit spatial volume per unit volume in 3-momentum space $Z^\perp - \{0\}$." Or it is "photon occupation number." Moreover, whenever the term "per unit spatial volume" occurs in the interpretations, it can be replaced by "per unit perpendicular area per unit time." The idea then is that, since photons travel at unit speed (Section 3.1), all the photons in a

5 Photons

unit spatial cube cross a unit spatial 2-area perpendicular to their travel direction in unit time (provided they are not destroyed in the interim).

EXERCISE 5.5.7

Suppose there is a very large, finite collection $\{(P_A, \eta_A)\}$ of photon beams that model the same physical situation that F_Z describes at z. Then, roughly, F_Z is world density per unit volume in 3-momentum space (cf. Exercise 5.5.6). More specifically, let $\mathcal{K} \subset \mathcal{S}_z \times (0, \infty)$ be an integration region and define $K = \{A \mid P_A z \in \pi_Z \mathcal{K}\}$, where π_Z is defined in Exercise 5.5.6a. The following approximate equalities cannot be proved since our two intuitively related models are mathematically independent; trying to demand exact equality $\forall \mathcal{K}$ would be nonsense. But try to make both approximate equalities plausible to yourself using Definition 3.2.1, Section 5.2.2, and some pictures:

(a) $$\int F_Z \pi_Z \cong - \sum_{A \in K} \eta_A z \, g(P_A, Z);$$

(b) $$\int e F_Z \pi_Z \cong \sum_{A \in K} [g(P_A, Z)]^2 \eta_A z = \sum_{A \in K} \hat{T}_A(Z, Z)$$

where $\hat{T}_A$ is the stress-energy tensor of (P_A, η_A).

5.6 Integration on lightcones

Suppose F_Z is a photon distribution function for the instantaneous observer (z, Z). F_Z describes only those photons near z. Worse, we do not yet know even how the photon distribution function for a different instantaneous observer (z, Z') at z is related to F_Z. We need an intrinsic, genuinely relativistic model, which does not single out any one instantaneous observer. This section sets up the formalism. The next gives the model. Both sections use the notation of the previous one. Thus if (x, X) is an instantaneous observer, P_X denotes his direction-energy space, π_X is the natural volume element on P_X (Section 5.5.1c), and $\pi_X: P_X \to \mathcal{L}_x^+$ is the natural diffeomorphism onto the future lightcone $\mathcal{L}_x^+$ at x (Exercise 5.5.6a).

We begin by analyzing intrinsic volume elements on lightcones. Let $(\mathbb{R}^4, g)$ be Minkowski space. We regard the origin, but not the axes, as distinguished since we actually have in mind the tangent space to a point in a general spacetime. Let Ω be the metric volume element $du^1 \wedge \cdots \wedge du^4$. Let ω be the "radial" 1-form $u^1 \, du^1 + u^2 \, du^2 + u^3 \, du^3 - u^4 \, du^4$; ω is intrinsic (Section 5.0.5). Let $\mathcal{L}_0^+$ be the set of all points to which the origin can send a light signal—that is,

$$\mathcal{L}_0^+ = \left\{ (u^1, \ldots, u^4) \in \mathbb{R}^4 \,\middle|\, u^4 > 0, \sum_{\mu=1}^{3} (u^\mu)^2 = (u^4)^2 \right\}.$$

Let $\iota: \mathcal{L}_0^+ \to \mathbb{R}^4$ be the inclusion map. Thus ι is an imbedding (Proposition 1.1.7) and $u^4 \circ \iota > 0$.

Proposition 5.6.1. *The 3-form $\Lambda_0 = \iota^*[(1/u^4)\,du^1 \wedge du^2 \wedge du^3]$ on $\mathscr{L}_0^+$ can be intrinsically characterized as follows. $\forall$ 3-form χ over ι such that $(\omega \circ \iota) \wedge \chi = \Omega \circ \iota$, $\Lambda_0 w = \iota^*(\chi w) \; \forall w \in \mathscr{L}_0^+$.*

Proof. There exists at least one such χ; for example, the 3-form
$$\chi = [(1/u^4)\,du^1 \wedge du^2 \wedge du^3] \circ \iota$$
works by our above expressions for ω, Λ_0, and Ω. We must show uniqueness of Λ_0. Let χ, χ' be 3-forms over ι with $(\omega \circ \iota) \wedge \chi = \Omega \circ \iota$. Then $(\omega \circ \iota) \wedge \chi' = \Omega \circ \iota$ iff $(\omega \circ \iota) \wedge (\chi - \chi') = 0$ iff $\chi - \chi' = (\omega \circ \iota) \wedge \eta$ for some 2-form η over ι (Bishop–Goldberg p. 95). But $\forall W$ tangent to $\mathscr{L}_0^+$ at $w \in \mathscr{L}_0$, $\omega(W) = 0$ (Section 5.0.5). Thus if χ and χ' both obey the stated conditions, $\iota^*(\chi w) = \iota^*(\chi' w) \; \forall w \in \mathscr{L}_0^+$. Thus Λ_0 is unique. □

We can now get an intrinsic volume element on the future lightcone at a point in any spacetime M. The factor e^{-1} in the following corollary corresponds to $(u^4)^{-1}$ in Proposition 5.6.1; the factor is essential in an observer independent treatment.

Corollary 5.6.2. *Suppose $x \in M$. Then there is a unique volume element Λ_x on $\mathscr{L}_x^+$ such that, $\forall$ instantaneous observer (x, X) at x, $\pi_X{}^*\Lambda_x = e^{-1}\pi_X$.*

PROOF. Let $\{X_1, \ldots, X_4\}$ be an orthonormal basis of M_x, and let $\tau: M_x \to \mathbb{R}^4$ be the diffeomorphism given by $\tau(\sum_i a^i X_i) = (a^1, \ldots, a^4)$. Then $\tau\mathscr{L}_x^+ = \mathscr{L}_0^+$; let $\tau_0: \mathscr{L}_x^+ \to \mathscr{L}_0^+$ denote the diffeomorphism induced by τ. The preceding proposition shows that $\tau_0^*\Lambda_0$ is a volume element on $\mathscr{L}_x^+$ independent of the orthonormal basis chosen. Write $\Lambda_x = -\tau_0^*\Lambda_0$. Observe that if $\nu: \mathscr{L}_0^+ \to \mathbb{R}^3$ denotes the restriction to $\mathscr{L}_0^+$ of the canonical projection $\mathbb{R}^4 \to \mathbb{R}^3$ such that $(a^1, \ldots, a^4) \to (a^1, a^2, a^3)$, then $\Lambda_0 = \nu^*(f\,du^1 \wedge du^2 \wedge du^3)$, where $f = [(u^1)^2 + (u^2)^2 + (u^3)^2]^{-1/2}$.

Now if (x, X) is given, we may assume $X = X_4$. Let $\rho: P_X \to \mathbb{R}^3$ be the mapping of Exercise 5.5.6b. Then $\rho = \nu \circ \tau_0 \circ \pi_X$. Exercise 5.5.6c then gives $\pi_X{}^*\Lambda_x = -\rho^*(f\,du^1 \wedge du^2 \wedge du^3) = e^{-1}\pi_X$. □

$\mathscr{L}_x^+$ is not compact and improper integrals, analogous to the limit in Section 5.5.3a, will arise. We give some machinery for dealing with them. As in Section 5.5.1a, let Λ be a volume element on a manifold N. Recall that $\mathbb{Z}^+$ denotes the positive integers. Define a sequence $\mathscr{K}_1, \mathscr{K}_2, \ldots$ of integration regions in N to be *exhaustive* iff $\mathscr{K}_i \subset \mathscr{K}_{i+1}^0$ (the interior of $\mathscr{K}_{i+1}$) $\forall i \in \mathbb{Z}^+$ and $\bigcup_i \{\mathscr{K}_i \mid i \in \mathbb{Z}^+\} = N$. For example, suppose (z, Z) is an instantaneous observer and $N = \mathscr{L}_z^+$. Then $\mathscr{K}_i = \{Y \in \mathscr{L}_z^+ \mid i \geq |g(Y, Z)| \geq (i + 1)^{-1}\}$ determines an exhaustive sequence of integration regions in N.

> Since N is paracompact (Section 0.0.4) there always is an exhaustive sequence for N. The simplest way to see this is to invoke the Whitney imbedding theorem and standard Morse theory to prove the existence of a C^∞ function $h: N \to [0, \infty)$ such that $h^{-1}[0, i]$ is compact $\forall i \geq 0$

5 Photons

and such that the critical points of h (= the points at which $dh = 0$) form a discrete set (cf. Milnor [1], p. 36). We may assume $0 \in hN$. Now define $\mathcal{K}_i = h^{-1}[0, i]$. The sequence $\{\mathcal{K}_i\}$ is then an exhaustive sequence. In fact, in this case each $\partial \mathcal{K}_i$ is an imbedded C^∞ submanifold of N except at a finite number of points.

Suppose $f: N \to \mathbb{R}$ is continuous. Suppose that for some a, $-\infty \leq a \leq \infty$, $\lim_{i \to \infty} \int_{\mathcal{K}_i} f\Lambda = a$ $\forall$ exhaustive sequence $\{\mathcal{K}_i\}$ of integration regions in N. Then by definition $\int_N f\Lambda$ exists and equals a, and we write $\int_N f\Lambda = a$. For example, suppose N is Minkowski space, Ω is the metric volume form and $f = 1$. Then $\int_M f\Omega$ exists and equals ∞. As another example, suppose f vanishes outside some compact set $\mathcal{K} \subset N$. Then $\forall$ such exhaustive sequence $\{\mathcal{K}_i\}$, there is an $i \in \mathbb{Z}^+$ such that $\mathcal{K} \subset \mathcal{K}_i$, since $\mathcal{K}_{i+1}^0 \supset \mathcal{K}_i$ $\forall i \in \mathbb{Z}^+$ and $\bigcup_i \mathcal{K}_i = N$. This implies that $\int_N f\Lambda$ exists and is a finite number. More generally, we have:

Proposition 5.6.3. *Suppose f keeps sign—that is, f is either nonnegative or nonpositive—then $\int_N f\Lambda$ exists.*

PROOF. We may assume f nonnegative. Let $\{\mathcal{K}_i\}$ be an exhaustive sequence. Then $\int_{\mathcal{K}_1} f\Lambda, \int_{\mathcal{K}_2} f\Lambda, \ldots$ is a nondecreasing sequence of nonnegative numbers and hence must approach a limit, say $a \in [0, \infty]$. Let $\{\mathcal{H}_i\}$ be a second exhaustive sequence; we must show $\int_{\mathcal{H}_i} f\Lambda \to a$.

Define a map $\nu: \mathbb{Z}^+ \to \mathbb{Z}^+$ by induction as follows. $\nu(1) = j$, where j is such that $\mathcal{K}_j \supset \mathcal{H}_1$; $\forall i = 2, 3, \ldots$, take $\nu(i)$ such that $\mathcal{K}_{\nu(i)} \supset \mathcal{H}_i$ and that $\nu(i) > \nu(k)$ $\forall k = 1, \ldots, i-1$. $f \geq 0$ now implies $\int_{\mathcal{H}_i} f\Lambda \leq \int_{\mathcal{K}_{\nu(i)}} f\Lambda$ and thus $\lim_{i \to \infty} \int_{\mathcal{H}_i} f\Lambda \leq a$. Reversing the roles of $\{\mathcal{K}_i\}$ and $\{\mathcal{H}_i\}$ in the argument, we have $a \leq \lim_{i \to \infty} \int_{\mathcal{H}_i} f\Lambda$. □

Proposition 5.6.3 is all we shall really need, and the above approach to improper integrals is the shortest for our purposes. It is by no means the most natural. The most natural approach begins with the definition of $\int_\mathcal{U} f\Lambda$ for a function f defined on a coordinate neighborhood $\mathcal{U}$ and having compact support. Then one can introduce a Borel measure on N and take over the whole machinery of measure theory. No detailed elementary exposition of this approach seems to be available, but Helgason [1], pp. 361–5, gives some of the main ideas.

We next specify the functions of interest. Suppose $f: \mathscr{L}_x^+ \to [0, \infty)$ is a function; thus f is C^∞ (Section 0.0.4). Recall that each $\omega \in M_x^*$ determines a function $\tilde{\omega}: \mathscr{L}_x^+ \to \mathbb{R}$ via its natural action on M_x (Exercise 0.0.10). The wrong way Schwarz inequality shows that if ω is causal, $\tilde{\omega}$ keeps sign. Thus $\forall$ nonnegative integer N, $\int_{\mathscr{L}_x^+} (\tilde{\omega})^N f\Lambda_x$ exists. We define f to be of *rapid decay* iff $\int_{\mathscr{L}_x^+} (\tilde{\omega})^N f\Lambda_x$ is finite $\forall$ such N and $\forall$ causal $\omega \in M_x^*$.

5.6 Integration on lightcones

EXAMPLE 5.6.4. Let (x, X) be an instantaneous observer and define $f: \mathscr{L}_x^+ \to (0, \infty)$ by $f(Y) = \exp(-u)/u$, where $u = -g(X, Y)$. We claim f is of rapid decay. Note that $f \circ \pi_x = e^{-1} \exp(-o)$. To estimate integrals involving $(\tilde{\omega})^N f$ we need bounds on $\tilde{\omega}$. $\forall Y \in \mathscr{L}_x^+$ we have $Y = e(X - U)$, $e \in (0, \infty)$, $U \in \mathscr{S}_x \subset X^\perp$. Suppose $\omega \in M_x^*$ is causal and decompose ω similarly—that is, $\omega = a\chi + \alpha$, where $a \in \mathbb{R}$, $\chi \in M_x^*$ is physically equivalent to X and $\alpha \in M_x^*$ with $\alpha(X) = 0$. Then $\forall Y \in \mathscr{L}_x^+$ $\tilde{\omega}Y = \omega(Y) = -e \times [a + \alpha(U)]$. Moreover, $[\alpha(U)]^2 \leq |\alpha|^2 |U|^2 = |\alpha|^2$ by the usual Schwarz inequality applied to $X^\perp$. Thus setting $b = |a| + |\alpha|$ we have $|\tilde{\omega}| \leq eb$ on all of $\mathscr{L}_x^+$. Thus with N a nonnegative integer

$$b^{-N} \left| \int_{\mathscr{L}_x^+} \tilde{\omega}^N f \Lambda_x \right| = \left| \int_{P_x} (\tilde{\omega} \circ \pi_x)^N e^{-2} \exp(-e) \pi_x \right| b^{-N}$$

$$\leq \int_{\mathscr{S}_x} \zeta \int_0^\infty e^{N-2} [\exp(-e)] e^2 \, de$$

$$= 4\pi \int_0^\infty e^N \exp(-e) \, de < \infty.$$

Thus $\int_{\mathscr{L}_x^+} \tilde{\omega}^N f \Lambda_x$ not only exists but is also finite. ω was any causal 1-form at x and N was any nonnegative integer, so f is of rapid decay, as claimed.

The generalization of these pointwise considerations to all of M will here require only one simple result on the tangent bundle TM. Define the *future lightcone in TM* as $\mathscr{L}^+ = \{(y, Y) \in TM \mid g(Y, Y) = 0 \text{ and } Y \text{ is future pointing}\}$. Define the *past lightcone* $\mathscr{L}^-$ in TM dually and take $\mathscr{L} = \mathscr{L}^+ \cup \mathscr{L}^-$. $\mathscr{L}^+$ is an open subset of $\mathscr{L}$ in the topology induced on $\mathscr{L}$ by that of TM since for any reference frame Z, $\mathscr{L}^+ = H^{-1}(-\infty, 0)$, where $H: \mathscr{L} \to \mathbb{R}$ is the (C^∞) function determined by $H(Y) = g(Z, Y)$.

Proposition 5.6.5. *The future lightcone $\mathscr{L}^+$ in TM is a 7-dimensional imbedded submanifold.*

Proof. As usual, let $K: TM \to \mathbb{R}$ be defined by $K(x, X) = g(X, X)$. Let $TM' = TM - \{(x, 0) \in TM\}$. TM' is an open submanifold of TM and dK is nowhere zero on TM' (Exercise 0.0.10). Since $\mathscr{L}$ is the level set $K = 0$ in TM', $\mathscr{L}$ is an imbedded submanifold of TM' and hence of TM. $\mathscr{L}^+$ being open in $\mathscr{L}$, the proposition follows. □

Let $F: \mathscr{L}^+ \to [0, \infty)$ be a function. We define F to be of *rapid vertical decay* iff both the following conditions hold. (a) $\forall x \in M$, the restriction of F to $\mathscr{L}_x^+$ is of rapid decay. (b) $\forall$ causal 1-form ω on M and each nonnegative integer N, the function on M defined by $x \to \int_{\mathscr{L}_x^+} [(\omega x)^\sim]^N F \Lambda_x$ is C^∞, where $(\omega x)^\sim$ denotes the function on M_x induced by ωx in the sense of Exercise 0.0.10.

Here "vertical" refers to the restriction to a fibre of the tangent bundle.

5 Photons

EXERCISE 5.6.6. FUNCTIONS OF RAPID DECAY

Let (x, X) be an instantaneous observer, $\chi \in M_x^*$ be physically equivalent to X, $\pi_X: P_X \to \mathscr{L}_x^+$, and Λ_x be as in this section. (a) Suppose $\ell: (0, \infty) \to (0, \infty)$ is a function such that, for all non-negative integers L, N, $\int_0^\infty e^N |d^L(e\ell)/de^L|\, de < \infty$. With $a, b \in (0, \infty)$, define $f = a\ell \circ (-b\tilde{\chi}): \mathscr{L}_x^+ \to (0, \infty)$; thus $f(Y) = a\ell(-bg(X, Y))$ $\forall Y \in \mathscr{L}_x^+$. Show f is a function of rapid decay. (b) Let F_X be a photon distribution function for (x, X); suppose F_X is Planck. Show from (a) that $F_X \circ \pi_X^{-1}$ is a function of rapid decay. (c) Let $f: \mathscr{L}_x^+ \to [0, \infty)$ be a function of rapid decay and suppose $\omega \in M_x^*$. Show that ω can be written as the sum of two causal 1-forms and then that $\int_{\mathscr{L}_x^+} \tilde{\omega} f \Lambda_x$ exists and is finite. (d) In (c) show $f \circ \pi_X$ is a photon distribution function for (x, X) and that the total energy density $\lim_{a \to 0, b \to \infty} u[a, b]$ in Section 5.5.3a is finite. (e) In (c) suppose $\eta \in M_x^*$. Show $\int_{\mathscr{L}_x^+} \tilde{\omega}\tilde{\eta} f \Lambda_x$ exists and is finite. (f) In (e) suppose f is not identically zero on $\mathscr{L}_x^+$ and both ω and η are future pointing. Show $\int_{\mathscr{L}_x^+} \tilde{\omega}\tilde{\eta} f \Lambda_x > 0$ (roughly: $\tilde{\omega}\tilde{\eta}$ is zero only on a set of measure zero).

EXERCISE 5.6.7. FUNCTIONS OF RAPID VERTICAL DECAY

Suppose X is a reference frame and $g: M \to (0, \infty)$, $j: M \to (0, \infty)$ are functions. Define a function $F: \mathscr{L}^+ \to (0, \infty)$ by requiring the restriction of F to $\mathscr{L}_x^+$ to equal f, where f is as in Exercise 5.6.6a with $X = Xx$, $a = gx$, and $b = jx$. Show F is of rapid vertical decay. (b) Let $H: \mathscr{L}^+ \to [0, \infty)$ be of rapid vertical decay, and let ω and η be 1-forms on M. Show there exists a function $h: M \to \mathbb{R}$ defined pointwise by $hx = \int_{\mathscr{L}_x^+} (\omega x)\tilde{}\, (\eta x)\tilde{}\, H\Lambda_x$ (what is required is to show h is C^∞).

5.7 A photon gas

We can now define the model often used to analyze such physical processes as the scattering of light by our atmosphere, the reflection of light by the moon and planets, the emission of light by the sun, the absorption of light by large gas clouds in our galaxy, the transmission of radio waves from a distant-early galaxy to here-now through the intervening spacetime, the time evolution of the microwave radiation (which seems to be a relic from an early epoch in the history of our universe; cf. Section 6.5), and so on.

Let $\mathscr{L}^+ \subset TM$ be the future lightcone; thus $\mathscr{L}^+$ is a 7-manifold (Proposition 5.6.5). A *photon gas on* M is a function $F: \mathscr{L}^+ \to [0, \infty)$ of rapid vertical decay. Throughout this section F is a photon gas on M. Roughly speaking, $F(y, Y)$ represents the probability of finding a photon with energy-momentum Y at $y \in M$ $\forall (y, Y) \in \mathscr{L}^+$; hence the requirement that F be non-negative. Requiring F to be of rapid vertical decay guarantees, *inter alia*, that a photon gas has a well-defined stress-energy tensor; to clarify this, we now make our interpretation more specific.

In more advanced treatments one imposes additional restrictions on F. These are stated and applied in the Optional exercises (Section 8.6).

The reader familiar with nonrelativistic one-particle kinetic theory will find that the discussion that follows merely adapts the ideas of that theory to the case of photons in general relativity.

The reader familiar with the geometric formulation of Hamiltonian mechanics, as described for example in Chapter 6 of Bishop–Goldberg [1], may find it useful to transcribe the discussion that follows into the language of that theory. We mention three mildly tricky features that then arise. (a) We are working on a submanifold $\mathscr{L}^+$ of TM, rather than working on all of T^*M. (b) F is not interpreted as a dynamic variable associated with a single photon; rather, F corresponds to the "density of photons in phase space." (c) Let K be the geometric energy function (Section 5.0.5). Thus $K = 0$ on $\mathscr{L}^+$, and K is not interpreted physically as energy (Section 5.0.5). Nevertheless the restriction of dK to $\mathscr{L}^+$ plays much the same role that the exterior derivative of the Hamiltonian plays in the ordinary geometric theory of Hamiltonian mechanics.

5.7.1 Physical interpretations

Let (x, X) be an instantaneous observer and let $\pi_X: P_X \to \mathscr{L}_x^+$ be the natural diffeomorphism of Exercise 5.5.6a. Then $F_X = F \circ \pi_X$ is a photon distribution function for (x, X) since $F\mathscr{L}^+ \subset [0, \infty)$.

(a) The physical interpretation of F is that, no matter which instantaneous observer (x, X) we consider, the operational definition indicated in Section 5.5.2 applies to $F_X = F \circ \pi_X$. Thus the other interpretations of Section 5.5 also apply. For example, F is defined as *Planck with temperature* T *for the instantaneous observer* (z, Z) iff $F \circ \pi_z$ is Planck with temperature T.

> Here the truly remarkable feature is that a single function can summarize the observations made by all instantaneous observers. If (x, X') is an instantaneous observer whizzing past (x, X) at half the speed of light, he does not agree with (x, X) on measured photon energies; nor do they even agree on rest spaces, let alone on measured spatial directions for photons. But they measure the same value for a given photon energy-momentum Y, namely, F(x, Y). This observer-independence of F is best regarded as a law of nature (cf. Section 3.3). However, it can be made plausible by comparing a photon gas to a mathematically independent, intuitively related model consisting of a large collection of photons (Ehlers [1]). Exercise 3.0.5 for the case $a = 1$ outlines the plausibility argument in its simplest form.

(b) Much as in Section 3.3, our interpretation of F leads to a stress-energy tensor for the photons described by F. By our assumption that F restricted to $\mathscr{L}_x^+$ is of rapid decay, the energy density (x, X) measures for the photons—namely, $u = \int_{P_X} eF_X\pi_X$ (Section 5.5.3a), exists and is finite (Exercise 5.6.6d). The following proposition shows more explicitly how u is determined by F; compare Exercise 5.5.7b.

5 Photons

Proposition 5.7.2. *Notation as in Sections 3.0.3 and 5.6. There exists a unique symmetric (2, 0)-tensor field $\hat{T}$ on M such that $\hat{T}(\omega, \omega) = \int_{\mathscr{L}_x^+} \tilde{\omega}^2 F \Lambda_x$ $\forall \omega \in M_x^*$ $\forall x \in M$. Moreover, the following properties hold:*

(a) *$\hat{T}$ is a stress-energy tensor;*
(b) *$\hat{T}x$ is normal unless $F \equiv 0$ on $\mathscr{L}_x^+$;*
(c) *trace $\hat{T} = 0$; and*
(d) *$\forall$ instantaneous observer (x, X), $T(X, X) = \int_{P_X} e(F \circ \pi_X)\pi_X = u$.*

Proof. Suppose $x \in M$. Since F restricted to $\mathscr{L}_x^+$ is of rapid decay, there exists a function $T: M_x^* \to \mathbb{R}$ defined by $\omega \to \int \tilde{\omega}^2 F \Lambda_x$ (Exercise 5.6.6e); here, and throughout this proof, the domain of integration will be $\mathscr{L}_x^+$ unless indicated otherwise. Since T is a quadratic function, it determines a unique symmetric tensor $\hat{T}_x \in T_0^2(M_x)$ such that

$$\hat{T}_x(\omega, \eta) = \int \tilde{\omega}\tilde{\eta} F \Lambda_x.$$

We note the following properties. (a') $\hat{T}_x(\omega, \omega) \geq 0 \forall \omega \in M_x^*$, since F is nonnegative. (b') If F is not identically zero on $\mathscr{L}_x^+$ and $W \in M_x$ is causal, then $\tilde{T}_x W$ is timelike. In fact suppose $\omega, \eta \in M_x^*$ are future pointing. Then $\hat{T}_x(\omega, \eta) > 0$ (Exercise 5.6.6). Thus if ω is physically equivalent to W, $\eta(\tilde{T}_x W) = \hat{T}_x(\eta, \omega) > 0$. Thus $\tilde{T}_x W$ is orthogonal to no causal vector in M_x and hence is timelike. (c') Trace $\hat{T}_x = 0$. For, let K be the geometric energy function and let $\{\omega^i\}$ be an orthonormal basis for M_x^*. We have:

$$\text{trace } \hat{T}_x = \sum_{\mu=1}^{3} \hat{T}_x(\omega^\mu, \omega^\mu) - \hat{T}_x(\omega^4, \omega^4)$$

$$= \int \left[\sum_{\mu=1}^{3} (\tilde{\omega}^\mu)^2 - (\tilde{\omega}^4)^2 \right] F \Lambda_x$$

$$= \int K F \Lambda_x = 0,$$

where the last equality is because K is zero on $\mathscr{L}_x^+$. (d') Let (x, X) be an instantaneous observer, χ be physically equivalent to X. Then

$$T_x(X, X) = \hat{T}_x(\chi, \chi) = \int \tilde{\chi}^2 F \Lambda_x$$

$$= \int_{P_X} e^2(F \circ \pi_X)(\pi_X^* \Lambda_x)$$

$$= \int_{P_X} eF_X \pi_X.$$

Now since F is of rapid vertical decay there exists a (C^∞) tensor field $\hat{T}$ on M defined pointwise by $\hat{T}x = \hat{T}_x$, with $\hat{T}_x$ as above. $\hat{T}$ then obeys (a)–(d) by virtue of (a')–(d'). □

$\hat{T}$ in Proposition 5.7.2 is defined as the *stress-energy tensor of the photon gas* F. We now examine the behavior of $\hat{T}$ in a special case which will play an important role in our later discussion of cosmology. Let (z, Z) be an instantaneous observer, $\mathcal{O}^3$ be the group of isometries from M_z onto itself leaving Z fixed, as defined in Section 2.1.7. F is defined as *spatially isotropic* for (z, Z) iff $F(z, \psi Y) = F(z, Y) \forall \psi \in \mathcal{O}^3$ and $\forall Y \in \mathcal{L}_z^+$. For example, the reader may check that if F is Planck for (z, Z) then F is spatially isotropic for (z, Z).

Proposition 5.7.3. *If F is spatially isotropic for (z, Z) the stress-energy tensor of F is spatially isotropic for (z, Z).*

Proof. Suppose $\sigma \in \mathcal{O}^3$ and denote the extension of σ to (r, s) tensors by $\sigma_s^r : T_s^r(M_z) \to T_s^r(M_z)$ (Exercise 0.0.14). Then $\sigma_2^0(gz) = gz$ and σ preserves the time orientation (Section 2.1.7). σ is therefore a diffeomorphism of the future lightcone $\mathcal{L}_z^+$ onto itself. By Proposition 5.6.1 the natural volume element Λ_z of $\mathcal{L}_z^+$ must obey $(\sigma^{-1})^* \Lambda_z = \Lambda_z$. Thus $\forall \omega \in M_z^*$, we have (integration over $\mathcal{L}_z^+$ will be understood in the following):

$$(\sigma_0^2 \hat{T})(\omega, \omega) = \hat{T}(\sigma_1^0 \omega, \sigma_1^0 \omega)$$

$$= \int [(\sigma_1^0 \omega)^\sim]^2 F \Lambda_z$$

$$= \int \tilde{\omega}^2 (F \circ \sigma^{-1})[(\sigma^{-1})^* \Lambda_z]$$

$$= \int \tilde{\omega}^2 F \Lambda_z = \hat{T}(\omega, \omega).$$

Thus $\sigma_0^2 \hat{T} = \hat{T}$—that is, $\hat{T}$ is spatially isotropic for (z, Z). □

Corollary 5.7.4. *Suppose there exists a reference frame Z such that F is spatially isotropic for $(x, Zx) \, \forall x \in M$. Then the stress-energy tensor of F is $\hat{T} = (\rho/3)(\hat{g} + 4Z \otimes Z)$, where $\rho = T(Z, Z)$.*

Proof. If F is spatially isotropic for $(x, Zx) \, \forall x \in M$, the same then holds for $\hat{T}$. So we have $\hat{T} = pZ \otimes Z + p(\hat{g} + Z \otimes Z)$ with p a function on M and $\rho = T(Z, Z)$ (Exercise 3.15.4). Trace $\hat{T} = 0$ now gives $p = \rho/3$. □

EXAMPLE 5.7.5. Let $F : \mathcal{L}^+ \to (0, \infty)$ be defined as in Exercise 5.6.7a in terms of functions $g, j : M \to (0, \infty)$, a function ℓ, and a reference frame X. F is a photon gas; we compute its stress-energy tensor. F is spatially isotropic for each instantaneous observer (x, Xx), so by Corollary 5.7.4, it suffices to find $\rho x = T(Xx, Xx) \forall x \subset M$. Write $X = Xx$. Then Proposition 5.7.2d gives

$$\rho x = \int_{P_X} e(gx)\ell[(jx)e] \pi_X = (gx) \int_{\mathscr{S}_X} \zeta \int_0^\infty e^3 \ell[(jx)e] \, de$$

$$= 4\pi(gx)(jx)^{-4} \int_0^\infty u^3 \ell(u) \, du.$$

Thus $\hat{T} = (\rho/3)(\hat{g} + 4X \otimes X)$, where $\rho = Agj^{-4}$ with $A = 4\pi \int_0^\infty u^3 \ell(u) \, du \in (0, \infty)$.

5 Photons

In particular, suppose there is a function $T: M \to (0, \infty)$ such that F is Planck for (x, Xx) with temperature Tx $\forall x \in M$. Taking $g = 1$, $j = (kT)^{-1}$ and $\ell = P$, the Planck function, we get $\rho = a_0 T^4$, where a_0 is the blackbody constant (Example 5.5.4b).

A photon gas is a matter model; we now define the simplest matter equations used for it. Let $\lambda: \mathscr{E} \to M$ be a lightlike geodesic. Then λ determines a curve $\tilde{\lambda}: \mathscr{E} \to \mathscr{L}^+$ via $\tilde{\lambda}(u) = (\lambda u, \lambda_* u)$ $\forall u \in \mathscr{E}$. Suppose $\mathscr{A} \subset M$ is a subset. A photon gas F is defined as *conserved* on $\mathscr{A}$ iff $F \circ \tilde{\lambda}$ is a constant $\forall$ lightlike geodesic λ whose image lies wholly in $\mathscr{A}$. Roughly, this means collisions in $\mathscr{A}$ do not affect F. We now give somewhat more specific interpretations and indicate in what sense the requirement that F be conserved is a matter equation.

> Let L be that vector field on $\mathscr{L}^+$ obtained by restricting the geodesic spray of the tangent bundle (cf. Bishop–Goldberg, Section 5.12) to $\mathscr{L}^+$. F is conserved on M iff $LF = 0$. Compare Section 8.6, which also gives more general—and more convincing—motivations and interpretations for this matter equation than does the next subsection.

5.7.6 Conserved photon gases

(a) Suppose F models a situation where no collisions involving photons occur: no emission, absorption, scattering, and so on. Then one demands F be conserved on M. More generally, suppose that collisions do not affect the average behavior of the photons—for example, that, on the average, whenever a photon is destroyed a photon of the same energy-momentum is created at the same location. Then one again assumes F is conserved on M. We now give some examples that illustrate the interpretations.

(b) Let M be Minkowski space, X be a parallel reference frame on M (e.g., $X = \partial_4$) and $j: M \to (0, \infty)$ be a function. Define a photon gas F on M by $F(y, Y) = \exp[(jy)g(X, Y)]$ (cf. Exercise 5.6.7). Suppose F is conserved on M; we will show j is a constant and then discuss in what sense $j = $ constant is consistent with the interpretations in (a).

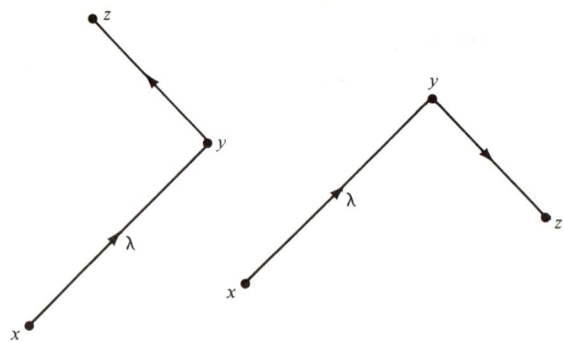

5.7 A photon gas

Suppose $x, z \in M$. There exists a $y \in M$ such that there is a lightlike geodesic λ from x to y and a lightlike geodesic from y to z (Exercise 5.2.7; the sketches give two examples). Since X is timelike and Killing, $|g(X, \lambda_*)|$ is a positive constant, say a (Section 3.6.3e). Since F is conserved on M, $\exp[-a(jx)] = \exp[-a(jy)]$—that is, $jx = jy$. Similarly, $jy = jz$. Since x and z were arbitrary, j is a constant.

Now the stress-energy tensor of F is $\hat{T} = (8\pi)j^{-4}(\hat{g} + 4X \otimes X)$ (Corollary 5.7.4). Since X is parallel, j = constant implies div $\hat{T} = 0$. The discussion of the equation div $\hat{T} = 0$ in Section 3.10 therefore lends weight to the terminology (that F is conserved) as well as to the interpretation given in (a).

On physical grounds, one always expects the stress-energy tensor $\hat{T}$ of a photon gas F to satisfy div $\hat{T} = 0$ whenever F is conserved. Indeed, the electric charge of a photon is zero (Section 5.1), so a photon gas does not exchange energy-momentum directly with any electromagnetic field (cf. Section 3.10). Now if, for example, no collision occur, F does not directly exchange energy-momentum with other forms of matter either, so one should find div $\hat{T} = 0$ (Sections 3.10 and 3.12). This heuristic argument actually leads to a correct mathematical result: under quite general conditions, F conserved $\Rightarrow$ div $\hat{T} = 0$ (Section 8.6). The converse is however not true (Exercise 5.7.8).

(c) Suppose there exists a subset $\mathscr{B} \subset M$ such that each inextendible lightlike geodesic intersects $\mathscr{B}$ exactly once. A level surface of u^4 in Minkowski space or in Einstein–de Sitter space is an example. Then the following uniqueness result holds. Given a function $F_0: \mathscr{L}^+ \cap (\Pi^{-1}\mathscr{B}) \to [0, \infty)$, there is at most one photon gas that is conserved on M and coincides with F_0 on $\mathscr{L}^+ \cap (\Pi^{-1}\mathscr{B})$.

Indeed, suppose $(y, Y) \in \mathscr{L}^+$. Let λ be the inextendible lightlike geodesic determined by $(\lambda 0, \lambda_* 0) = (y, Y)$. By assumption, there is a $u \in \mathbb{R}$ such that $\lambda u \in \mathscr{B}$. If there exists a photon gas F such that our requirements hold, $F(y, Y) = F_0(\lambda u, \lambda_* u)$. Since (y, Y) was any point in $\mathscr{L}^+$, there exists at most one such F. Thus, in this case, one has a "present determines the future" type of result, as one would expect of appropriate matter equations (Section 3.12).

EXERCISE 5.7.7. NUMBER DENSITY

A good way to review our results on $\hat{T}$ is to derive very similar results for a slightly simpler, slightly less important quantity, as follows. (a) Show there is a unique vector field N on M such that $N(\omega) = \int_{\mathscr{L}_x^+} \tilde{\omega} F \Lambda_x \; \forall \omega \in M_x^*, \forall x \in M$. (b) Show that for each instantaneous observer (x, X), $g(N, X) = -\int_{P_X} (F \circ \pi_X) \pi_X$. In view of this result and the interpretations in Sections 5.5.2a and 5.7.1a, N is defined as the *number density* of the photon gas F; if $\mathscr{B} \subset M$ is a space-section and Ω is the volume form of M, $|\int_{\mathscr{B}} i(N)\Omega|$ is defined as the *total number of photons in $\mathscr{B}$*. (c) Show Nx is future pointing timelike unless F is zero on $\mathscr{L}_x^+$. (d) Show that if F is spatially isotropic for (z, Z), N is also. (e) Suppose F is

5 Photons

Planck for (z, Z) with temperature T. Show $Nz = bT^3 Z$, where b is a universal constant. (f) In Section 5.7.6b, show div $N = 0$.

EXERCISE 5.7.8

Generally speaking, a photon gas F has many properties in addition to those properties described by its stress-energy tensor $\hat{T}$ and number density N. You are asked to construct two examples, as follows. (a) Find photon gases F, F' on Minkowski space such that F ≠ F', $\hat{T} = \hat{T}'$, and $N = N'$. (b) Find a photon gas F on Minkowski space M such that div $\hat{T} = 0 = $ div N but F is not conserved on M.

EXERCISE 5.7.9

Suppose that, $\forall x \in M$, the restriction of F to $\mathscr{L}_x^+$ is somewhere nonzero. (a) Show F is spatially isotropic for (z, Z) iff F ∘ π_Z is independent of the direction U in $\mathscr{S}_z$ and that then F is not spatially isotropic for (z, Z') if $Z' \neq Z$. (b) Show $\hat{T}$ obeys the timelike convergence condition. (c) Suppose there is a reference frame X such that, $\forall x \in M$, F is spatially isotropic for (x, Xx); define $\rho = T(X, X)$. Show $(\rho, \rho/3, X)$ is a rest-mass zero perfect fluid with the same stress-energy tensor as F.

> Suppose one wants to describe light. Then, depending on the situation, any one, or any combination, of the following mathematically independent but intuitively related models is used. A collection of photons (Section 3.8); a finite superposition of photon beams (Sections 3.2 and 3.13); a photon gas; a finite superposition of rest-mass zero perfect fluids; an electromagnetic field (Section 3.4); a "statistical superposition" of such fields; a quantum electrodynamics model (Messiah [1]); or a verbal description. No wonder a mathematician, accustomed to regarding a mathematical framework as primary, finds the plethora of alternatives confusing. The physics view is that nature is primary, and that even the most careful, mathematically rigorous mathematical models are at best approximations to nature. So the physicist finds nothing odd in having to switch models occasionally, and indeed revels in the intuitive arguments needed to decide which model is least inaccurate physically.

Cosmology 6

The universe in the large is fascinating. Its study will here supply an example of how the basic assumptions of macrophysics, especially the Einstein field equation, are actually applied.

Some skepticism is called for. Cosmology is even further from being exact than is most of physics. In discussing it, such conceptual problems as the "mathematics vs. physics" or "logic vs. history" ones mentioned in Section 2.1 are particularly acute. At present, all our cosmological models have severe limitations. Cranks to the contrary, a fully satisfactory model is not yet in sight. So the game is to use intentionally oversimplified models to understand qualitatively which physical effects are dominant, then gradually sneak up on the actual universe by considering ever more detailed and sophisticated models.

Section 6.0 reviews Einstein–de Sitter spacetime. Section 6.1 outlines some relevant empirical facts, using as few theoretical assumptions as possible. Broadly speaking, the data suggests two things: (a) near here-now the universe seems to be simpler than was thought likely ten years ago, so the classical cosmological spacetimes are probably better approximations than more sophisticated modern alternatives; (b) however, there probably is a hot, dense, high-curvature region in the history of the universe, "near the big bang," where rather sophisticated matter models are needed.

Though the universe's apparent predilection for simple spacetimes and complicated matter strikes geometers as misguided, Sections 6.2 to 6.5 present some models that respect it. Section 6.2 outlines the assumptions of nonquantum general relativistic cosmology and then defines the most naive—and most important—special case: the Einstein–de Sitter model, constructed from the Einstein–de Sitter spacetime. In Section 6.3 we shall find that this model fits the data better than such a crude model has any right

6 Cosmology

to, though *not* to high precision. Section 6.4 discusses one of the model's worst drawbacks: it predicts matter was very dense in the early universe but is itself not applicable to spacetime regions where the density is more than about a billion times the average cosmological density near here-now. We next show how this disease can be partially cured by using a more detailed matter model. Then in Section 6.5 we can outline current opinion on the history of our cosmos, from such times as a few minutes after a big bang, which recent data and theories actually seem to have brought within our grasp, to about ten billion years later. The chapter concludes with a few comments on the things we must learn before we can understand more deeply the profound drama the heavens present to our view.

Throughout this chapter and the next we adopt the attitude that if the underlying physical ideas are clearly explained using the simplest available models, our mathematician reader will have little trouble with mathematically more complicated models in the literature.

6.0 Review, notation and mathematical preliminaries

Throughout the chapter, the notation summarized in Section 4.0 holds. In particular, (M, g) is a spacetime with Einstein tensor G and scalar curvature S, $\hat{T}$ is the stress energy tensor of a matter model $\mathcal{M}$ on M, and T is the (0, 2)-tensor field physically equivalent to $\hat{T}$.

6.0.1 Einstein–de Sitter spacetime

This section reviews the key example, Einstein–de Sitter spacetime. Thus let $M = \mathbb{R}^3 \times (0, \infty)$ and let $g = -du^4 \otimes du^4 + (R^2 \circ u^4) \sum_{\mu=1}^{3} du^\mu \otimes du^\mu$, where $Ru = u^{2/3}$ throughout the rest of the section (cf. Example 1.4.3). (M, g) is maximal (Exercise 5.2.7), but not geodesically complete (cf. Section 1.4 and Example 5.2.2). Suppose $x \in M$. Since $M = \mathbb{R}^3 \times (0, \infty)$ we will sometimes write, for example, $x = (w, t)$ with $w \in \mathbb{R}^3$ and $t = u^4 x \in (0, \infty)$; on rare occasions we will write $x = (x^1, x^2, x^3, x^4)$, with $x^i = u^i x$, $i = 1, 2, 3, 4$. Times, and thus also distances, will be measured in years, u^4 will be assigned the dimension of time.

6.0.2 The comoving reference frame

(a) ∂_4 is a reference frame on (M, g). It can be canonically distinguished from other reference frames by either of the following criteria: (1) for each $x \in M$, $\partial_4 x$ is an eigenvector of the Einstein tensor (Corollary 1.4.5); (2) the 1-form physically equivalent to ∂_4 is proportional to dS (Section 1.4). Recall that ∂_4 is designated the *comoving reference frame* and, $\forall x \in M$, $(x, \partial_4 x)$ is the *comoving instantaneous observer at x*.

In addition to its purely geometric meaning above, the term "comoving" has important, slightly subtle empirical connotations. These will be discussed in Section 6.2, after we have discussed the observations and introduced a

6.0 Review, notation and mathematical preliminaries

matter model. For the moment, the reader may think of ∂_4 as, roughly speaking, comoving with the centers of galaxies.

> One should not regard ∂_4 as comoving with all the individual particles in a nearly rigid body such as a rock. If, in a model, the center of the rock follows an integral curve of ∂_4, then a grain on the outside would not quite follow such an integral curve, though the deviation would not be very significant for a billion years or so. Similarly, a star, the solar system, and our own galaxy do not share in the expansion mentioned below, which refers to the running apart of different galaxies.

(b) The comoving reference frame ∂_4 is geodesic and proper time synchronizable while u^4 is a proper time function for ∂_4 (Proposition 2.3.7). ∂_4 is irrotational (Proposition 2.3.5). It is not rigid anywhere (Section 2.3). It is expanding in the sense of Proposition 2.3.7c and the expansion is slowing down in the sense of Proposition 2.3.7d.

6.0.3 Cosmological time

As was shown in the course of proving Proposition 1.4.4, u^4 is canonically defined as the function $u^4 = (4/(3S))^{1/2}$; other characterizations are outlined in Exercise 6.0.14. u^4 is defined as the *cosmological time* for M. $u^4 \to 0$ is referred to as *approaching the big bang* since then $S \to \infty$. If $z \in M$ models here-now, we use "near the big bang" or "early universe" to mean $u^4 \ll u^4 z$.

6.0.4 Space slices

Each level surface of cosmological time u^4 is called a *space slice*. From the form of (M, g) the level surface $u^4 = a$ is a spacelike 3-submanifold $\mathbb{R}^3 \subset M$ with induced metric $a^{4/3}[du^1 \otimes du^1 + du^2 \otimes du^2 + du^3 \otimes du^3]$ (cf. Exercise 1.1.12). Thus such a level surface is isometric to Euclidean 3-space $(\mathbb{R}^3, \sum_{\rho=1}^{3} du^\rho \otimes du^\rho)$, via $u^\rho \to a^{2/3} u^\rho$.

It is drastically misleading to focus attention, as many popularizations do, on the properties of space slices. Indeed, a space slice cannot be observed physically (cf. the figure in Section 6.0.11). Moreover, M has a nonzero curvature which tends to focus causal geodesics (cf. Section 4.3 and Section 6.3.9); the fact that each space slice is flat has almost no relevance to this crucial focusing effect.

6.0.5 Isometries

We now analyze the isometries of Einstein–de Sitter spacetime more systematically than was done in Exercise 5.2.5. Roughly speaking, the isometries are simply the translations, rotations, and reflections of Euclidean 3-space. More specifically, let $\psi: M \to M$ be an isometry. Then: (a) $u^4 \circ \psi = u^4$ since $S \circ \psi = S$ (Exercise 1.0.4); (b) $\psi^* du^4 = du^4$ by (a); (c) ψ preserves the time orientation by (b); (d) thus $\psi_* \partial_4 = \partial_4$ by the characterizations of ∂_4 in Section 6.0.2; (e) thus γ is an integral curve of ∂_4 iff $\psi \circ \gamma$ is. It now follows that the obvious isometries are the only ones, in the sense of the following proposition.

6 Cosmology

Proposition 6.0.5f. $\psi\colon M \to M$ *is an isometry iff there is a Euclidean isometry* $\tilde{\psi}\colon \mathbb{R}^3 \to \mathbb{R}^3$ *such that* $\psi(w, t) = (\tilde{\psi}w, t) \;\forall (w, t) \in M$ *(cf. Section 6.0.1 for the notation).*

Proof. Let $\tilde{\psi}$ be a Euclidean isometry from $\mathbb{R}^3$ to $\mathbb{R}^3$ and define ψ as in the proposition. Clearly ψ is then a diffeomorphism. With $h = \sum_{\rho=1}^{3} du^\rho \otimes du^\rho$ on $\mathbb{R}^3$ we have $\tilde{\psi}^*h = h$; thus $\psi^*h = h$, where h is now regarded as a tensor on M. Moreover, $\psi^*du^4 = du^4$. Thus $\psi^*g = g$ and ψ is an isometry.

Conversely, let $\chi\colon M \to M$ be an isometry and designate the level surface $u^4 = 1$ by $\mathscr{B}$. We may regard $\mathscr{B}$ together with the Riemannian metric induced on $\mathscr{B}$ by g as $(\mathbb{R}^3, h)$. Moreover, if we denote the restriction of χ to $\mathscr{B}$ by $\tilde{\psi}$, $\tilde{\psi}$ is a diffeomorphism from $\mathscr{B}$ to $\mathscr{B}$, by (a) above. Since χ is an isometry, $(\tilde{\psi})^*h = h$ and $\tilde{\psi}$ is a Euclidean isometry on $\mathscr{B}$. Define ψ as before, then χ and ψ are isometries of M which agree on $\mathscr{B}$. We will show they agree everywhere, thus completing the proof.

Let γ be an inextendible integral curve of ∂_4 which intersects $\mathscr{B}$ at x. Then both $\chi \circ \gamma$ and $\psi \circ \gamma$ are inextendible integral curves of ∂_4 by (e) and they intersect $\mathscr{B}$ at the same point $\chi x = \psi x$ and at the same parameter value. Thus $\chi \circ \gamma = \psi \circ \gamma$. But $\forall x = (w, t) \in M$, $w \in \mathbb{R}^3$ is uniquely determined by the inextendible integral curve of ∂_4 on which x lies. Thus the results $\chi \circ \gamma = \psi \circ \gamma \;\forall \gamma$ as above together with $u^4 \circ \chi = u^4 = u^4 \circ \psi$ imply $\chi = \psi$. $\square$

(g) For $x \in M$, denote by $\mathcal{O}_x^{\,3}$ that subgroup of the isometry group $\mathscr{G}M$ which leaves x fixed—that is, $\psi x = x$ for each $\psi \in \mathcal{O}_x^{\,3}$. By Proposition 6.0.5f, $\mathcal{O}_x^{\,3}$ is isomorphic to the ordinary rotation group of $\mathbb{R}^3$ and coincides with the group $\mathcal{O}^3$, discussed in Section 2.1.7 which guarantees Einstein–de Sitter spacetime is spatially isotropic for every instantaneous comoving observer. We will call ψ a *spatial rotation around* x iff $\psi \in \mathcal{O}_x^{\,3}$. (h) We give some examples of isometries. Suppose $z = (z^1, z^2, z^3, z^4) \in M$. Then for any $a \in \mathbb{R}$, there is an isometry ψ such that $\psi z = (0, 0, a, z^4)$, obtained by choosing an appropriate Euclidean translation ψ. Thus given any pair x and z there exists an isometry ψ such that $\psi z = (0, 0, 0, z^4)$ and

$$\psi x = (0, 0, [(a^1)^2 + (a^2)^2 + (a^3)^2]^{1/2}, x^4),$$

where $a^\rho = z^\rho - x^\rho$. (i) Note that for $x, y \in M$, $u^4 x = u^4 y$ iff there is an isometry ψ such that $\psi x = y$.

We now turn to the quantities that are of prime importance in analyzing actual data: photons, light signals, frequency ratios, and the causality relation.

6.0.6 Photons and light signals

(a) Suppose $x, z \in M$. There exists a freely falling photon from x to z iff there is a unique light signal from x to z (cf. Exercise 5.1.6 and Section 5.2). (b) The standard photon λ is given by: $\lambda u = (0, 0, 3u^{1/5}, u^{3/5}) \;\forall u \in (0, \infty)$. It has the property that $u \to u^4 \lambda u$ is a monotonically increasing function from $(0, \infty)$ to $(0, \infty)$. (c) Any freely falling photon can be obtained from the standard photon by using isometries, reparametrizations, and restrictions (Exercise 5.2.5).

With x and z assumed to lie on the standard photon, we sketch the situation in (a) above using a spacetime diagram. Here, and subsequently, it will sometimes be convenient to add "boundaries" pictorially to our spacetime diagrams. Thus the u^3 axis in the picture does not really represent the level surface $u^4 = 0$ since M contains no such surface; indeed $S \to \infty$ as $u^4 \to 0$ (Section 1.4).

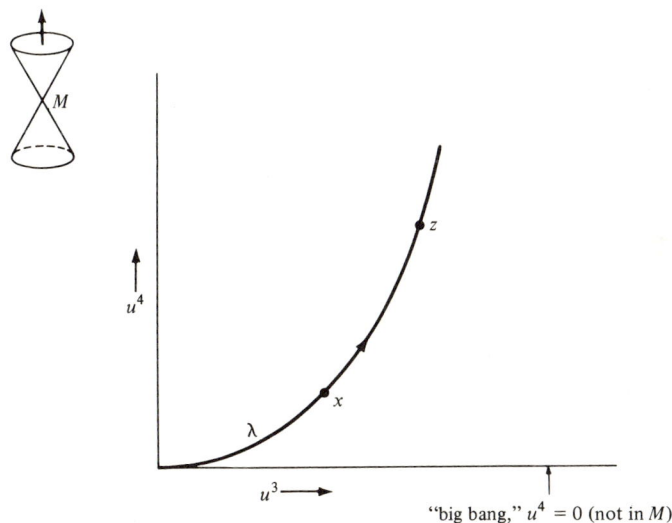

"big bang," $u^4 = 0$ (not in M)

However, there is a systematic way to assign such boundaries, the causal boundary method, whose main ideas are outlined in Hawking and Ellis [1]. The boundaries we add pictorially will be consistent with these causal boundaries. For Einstein–de Sitter spacetime the causal boundary construction gives a space homeomorphic to $\mathbb{R}^3$ for the "big bang, $u^4 = 0$" part of the boundary, as our spacetime diagram suggests.

In particular, the spacetime diagram suggests that the "big bang" in the Einstein–de Sitter spacetime is not merely a single point. This is intentional. Popularizations to the contrary, it is not at all useful to regard the big bang, in this and similar models, as one point. One reason will crop up in Section 6.3.12, which will indicate that, roughly speaking, only part of the big bang is visible to any one instantaneous observer.

6.0.7 The cosmological redshift

Suppose $x, z \in M$ and there exists a light signal, which we shall designate $[\lambda]$, from x to z. Then the frequency ratio $\imath$ for $([\lambda], \partial_4 x, \partial_4 z)$ is defined as the *cosmological frequency ratio for (x, z)*. If z represents here-now, $\imath$ may usually

6 Cosmology

be regarded as directly measurable (cf. Section 5.5.3d). The term *cosmological* refers mainly to the fact that [λ] and the instantaneous comoving observers $\partial_4 x$, $\partial_4 z$ are canonically determined given x and z (cf. Sections 6.0.1 and 6.0.6a). Thus an instantaneous observer at z need not always regard $\imath$ as giving primarily information about the world velocity of the emitter as would normally be the case; he may assume this world velocity is $\partial_4 x$ and regard $\imath$ as giving cosmological information on the location of x and the properties of the intervening spacetime. We will often write $\imath(x, z)$ for the cosmological frequency ratio for (x, z).

Define the *cosmological redshift for* (x, z) to be $z = \imath - 1$. Here "red" refers to the fact that, as the next proposition shows, $z > 0$ and thus $\imath > 1$; $\imath > 1$ corresponds to redder light at reception than at emission (cf. Section 5.1).

Proposition 6.0.8. *Suppose* $x, z \in M$ *and there exists a light signal from x to z. Then the cosmological frequency ratio for (x, z) is* $\imath(x, z) = R(u^4 z)/R(u^4 x) = (u^4 z/u^4 x)^{2/3} > 1$.

Remarks. The idea of the proof is to use an isometry to simplify the calculation; this idea will often be used later. We therefore isolate two lemmas. Incidentally, there is a slicker proof (Exercise 6.0.17d).

Lemma 6.0.9. *If* $\psi: M \to M$ *is an isometry, then* $\imath(\psi x, \psi z) = \imath(x, z)$.

Lemma 6.0.10. *If* λ *is the standard photon and* $a \in (0, \infty)$ *then* $\imath(\lambda u, \lambda a) = (u^4 \lambda a/u^4 \lambda u)^{2/3}$ $\forall u \in (0, a)$. *Furthermore, for* $u \in (0, a)$, $u \to \imath(\lambda u, \lambda a)$ *defines a monotonically decreasing function onto* $(1, \infty)$.

Proof of Lemma 6.0.9. Let $\lambda: [a, b] \to M$ be a freely falling photon from x to z. Since ψ is an isometry it preserves the time orientation (Section 6.0.5) and thus $\psi \circ \lambda$ is a freely falling photon from ψx to ψz. Now we use the definition of $\imath(x, z)$ in Section 6.0.7, the definition of frequency ratio in Section 5.4, and the properties in Section 6.0.5 of ψ to get: $\imath(\psi x, \psi z) = g((\psi \circ \lambda)_* b, \partial_4 \psi z)/g((\psi \circ \lambda)_* a, \partial_4 \psi x) = g(\psi_* \lambda_* b, (\psi_* \partial_4)\psi z)/g(\psi_* \lambda_* a, (\psi_* \partial_4)\psi x) = \psi^* g(\lambda_* b, \partial_4 z)/\psi^* g(\lambda_* a, \partial_4 x) = g(\lambda_* b, \partial_4 z)/g(\lambda_* a, \partial_4 x) = \imath(x, z)$. □

Proof of Lemma 6.0.10. $\imath(\lambda u, \lambda a) = (u^4 \lambda a/u^4 \lambda u)^{2/3}$ by Example 5.4.3. Now $u \to u^4 \lambda u$ is monotonically increasing (Section 6.6.6b) so $u \to \imath(\lambda u, \lambda a)$ is monotonically decreasing. Moreover, we have $u^4 \lambda a > u^4 \lambda u > 0$, and $u^4 \lambda u \to 0$ as $u \to 0$ by Example 5.4.3, so the image of $(0, a)$ is $(1, \infty)$. □

Proof of Proposition 6.0.8. Let $[\lambda_0]$ be the light signal from x to z. There exists an isometry ψ such that $[\psi \lambda_0]$ is the standard light signal (Exercise 5.2.5). Thus

$$\imath(x, z) = \imath(\psi x, \psi z) = (u^4 z/u^4 x)^{2/3} > 1$$

by Section 6.0.5a and the two preceding lemmas. □

6.0.11 The causal past 2-dimensionally

For vividness, suppose $z \in M$ models here-now. Then the spacetime region from which we can receive signals, that is, particles, is the causal past of z (cf. Section 5.0.1 and Exercise 6.0.15c). This is of course, the region of main interest—the "observable universe." Assuming for the moment z lies in the standard photon, and temporarily focusing attention on the (u^3, u^4)-plane we sketch the result of Proposition 6.0.13 below on a spacetime diagram. Here the image of the standard photon is shown as the curve $u^4 = (u^3/3)^3$, $u^3 > 0$ (Section 6.0.6b). The other photon shown is given by reflection of the standard photon; specifically $\mu u = (0, 0, 2u^3z - 3u^{1/5}, u^{3/5})$. The dotted region labelled "our causal past" is determined by: $3(u^4)^{1/3} \leq u^3 \leq 2u^3z - 3(u^4)^{1/3}$.

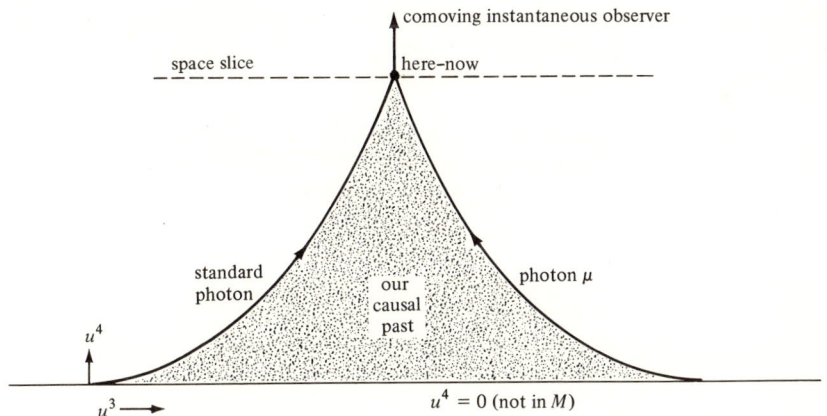

6.0.12 An auxiliary function

To analyze 4-dimensionally the causal past of $z \in M$, we will need an auxiliary function. For each pair $x, z \in M$, define $\delta(x, z) = \{\sum_{\mu=1}^{3} (u^\mu z - u^\mu x)^2\}^{1/2}$. Thus $\delta(x, z) = \delta(z, x) \geq 0$. Note that: (a) If $\psi: M \to M$ is an isometry $\delta(\psi x, \psi z) = \delta(x, z)$ (Section 6.0.5). (b) Denote the level surface $u^4 = 1$ by $\mathscr{B}$. Let x' be the point at which the inextendible integral curve of ∂_4 through x intersects $\mathscr{B}$. Define z' similarly. Then $\delta(x, z)$ is the distance, within $\mathscr{B}$, between x' and z'.

Despite its formal similarity to the distance function of $\mathbb{R}^3$, δ has no important physical interpretation: (b) above is not only clumsy but also refers to a distance that cannot be measured directly. Thus δ neither deserves nor gets a special name in the physics literature. Formally, however, its invariance property (a) makes it quite useful.

We now state and prove the generalization of the result sketched pictorially in Section 6.0.11.

Proposition 6.0.13. *Suppose $x, z \in M$. Then x causally precedes z iff $\delta(x, z) \leq 3(u^4z)^{1/3} - 3(u^4x)^{1/3}$.*

Proof. Define the function $u = 3(u^4)^{1/3}$ on M. Then $du = (u^4)^{-2/3}du^4$, so $g = (u^4)^{4/3}\{-du \otimes du + \sum_{\mu=1}^{3} du^\mu \otimes du^\mu\}$. In view of our comments on conformal invariance in Section 5.0.1 and the behavior of the causality relation for Minkowski space (cf. Exercise 5.0.7), this gives $x \leq z$ iff $0 \leq \delta(x, z) \leq uz - ux$. Thus $x \leq z$ iff $3\{(u^4z)^{1/3} - (u^4x)^{1/3}\} \geq \delta(x, z)$. □

EXERCISE 6.0.14. CHARACTERIZATIONS

In addition to reviewing Proposition 6.0.5, this exercise gives some characterizations that, as will be seen in later sections, are closer to observations than the geometric characterizations in Sections 6.0.2 and 6.0.3. (a) Extend Exercise 2.1.11 by showing Einstein–de Sitter spacetime is spatially isotropic for an instantaneous observer (x, X) iff $X = \partial_4 x$. (b) Show the Einstein tensor Gx is spatially isotropic for (x, X) iff $X = \partial_4 x$. (c) Show u^4 is the unique time function for ∂_4 whose image is $(0, \infty)$. (d) For $z \in M$, show $u^4 z$ is the "longest timelike distance to the big bang" in the following sense. No future-pointing timelike curve with future endpoint z has arclength less than $u^4 z$ but the past endless ∂_4 integral curve with future endpoint z has arclength $u^4 z$.

(d) will be quite useful. The reversal of "longest" and "shortest" here, so anti-intuitive for a Riemannian geometer, of course stems from the wrong-way character of the wrong-way triangle inequality (Exercise 1.2.4).

EXERCISE 6.0.15. CHRONOLOGY AND CAUSALITY

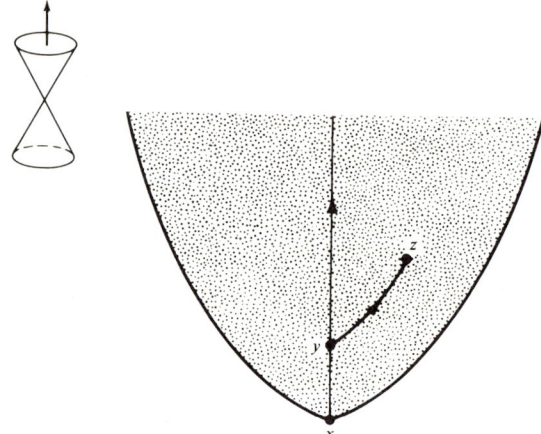

Suppose $x, z \in M$. Using the methods of Exercise 5.1.6 and the proof of Proposition 6.0.13, show the following properties of causality and chronology (Section 5.0.1). (a) $x \ll z$ iff $\delta(x, z) < 3[(u^4z)^{1/3} - (u^4x)^{1/3}]$. (b) Let γ be the future endless integral curve of ∂_4 with past endpoint x. There exists a light signal to z from some point y on γ other than x iff z lies in the chronological future $\mathscr{U}$ of

x with the image of γ deleted (figure). Moreover $\mathcal{U}$ is open. (c) $x \leq z$ iff $x = z$ or there is a particle from x to z. (d) $x \leq z$ iff $x = z$ or there is a freely falling particle from x to z.

(c) is true for any pair of points in any spacetime (Penrose [1]). However (d) is very special; for example, by removing one point from Minkowski space one gets a spacetime where (d) is false.

EXERCISE 6.0.16

Suppose $x, z \in M$ and $\delta(x, z) \neq 0$ (Section 6.0.12). With $\mathcal{O}_z^3$ the group (Section 6.0.5g) of spatial rotations about z, denote by $\mathscr{S}$ the set $\{y \mid y = \psi x$ for some $\psi \in \mathcal{O}_z^3\}$. (a) Show $\mathscr{S}$ is a spacelike 2-submanifold, diffeomorphic to $\mathscr{S}^2$ and given explicitly by: $y \in \mathscr{S}$ iff $u^4 y = u^4 x$ and $\delta(x, z) = \delta(y, z)$. (b) Show the induced metric on $\mathscr{S}$ is that for a sphere of radius $(u^4 x)^{2/3} \delta(x, z)$ in $\mathbb{R}^3$. (Hint: regard $\mathscr{S}$ as a submanifold of the level surface $u^4 = u^4 x$).

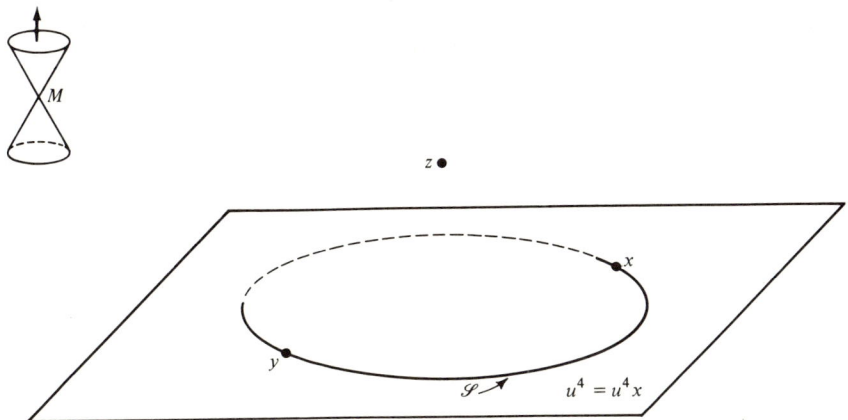

EXERCISE 6.0.17

Let (N, h) be a simple cosmological spacetime (Example 1.4.3). (a) Show that each Euclidean isometry $\tilde{\psi}$ gives rise to an isometry $\psi: N \to N$ just as in Proposition 6.0.5f. (b) Find an example where $\mathscr{G}N$ is larger than the Euclidean isometry group of $\mathbb{R}^3$. (c) Show from (a) and Section 3.6.3a that ∂_μ is a Killing vector field $\forall \mu \in (1, 2, 3)$. (d) Suppose there exists a light signal $[\lambda]$ from $x \in N$ to $z \in N$. Show that the frequency ratio for $([\lambda], \partial_4 x, \partial_4 z)$ is $\imath = Ru^4 z / Ru^4 x$. (Hint: use (c), Section 3.6.3e, $g(\lambda_*, \lambda_*) = 0$, and algebra; you do not have to find λ explicitly by integrating.)

EXERCISE 6.0.18

Let (M, g) and (M', g') be simple cosmological spacetimes; thus M is determined by $R: \mathscr{F} \to (0, \infty)$, M' by $R': \mathscr{F}' \to (0, \infty)$. (a) Suppose $\mathscr{F} = \mathscr{F}'$ and $R = aR'$, $a \in (0, \infty)$. Use the diffeomorphism determined by $u^4 \to u^4$ and $u^\mu \to$

6 Cosmology

$au^\mu\ \forall \mu \in (1, 2, 3)$ to show (M, g) is isometric to (M', g'). (b) Suppose the transformation τ given by $u \to u + a$, $a \in (0, \infty)$ carries $\mathscr{F}$ onto $\mathscr{F}'$ and that $R \circ \tau = R'$. Show again that both spacetimes are isometric. [In both cases (M, g) and (M', g') belong to the same spacetime equivalence class. Later, when (M, g) is supplied with a matter model and other structures, there will of course be a corresponding equivalence relation, with equivalent models representing the same physical situation.]

6.1 Data

This section summarizes some results of observational cosmology. Most of the data is low precision, but so much is now available, owing mainly to work during the last 15 years, that at least the spacetime region near here-now may be considered reasonably well understood. The most important, and most nearly precise observational results are the Hubble law (Section 6.1.7) and the isotropy of the microwave radiation (Section 6.1.9). Both suggests a surprising overall simplicity. The order-of-magnitude agreement between various independently measured time scales for the universe is encouraging (Section 6.1.8). The most annoying of the many observational uncertainties is that we are not sure what forms of matter are present in significant amounts (cf. Peebles [1] and Section 6.1.10). On the other hand, the possibility exists that the observed helium content (Section 6.1.2) together with the microwave photons (Section 6.1.9) may be giving us rather detailed information about the state of the universe a few minutes after the big bang. For more details on the observational data than is given here, see, for example, Weinberg [1], Peebles [1], or Longair and Rees [1].

6.1.1 Circular reasoning

Cosmology (like the rest of physics) is circular reasoning in the following sense: one cannot really discuss the empirical data coherently without using, explicitly or implicitly, some tentative theoretical model; one cannot sensibly choose even a tentative theoretical model without some reference to the empirical data. For the moment, we shall break into this circle mainly by using various terms rather intuitively and broadly (e.g., "observable universe," "spatial isotropy," etc.) without assuming a specific model. On the other hand, we will assume standard general relativity, including its Newtonian limit, throughout this section.

6.1.2 Galaxies

(a) The observable universe contains about 10^{11} galaxies. Naively, imagine all these now distributed within a big sphere having here as center and radius of about 10^{10} years (i.e., 10^{10} light years). Imagine them running away from each other. There is some clumping. The biggest clumps of galaxies may be about 10^8 years across and contain perhaps a million galaxies. Our own galaxy is part of a small local group which in turn is part of a big clump.

However, on scales still larger than 10^8 years the distribution seems to be rather uniform (approximate "spatial homogeneity"). (b) Though some idealizations are involved, it makes sense to talk of the random 3-velocities of galaxies (cf. Exercises 3.8.8 and 3.15.6 and the comments in Section 2.1.2 on replacing a small part of a spacetime by part of a tangent space). The measured random 3-velocities are quite small in magnitude compared to the speed of light: usually considerably less than 0.01. Thus we shall ignore these random 3-velocities throughout. (c) There are various kinds of galaxies, identified by their shape and their color (spectrum) (cf. Section 5.5). Our own is a spiral galaxy. Giant elliptical galaxies are among the most useful in cosmological observations, not only because they are large and bright but also because the brightest giant elliptic galaxies all seem to be similar. Thus if an astronomer observes a very distant cluster of galaxies, he would feel comparatively confident that the intrinsic brightness of the biggest giant elliptical galaxy in the cluster is comparable to the intrinsic brightness of the biggest relatively close giant elliptic galaxies. The latter brightness can be estimated by a beautiful series of steps known as the cosmological distance ladder (Weinberg [1], Tamman [1]). So the intrinsic brightness of the very distant giant elliptical galaxy can be regarded as approximately known. Then its apparent luminosity gives us a handle on how distant-early it is. A similar discussion applies to the intrinsic and apparent sizes. Some galaxies are called radio galaxies because they send out very strong radio signals in addition to visible light. Quasars, which are probably very dense galaxies, also emit both visible light and radio signals. (d) The most important physical property of an individual galaxy is beauty; the reader should look at some color slides or photographs. A typical galaxy has a rest-mass of the order of $(1/10)$ year or less in our units (Section 0.1.4), and a diameter of perhaps 50,000 years. It contains several billion stars, some gas, some dust, and other constituents that are minor in the sense that their contribution to the total rest-mass is small. Whether a significant number of black holes is present in addition is not known. Hydrogen is the predominant element. But about 30% of the rest-mass is in helium, which seems to be rather uniformly distributed, and traces of most elements are present.

Typically, the center of a galaxy is very dense, small, and easy to identify observationally.

6.1.3 Idealizations

The numbers quoted in Section 6.1.2 suggest several convenient idealizations. (a) Since an individual galaxy is much smaller than the observable universe, we will often regard a galaxy as a (point) particle—that is, model its history as a curve (Definition 3.1.1). (b) Even when we consider a galaxy as an extended body we will model the center of the galaxy as a particle. Moreover, an observed galaxy history is also small compared to the spacetime region over which spacetime curvature becomes important (cf. Section 2.1.2). Thus, when considering the intrinsic properties of a single galaxy we will often have in mind an (hypothetical) instantaneous observer (x, X) in the galaxy center with X tangent to the history of the center, and a model of the galaxy on M_x. Then various Newtonian, Euclidean, and special relativistic concepts become applicable: the galaxy history intersects $X^\perp$ in a region with a certain Euclidean volume and the plane cross-sections of that region have a well-defined Euclidean area; the Newtonian speed measured by (x, X) for a star in the galaxy is well-defined (Section 2.1.3); and so on. (c) These idealizations apply to our own galaxy. Moreover, the speed of the earth relative to the center of our galaxy is less than 0.001. Thus we can and shall idealize as follows: (z, Z) will denote an instantaneous observer on spacetime M, with z interpreted not only as here-now but also as an appropriate point on the history of the center of our galaxy; Z will be interpreted not only as tangent to the history of an actual telescope but also to the history of the center of our galaxy. In the remainder of Section 6.1 we use the symbol (z, Z) or the phrase "actual observer" only if we have this interpretation in mind. Roughly, (z, Z) is "at rest" with respect to our galaxy.

6.1.4 Local physics there-then

Let x represent a moderately distant-early point in the observable universe— for example, x is halfway or less of the way back in time toward the big bang in the sense of the figure in Section 6.0.11 or of Exercise 6.0.15. There is considerable empirical evidence that the basic laws of local physics at x are the same as those at z (cf. Section 5.5.3d); general relativity assumes this, as indicated by the fact that in stating the laws we have never referred to a distinguished spacetime point; we assume it throughout.

6.1.5 Redshifts

The observed frequency ratio $\imath$ of galaxies and quasars, determined by the method mentioned in Section 5.5.3d, is greater than 1. The only exceptions are a few comparatively very nearly galaxies whose random velocities happen to mask this systematic effect, the latter being small for nearby galaxies (Section 6.1.7). We henceforth leave the exceptions out of all discussions. Mainly for historical reasons one again introduces the *observed cosmological red shift* $z = \imath - 1$; the interpretations and comments in Section 6.0.7 apply. Many texts even introduce another quantity $v = (\imath^2 - 1)/(\imath^2 + 1)$. v is then called the *observed recession velocity* since it would measure an honest recession speed in the case of 2-dimensional Minkowski spacetime (Example 5.4.2). Note that for $0 < z \ll 1$, $v \cong z$.

> In intuitive discussions, when no particular model is explicitly assumed, observational astronomers normally identify v or even z with outward speed. When analyzing galaxies in a sufficiently small neighborhood of here-now this is a legitimate and useful way to think. But we shall soft-pedal it since it has the misleading features mentioned in the fine print comment below Example 5.4.3.

Until quite recently the largest redshift observed for a galaxy was $z = 0.465$, observed by R. Minkowski in 1960 for the radio galaxy 3C295 (i.e., number 295 in the third Cambridge catalogue of radio sources). Only during the past year have techniques for measuring the red shifts of galaxies for which $z > 0.5$ become available (Gunn and Oke [1]; Spinrad [1]). But for quasars, redshifts as large as 3.5 have now been measured (Lang et al. [1]). Even $z = 0.465$ corresponds to $v = 0.36 \times$ (the speed of light), and in typical models this corresponds to looking almost halfway back in time toward a big bang and very far out in space; $z = 3.5$ similarly corresponds to going about 95% of the way back in time; see the arrows marked "galaxy" and "quasar" in Figure 6.5.2b following.

6.1.6 Spatial distance

Suppose (z, Z) observes a distant-early galaxy. Typically it appears as a very small, dim patch on the night sky.

(a) Let ℓ be the apparent luminosity (z, Z) measures (Section 5.5.5b). Assume the absolute luminosity $L > 0$ is known (at least approximately; cf. Sections 6.1.2c and 6.1.4). Abstracting from the elementary result that the apparent brightness of a small distant light source is inversely proportional to the square of the distance, one may define a *luminosity distance* $d_L \in (0, \infty)$ operationally by $4\pi\ell = L/d_L^2$ (cf. Section 5.5.5b and Exercise 6.7.5).

(b) In the sense of Section 6.1.3b, let A be the intrinsic cross-sectional area of (the observable part of) the galaxy. Ignoring several serious technical problems (briefly described in Weinberg [1]), assume A is (approximately) known. Let $\Omega \in (0, 4\pi)$ be the apparent size of the galaxy for (z, Z). (Section 5.5.5c.) Assuming $\Omega \ll 4\pi$ as is in practice the case, we may again abstract, defining an *area-distance* d_A operationally by $\Omega = A/d_A^2$. Often d_A is called "distance by apparent size" in the literature, and the "distance by apparent angular diameter" of some texts is also in effect d_A.

(c) In principle, many further operationally defined spatial distance concepts are available—"parallax distance," "radar distance," and so on (Weinberg [1]). We shall not need these. Moreover, given a model, one can usually define even other spatial distance (e.g., $|W|$ in Section 6.3.3, d in Proposition 6.3.4, etc.) In general there are no simple interrelations among all these spatial distances; too much depends on the properties of the intervening spacetime. However, there are two exceptions as follows.

(d) Under rather general circumstances $d_L = z^2 d_A$, where z is the frequency ratio (cf. Ellis [1]).

We shall prove this only in a special case (Proposition 6.7.3). Physically, it is a rather deep result—somehow the properties of the intervening spacetime cleverly cancel out of the ratio d_L/d_A. The simplest formal proof is based on the assumption that a conserved photon gas models the light from the galaxy (cf. Section 5.7).

6 Cosmology

The observations are consistent with this result (Longair and Rees [1]).

(e) There exists a neighborhood $\mathcal{U}$ of z so small that, for (z, Z), spacetime curvature within $\mathcal{U}$ is negligible compared to empirical inaccuracies (Sections 2.1.2 and 6.1.3b). Within $\mathcal{U}$ one may normally identify d_L, d_A and the other spatial distance concepts mentioned, provided all relative speeds relevant to the measurements are negligible compared to 1. Then one can regard each as "Newtonian distance" d_N or as Euclidean distance within a flat 3-dimensional submanifold $\mathbb{R}^3 \subset M_z$, where $\mathbb{R}^3$ is parallel to $Z^\perp$. More specifically, if $d_L < 10^8$ years, we may take $d_L = d_A = d_N$ to well within the limits of empirical accuracy.

6.1.7 The Hubble law and Hubble time

Suppose we observe a distant galaxy, measuring a recession velocity v as in Section 6.1.5 and a distance—say, d_L to be specific—as in Section 6.1.6. Then we can also assign a time t via $d_L = vt$. Interpreted naively, t measures how long ago the distant galaxy was right on top of us.

The empirical *Hubble law* states there is some one time $t_H \in (0, \infty)$, the empirical *Hubble time*, such that $t = t_H$, to good approximation, for all galaxies; thus the observed t is spatially isotropic in the sense that it does not depend noticeably on the direction of observation. Moreover, t is also more or less independent of d_L, at least for a large range—for example, 5×10^6 years $< d_L < 2 \times 10^9$ years.

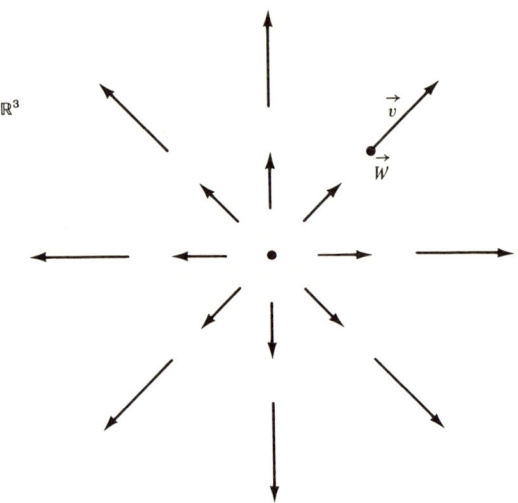

For technical reaons, it is easier to check that there is just one number t_H involved than to measure the actual numerical value. The presently accepted value is $t_H = 18$ billion years $\pm 15\%$ (Tamman [1]).

A Newtonian description of the Hubble law is the following. Our own galaxy is at rest at the origin of $\mathbb{R}^3$. If another galaxy is centered at $\vec{W} \in \mathbb{R}^3$, then it is running away from us with Newtonian velocity $\vec{v} = H\vec{W}$, where H is independent of which other galaxy is being considered. On regarding $|\vec{v}|$, $|\vec{W}|$, and H, respectively, as v, d_L, and $t_H{}^{-1}$, the description corresponds to the empirical Hubble law.

> t_H^{-1} is called the *Hubble constant* (sometimes, sarcastically, the Hubble variable since the empirical estimate has changed so often) in the literature. Contrary to what the picture may suggest, the law does not indicate a distinguished origin, as discussed in Exercise 6.1.12.

6.1.8 Other time scales

One can obtain time scales by other methods: radioactive dating of old rocks in the solar system; estimating the age of old stars in our galaxy; applying a dimensional argument to the observed stress-energy density discussed below; and in several other ways. Some of these measurements are difficult and controversial. However, each usually gives a time of somewhat more than 10^{10} years, so there is some kind of rough, overall consistency.

6.1.9 The microwave radiation

We observe many photons with measured wavelengths between 0.1 and 10 cm. These are called *microwave photons*. They have three remarkable properties, whose discovery, interpretation, and implications have been central in cosmology during the last decade.

First, they do not come from identifiable discrete sources such as stars or galaxies. Probably the ones we see were created slightly later than 10^5 years after a big bang. In this sense observing them probably involves looking backward in time almost 99.999% of the way and thus also almost to the very edge of the observable universe; compare Figure 6.5.2b following.

Second, the observed pattern is spatially isotropic to an accuracy of considerably better than 0.1%. This counts as extremely high precision in cosmology. In view of the first property, it seems to indicate a surprisingly high uniformity of the whole observable universe. Moreover, the observed spatial isotropy indicates not only a symmetry of nature but also that an actual observer (z, Z) is moving towards the future in a very special way (cf. Exercise 6.1.13). Roughly speaking, (z, Z) is "almost at rest with respect to the universe," assuming the microwave radiation gives some overall measure of the universe's average motion. That the observer at rest with respect to our galaxy (Section 6.1.3) should also be almost at rest with respect to the microwave radiation is really pretty, though not unexpected. Recent measurements are almost precise enough to detect residual effects due to the slight residual velocity presumably due to the sun's motion with respect to the center of our galaxy, that of our galaxy with respect to our local group, and so on; compare Sections 6.1.2a and 6.1.13c.

Third, the spectrum of the microwave radiation is, to good approximation, Planck ("thermal equilibrium," "blackbody"; cf. Example 5.5.4) for (z, Z). The temperature is slightly higher than 2.7 Kelvin. Near here-now, there seems to be no photon source sufficiently strong and close to thermal equilibrium to account for the observed characteristic Planck spectrum. However, big bang models can explain the spectrum in a reasonably plausible way (see Section 6.5). Thus the microwave data are generally regarded as the most nearly convincing of a number of observational results which indicates that something like a big bang actually occurred.

6.1.10 Stress energy

Before summarizing some observations on matter near here-now we give some background comments. (a) Let $\hat{T}$ be the stress-energy tensor of matter, let (z, Z) be the actual observer (Section 6.1.3c), and adopt the notation of Section 3.0.3 for tensors physically equivalent to a (2, 0) tensor. Let us take advantage of additivity (Example 3.13.3) to write $T = T_g + T_p + T'$, where T_g is due to the matter in galaxies, T_p is due to the microwave photons (Section 6.1.9), and T' includes the contribution of all other forms of matter—for example, of neutrinos (cf. Example 3.1.4), and of protons in intergalactic space. (b) In building models, one must, willy nilly, neglect some contributions to $T + E$ in the Einstein field equation $G = T + E$ for the influence of matter and electromagnetism on spacetime. If a cosmologist tried to include such small contributions as the stress-energy tensor of the electromagnetic field in your TV set, he would have a long row to hoe. (c) Near here-now we observe a galactic energy density $T_g(Z, Z)$ of roughly $(10^{10} \text{ years})^{-2}$. Here it is understood that the average value over a "very small" spacetime volume—say, 10^7 years across—has been taken. The observations, based on estimating how much matter is needed to produce the observed amount of light from a galaxy or alternately on estimating how much nearby galaxies pull on each other gravitationally, are difficult. $T_g(Z, Z)$ may be less than the value quoted above by a factor of 10 or even 30 (Longair and Rees [1]). In any case, near here-now T_g almost certainly dominates. $T_p(Z, Z)$ is about $10^{-4} \times (10^{10} \text{ years})^{-2}$ (Exercise 6.1.14); $E(Z, Z)$ is similarly negligible. $T'(Z, Z)$ is probably also negligible (Longair and Rees [1]). (d) Finally, we remark that even when a particular form of matter is being neglected in (b), one often includes this form in an overall model as "test matter," which "responds to, but does not influence, spacetime" (cf. Section 3.5).

A word of caution about (c). Some forms of matter are very difficult to detect directly, so one is not certain if $T'(Z, Z)$ is negligible (cf. Section 6.6.4); moreover, extrapolations suggest T_p and T' are very important near the big bang (Sections 6.4 and 6.5).

6.1.11 Other data

Many other measurements relevant to cosmology have been made. For example, one can try to count the number N of all observed galaxies whose luminosity distance is less than a certain value d_0. On a naive Newtonian model one would expect $N = (4\pi/3)nd_0^3$, where n is the number of galaxies per unit $\mathbb{R}^3$ volume near here-now. One job of a cosmological model is to make a corresponding prediction. Section 6.3 gives the canonical example and outlines the data.

As another example, various high-energy particles ("cosmic rays," including protons and high-energy photons such as x-rays or gamma rays) come from outside our galaxy to us. The information they carry must be fitted into any sensible model, at least in a qualitative way. We shall not discuss these particles or a host of other observed phenomena here. But we mention that the observations are indeed consistent, in a number of nontrivial ways, with the models we shall present (cf. Sciama [1] and Longair-Rees [1]).

EXERCISE 6.1.12. SPATIAL HOMOGENEITY

In Section 6.1.7, show that the Newtonian model for the Hubble law is spatially homogeneous in the following sense. If an observer on any other galaxy center regards herself as at rest, and thus uses locations $\vec{W}'$ and velocities $\vec{v}'$ relative to herself (Section 0.1.6) she will find $\vec{v}' = H\vec{W}'$ exactly as we do, with the same value of H.

EXERCISE 6.1.13. SPATIAL ISOTROPY AND PREFERRED INSTANTANEOUS OBSERVERS

Spatial isotropy not only indicates some physical or geometrical symmetry but normally, as in Proposition 3.15.1b, Exercise 5.7.9, and Exercise 6.0.14 a and b, singles out a preferred instantaneous observer at a point. This exercise gives another, simpler example of this kind of uniqueness. Suppose $(\mathbb{R}^2, g_0)$ is two-dimensional Minkowski space and $z \in \mathbb{R}^2$. Using the notation of Section 5.1, $Y^\pm = hf(\partial_2 \pm \partial_1)(z)$ determines two photon energy-momenta at z. (a) Show the instantaneous observer $(z, \partial_2 z)$ observes the same frequency, namely f, for both photons. (Thus, intuitively speaking, this instantaneous observer sees the same color sky whether he looks East or West.) (b) Show that no other instantaneous observer (z, X) at z sees East-West spatial isotropy as in (a). (c) Find the Newtonian speed (Section 2.1.3) of (z, X) observed by $(z, \partial_2 z)$ in terms of f^+/f^-, where $f^\pm$ is the frequency (z, X) measures for $Y^\pm$.

EXERCISE 6.1.14

Notation as in Section 6.1.10 and assume T_p is due to a photon gas Planck for (z, Z) with temperature 2.75 Kelvin (Section 6.1.9). (a) Show from Example 5.7.5 that $T_p(Z, Z) \cong 10^{-4} \times (10^{10} \text{ years})^{-2}$ as claimed in Section 6.1.10. (b) Assume $\hat{T}_g z = (10^{10} \text{ years})^{-2} Z \otimes Z$ [i.e., dust with $\rho z = 10^{-20} (\text{years})^{-2}$]. Show from the wrong-way Schwarz inequality that $T_p(X, X) \ll T_g(X, X)$ for all instantaneous observers (z, X), and not just for (z, Z).

6 Cosmology

(b) certainly suggests $T_p z$ is negligible and more detailed models (e.g., that of Proposition 6.4.5) bear this out.

EXERCISE 6.1.15

Using freshman optics (i.e., assuming Newtonian physics, Euclidean geometry, etc.) and the following figure, discuss the reason for the assumption $\Omega \ll 4\pi$ made in our definition (Section 6.1.6b) of area distance.

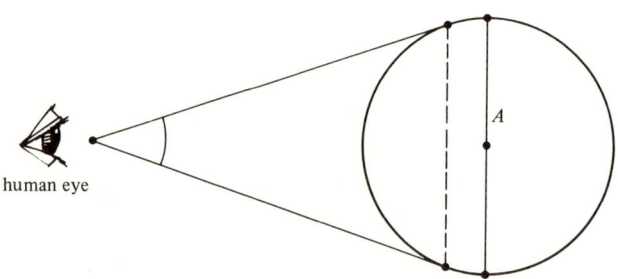

6.2 Cosmological models

To get a coherent picture of the universe, we need models that can unify the data in Section 6.1. After some general comments, we state our assumptions and then analyze a simple example that will concern us for some time.

6.2.1 General remarks

Since our treatment of cosmological models will be highly selective, we make some comments on our approach. A precise definition of a "cosmological model" will be given in Section 6.2.5. For the purpose of the informal discussion in this section, this term may be taken to mean a relativistic model consistent with the data in Section 6.1.

(a) In practice, the construction of a cosmological model goes roughly like this (cf. Sections 2.1 and 6.1.1). The theorist summarizes the known data, guesses at a systematic model that seems to be roughly consistent with the data, uses the model to analyze the data more closely, correspondingly refines his model, makes further adjustments if necessary when new data comes in, and so on. Sometimes a model must be discarded as beyond repair; often one is kept but only with a specific qualification as to its range of applicability. Then new models are sought, and the whole process begins afresh. In physics, this tentative, groping character of the models is a way of life. Therefore we shall not attempt the (hopeless) task of writing down the "most general" viable cosmological model. Instead, Sections 6.2 to 6.6 present canonical examples of this step-by-step procedure of refining models.

(b) As we proceed, and again in Section 6.6, we shall emphasize the limitations and diseases our models have. For example, all general relativistic models normally considered are self-defeating in the following sense: each predicts a big bang and simultaneously predicts that its own assumptions become physically unrealistic in the region sufficiently close to the big bang (cf. Proposition 6.4.2 and Section 6.6.5).

(c) By now, literally thousands of cosmological models have been rather intensively investigated. An introductory treatment, especially one for mathematicians who do not plan to specialize in cosmology, must thus proceed by example rather than by exhaustion, as was attempted in older texts and popularizations.

(d) In older discussions of cosmology, interest focused on the choice of a spacetime. Nowadays, as the data in Section 6.1 perhaps suggests, one is more interested in another problem: choosing matter models to represent the many constituents of the universe. Our treatment—for example, in Sections 6.4 and 6.5, will emphasize this point; there we shall only consider a simple cosmological spacetime (Example 1.4.3). For the analysis of the region near here-now, we will even confine attention to the Einstein–de Sitter spacetime (see Section 6.3).

(e) Popularizations of cosmology usually take the concept of "3-space" for granted and classify cosmological spacetimes into three categories according as "3-space" has constant positive curvature, zero curvature, or constant negative curvature (cf. Exercise 6.2.14). This famous and vivid popular trichotomy is obsolete: many current models belong to none of the three categories; focusing attention on "3-space" is rather silly from an observational standpoint (cf. Section 6.0.4); (a)–(d) above suggest quite a different approach; and so on. In fact, when one actually uses the models of Exercise 6.2.14 to analyze observable effects (Weinberg [1]), one finds all three types make surprisingly similar predictions rather than grossly different ones as the trichotomy suggests. Thus our decision to concentrate solely on simple cosmological spacetimes, whose space slices have zero curvature, will not result in excluding anything really essential.

(f) For similar reasons many other famous concepts will here get short shrift. We mention a few: "ultimate collapse"; "spatial" compactness; "Mach's principle"; the "cosmological constant" (always in effect taken as zero in this book); "tired light" cosmologies; torsion in cosmology; and nonquantum "unified field theory" cosmologies. Perhaps some of these have some interest. But we feel none has enough to warrant emphasis in an introductory treatment. Omitting them will leave room for a discussion of other topics more closely related to present observations—for example, the microwave radiation and the primordial fireball (Sections 6.4 and 6.5).

(g) Despite some differences in detail, our attitude towards models will thus be very similar to that expressed by Weinberg [1]:

6 Cosmology

"Of course the standard model may be partly or wholly wrong. However, its importance lies not in its certain truth but in the fact that it provides a common meeting ground for an enormous variety of observational data. By discussing these data in the context of a standard cosmological model, we begin to appreciate their cosmological relevance, whatever model ultimately proves correct."

In view of Section 6.2.1d, we begin with a proposition that indicates the matter models normally used and motivates the assumptions we shall presently make on matter.

Recall that if a relativistic model $(M, \mathcal{M}, F)$ is a finite superposition of relativistic models, then the matter model $\mathcal{M}$ is a collection of $N \geq 1$ components $\mathcal{M}_1, \ldots, \mathcal{M}_N$, each of which is itself a matter model on M.

Proposition 6.2.2. *Suppose $(M, \mathcal{M}, F)$ is a finite superposition of relativistic models and each component is either a particle flow, or a perfect fluid, or a photon gas. Suppose at least one of the following generality conditions holds for the superposition:* (a) *One component is a perfect fluid; or* (b) $\forall x \in M$ *one component is a photon gas not identically zero on $\mathcal{L}_x^+$; or* (c) $\forall x \in M$ *there is a component that is a particle flow (m, e, P, η) with $m\eta(x) \neq 0$. Then the stress-energy tensor $\hat{T}$ of the matter model $\mathcal{M}$ is normal and obeys the timelike convergence condition.*

Proof. We will prove $\hat{T}$ is normal. The proof that $\hat{T}$ obeys the timelike convergence condition is similar, easier and will be left as an exercise (Exercise 6.2.12a).

In the notation of Section 3.0.3, we have $\tilde{T} = \tilde{T}_1 + \cdots + \tilde{T}_N$, where $\tilde{T}_i$ is the stress-energy tensor of the ith component, and we must show TX is timelike for each causal $X \in M_x, \forall x \in M$. We first note the following:

(d) Suppose $W_1 \in \mathcal{T}_x^+$ and $W_2, \ldots, W_N$ lie in the closure of $\mathcal{T}_x^+$ — that is, W_1 is future pointing timelike and each of the others is either zero or future pointing causal. Then $W_1 + \cdots + W_N \in \mathcal{T}_x^+$ (Exercise 1.2.4).

Now fix $x \in M$ and let $X \in M_x$ be past pointing causal. Designate the component which obeys (a), (b) or (c) at x as $\mathcal{M}_1$. Setting $W_i = \tilde{T}_i X$, we will show $W_1 \in \mathcal{T}_x^+$ and $W_i \in$ closure $\mathcal{T}_x^+ \forall i$. In fact, suppose first $\mathcal{M}_i$ is a particle flow (m, e, P, η). Then $W_i = g(P, X)\eta(x)P$. $\eta(x) \geq 0$ and P is future pointing causal; by the usual argument, $g(P, X) > 0$, so $W_i \in$ closure $\mathcal{T}_x^+$. If $m \neq 0$, $g(P, X) \geq 0$ and $Px \in \mathcal{T}_x^+$; thus if in addition $\eta(x) \neq 0$, then $W_i \in \mathcal{T}_x^+$. Therefore $W_i \in$ closure $\mathcal{T}_x^+$ and $W_1 \in \mathcal{T}_x^+$ in case (c).

Similarly, if $\mathcal{M}_i$ is a perfect fluid then $W_i \in \mathcal{T}_x^+$; the argument that W_i is timelike repeats the proof of Proposition 3.15.1a verbatim, with $\rho z = a, pz = b$. W_i must be future pointing since $T_i(X, X) < 0$ is impossible (Definition 3.3.1). Finally if $\mathcal{M}_i$ is a photon gas, $W_1 \in$ closure $\mathcal{T}_x^+$ and $W_1 \in \mathcal{T}_x^+$ if (b) holds (Proposition 5.7.2b).

We have thus shown that in all cases (a) to (c), the conditions in (d) hold. Thus $\tilde{T}X$ is timelike for each future-pointing causal vector $X \in M_x$. Thus $\tilde{T}X$ is timelike for each causal $X \in M_x$. Thus $\hat{T}x$ is normal. This argument applies $\forall x \in M$, so $\hat{T}$ is normal. □

We can now make explicit some basic assumptions, interpretations, and motivations often taken for granted and left implicit in discussions of cosmology.

6.2.3 Nonquantum general relativistic cosmology

Formally, the basic assumptions are the following: (a) The universe is described by a relativistic model $(M, \mathcal{M}, F)$ which obeys Einstein's and Maxwell's equations. (b) The spacetime M is maximal. (c) The stress-energy tensor $\hat{T}$ of the matter model $\mathcal{M}$ is normal and obeys the timelike convergence condition.

By assumption (b) and Corollary 3.14.2 there is a unique eigenvector reference frame Z for $\hat{T}$. We shall call Z the *comoving reference frame* for the model. The final assumption now reads: (d) There exists a distinguished instantaneous observer (z, Z), designated the *actual observer* with z designated *here-now*, who enjoys the properties $Z = Zz$ and div $Z(z) > 0$.

Assumption (c) seems reasonable since there is a lot of matter wandering around the universe (Section 6.1) and any reasonably general nonquantum matter model has a normal stress-energy tensor which obeys the timelike convergence condition (Proposition 6.2.2). The name for Z and assumption (d) will both be motivated in Section 6.2.5.

Roughly speaking, assumptions (a) to (d) just mean that one is using nonquantum general relativity. This is a reasonable thing to try but one must expect that neglecting quantum effects may impose limitations on the applicability of the results (cf. Section 6.2.1a). And so it does (cf. Section 6.6.5).

6.2.4 A simplifying assumption

The observations indicate that near here-now the electromagnetic field F does not contribute significantly to the total stress-energy tensor. Except in very detailed, sophisticated models one usually takes $F = 0$ outright. Though this assumption is less fundamental than those in Section 6.2.3, it is convenient, and will be adopted here.

The formal assumptions above must be supplemented by various interpretation rules. The latter count as an essential part of a model. But one cannot even give a full list or state the rules in fully precise mathematical terms, much less prove them (cf. Einstein's comment quoted in Section 2.1.2). We now state and motivate the most important interpretation rule.

6.2.5 The comoving reference frame

The comoving reference frame Z in Section 6.2.3 is, intuitively speaking, comoving with the matter of the universe (cf. Exercises 3.8.8 and 3.15.6). Now, at least near here-now, galaxies seem to form the predominant form of matter (Section 6.1.10). Thus, taking advantage of the fact that galaxy random velocities are small (Section 6.1.2), one makes the interpretation rule that *the history of each galaxy is modelled by an integral curve of* Z. We add some comments and qualifications. (a) In some arguments a galaxy must be treated as an extended object. Then its center is still modelled by an integral curve of Z (cf. Section 6.1.3). (b) Of course not every integral curve of Z models some galaxy center (cf. Section 6.3.12 and the beginning of Section 6.4), nor is any of the relevant curves necessarily inextendible. In particular, if there is a hot, dense, violent region in the history of the universe "near a big bang," no galaxies are present in that region. (c) The interpretation rule (a) motivates the assumption $Z = Zz$ in Section 6.2.3d since (z, Z) is (to good approximation) "at rest" in our own galaxy (Section 6.1.3). (d) The assumption div $Z(z) > 0$ now merely corresponds to the fact that the empirical Hubble law (Section 6.1.7) indicates, at least qualitatively, expansion (cf. the interpretation of div below Lemma 4.3.3).

With the preceding interpretation rule and qualifications understood, we define a *cosmological model* to be a triple $(M, \mathscr{M}, z)$, where M is a spacetime, $z \in M$ and $\mathscr{M}$ is a matter model on M, such that: if F is taken to be zero and z is here-now, then all the assumptions in Section 6.2.3 are satisfied (cf. Section 6.2.4). The more correct appellation of "nonquantum general relativistic cosmological model with zero electromagnetic field" will not be employed here, for obvious reasons. A large number of such cosmological models are known. We now proceed to characterize, motivate, and indicate the limitations of the most naive one.

Recall that if (M, g) is a simple cosmological spacetime then $M = \mathbb{R}^3 \times \mathscr{F}$ and $g = R^2(u^4) \sum_{\mu=1}^{3} du^\mu \otimes du^\mu - du^4 \otimes du^4$, where $R > 0$. In this case denote by $\dot{R}$ and $\ddot{R}$ the first and second derivatives.

Lemma 6.2.6. *Let (M, g) be a simple cosmological spacetime.* (a) *The metric volume element is $\Omega = R^3(u^4)du^1 \wedge du^2 \wedge du^3 \wedge du^4$.* (b) *The Einstein tensor is $G = \rho_0 du^4 \otimes du^4 + p_0(g + du^4 \otimes du^4)$, where $\rho_0 = 3(\dot{R}/R)^2 \circ u^4$ and $p_0 = -[2(\ddot{R}/R) + (\dot{R}/R)^2] \circ u^4$.* (c) div $\partial_4 = 3(\dot{R}/R) \circ u^4$.

Here (a) follows directly from Section 3.0.1b since $(Rdu^1, Rdu^2, Rdu^3, du^4)$ is a (consistently oriented) orthonormal basis. The proof of (b) follows the proof of Proposition 1.4.4 almost verbatim and is left to the reader as per an earlier agreement. To prove (c) note that $i(\partial_4)\Omega = -(R^3 \circ u^4)du^1 \wedge du^2 \wedge du^3$ so $d[i(\partial_4)\Omega] = [3(\dot{R}/R) \circ u^4]\Omega$, that is, div $\partial_4 = 3(\dot{R}/R) \circ u^4$. □

Proposition 6.2.7. *Let $(M, \mathscr{M}, F)$ be a relativistic model, (z, Z) be an instantaneous observer on M. Suppose:* (a) *The nonquantum general relativistic cosmology assumptions Section 6.2.3 hold and $F = 0$;* (b) *M is a simple cosmological spacetime; and* (c) *$\mathscr{M}$ is a dust (Z, ρ) on M. Then, up to equivalence (Exercise 6.0.18), M is Einstein–de Sitter spacetime, $Z = \partial_4$, and $\rho = (4/3)(u^4)^{-2}$.*

6.2.8 Remarks

Before giving the proof we make some remarks.

The very specific assumptions (b) and (c) above are two-edged. On the one hand, (b) seems fairly reasonably *ab initio* in view of the correspondence between the isometries (Exercise 6.0.17a) of a simple cosmological spacetime and the observed, approximate spatial homogeneity and isotropy of the universe near here-now (cf. Sections 6.1.2, 6.1.7, 6.1.9, and 6.1.11 and Exercise 6.1.12, and the heuristic discussions of the "cosmological principle" in such standard texts as Weinberg [1]). Moreover, (c) seems to be a reasonable idealization: near here-now galaxies, are apparently the dominant form of matter (Section 6.1.10); they have small random velocities (Section 6.1.2b), thus corresponding to dust (Exercises 3.15.6 and 3.15.8). Therefore, choosing a dust to model the matter in galaxies (plus intergalactic matter which has small random velocities, if any is present in significant amounts) should be a good approximation, at least near here-now.

Indeed assumptions 6.2.7b,c lead to a powerful, convenient, and reasonably accurate though intentionally oversimplified model for the universe near here-now (Section 6.3). On the other hand, both are too naive to be more than zeroth order approximations, and the resulting model has at best a restricted domain of validity (cf. Sections 6.2.1, 6.4, 6.5, and 6.6).

PROOF OF PROPOSITION 6.2.7. Since $F = 0$, the Einstein field equation $G = T + E$ becomes $\rho_0 du^4 \otimes du^4 + p_0(g + du^4 \otimes du^4) = \rho\omega \otimes \omega$, where ω is the 1-form physically equivalent to Z and we have used Lemma 6.2.6b. At $x \in M$, choose a nonzero $X \in M_x$ such that $du^4(X) = 0 = \omega(X)$; X is spacelike. Then the preceding equation implies $p_0 g(X, X) = 0$. Since $g(X, X) \neq 0$, $p_0 = 0$.

We now have $\rho_0 du^4 \otimes du^4 = \rho\omega \otimes \omega$ with ρ nowhere zero since $\mathcal{M}$ is a dust. This implies $\omega = f du^4$ for some function f on M; since both ω and $-du^4$ are unit and future pointing, $f = -1$. Thus $\omega = -du^4$—that is, $Z = \partial_4$ as claimed. Moreover we also have $\rho = \rho_0 = 3(\dot R/R)^2$, by Lemma 6.2.6b again. To finish the proof, we return to the condition $p_0 = 0$.

This reads $2\ddot R = -\dot R^2/R$. For any constants A and B, $Ru = A(u - B)^{2/3}$ is a solution whenever $u \neq B$; by the standard uniqueness theorem and the fact that $R > 0$, each solution has this form. Without loss of generality, we may take $B = 0$ (Exercise 6.0.18b). Since $R > 0$ and M is maximal, we must have either: (i) $\mathscr{F} = (0, \infty)$ and $A > 0$, or (ii) $\mathscr{F} = (-\infty, 0)$ and $A < 0$. We exclude (ii) (which gives a relativistic model all right but a collapsing one) by using the assumption 6.2.3d that div ∂_4 is somewhere positive. By Lemma 6.2.6, div $\partial_4 = 3(\dot R/R) \circ u^4 = 2(u^4)^{-1}$. In case (ii), this is everywhere negative so we must have case (i). Since $A > 0$, we may take $A = 1$ without loss of generality (Exercise 6.0.18a). Now $Ru = u^{2/3}$, so M is Einstein–de Sitter spacetime and $\rho = 3(\dot R/R)^2 = (4/3)(u^4)^{-2}$. □

6.2.9 Einstein–de Sitter model

Let M be Einstein–de Sitter spacetime; thus M is maximal (Exercise 5.2.7). Define $\rho = (4/3)(u^4)^{-2}$, so that $\mathcal{M} = (\partial_4, \rho)$ is a dust on M. Choose any point $z \in M$. The triple $(M, \mathcal{M}, z)$ is then a cosmological model, as the reader

6 Cosmology

may check by a routine verification using the equations given in the proof of Proposition 6.2.7 and the definition in Section 6.2.5.

On comparing the direct data (Section 6.1.8) on the age of objects in our solar system or galaxy with the meaning (Exercise 6.0.14c,d) of u^4, it is reasonable to guess u^4z is a bit more than 10 billion years. Since the data are uncertain and the model intentionally oversimplified, no significance is attached to the exact value chosen. All one can really say is that in all likelihood $8 \times 10^9 < u^4z < 15 \times 10^9$ if the model applies at all.

With M, $\mathcal{M}$ as above and with z chosen so that $u^4z = 10^{10}$ years, we shall call $(M, \mathcal{M}, z)$ the *Einstein–de Sitter model*. $\mathcal{M}$ here is so naive that it is hard to tell the difference between the Einstein–de Sitter model and the Einstein–de Sitter spacetime; one may note nevertheless that the defining property "$u^4z = 10^{10}$ years" belongs to the model and not to the spacetime. In later models—for example, that of Proposition 6.4.5—$\mathcal{M}$ will be more complicated.

A more detailed analysis of the Einstein–de Sitter model will occupy us in Section 6.3.

> Using 11 or 12 billion years in the choice of u^4z would actually give a slightly better fit to almost all the data. But we prefer to use the rounded-off value of 10 billion, handicapping the model a little, to emphasize its intentional naivete. $u^4z = 1.2 \times 10^{10}$ years would suggest a precision that neither the model nor the data possesses.

6.2.10 Adjustable parameters

By adjustable parameters, we mean the indeterminate constants in a cosmological model. As an example, we note that in the Einstein–de Sitter model the numerical value of the cosmological time u^4z at here-now is not predetermined in any mathematically natural way. Next to vague philosophy, gratuitous adjustable parameters are the biggest curse of theoretical cosmology. One main reason the Einstein–de Sitter model is more fun than other cosmological models is that the others have adjustable parameters that observations do not, as yet, determine to reasonable accuracy.

EXERCISE 6.2.11. THE PURE RADIATION UNIVERSE

Replace assumption (c) of Proposition 6.2.7 by the assumption that $\mathcal{M}$ is a rest-mass zero perfect fluid (physics texts refer to this as "pure radiation") and show the following. (a) $\mathscr{F}$ has a finite lower bound but no finite upper bound. (b) Up to equivalence (Exercise 6.0.18): $\mathscr{F} = (0, \infty)$, $Ru = u^{1/2}$, $Z = \partial_4$, and $\rho = 3(\dot{R}/R)^2 \circ u^4 = \frac{3}{4}(u^4)^{-2}$. (c) With $\mathscr{F} = (0, \infty)$, $G_j{}^i G_i{}^j \to \infty$ as $u^4 \to 0$, where we use the index notation of Section 3.6.1f. Since $G_j{}^i G_i{}^j$ is a bona-fide function on M, this result again indicates the existence of a big bang.

> The above assumption on $\mathcal{M}$ is probably an excellent approximation very near the big bang—for example, for 1 second $< u^4 < 10^3$ years in the resulting model (cf. Sections 6.4–6.6). Moreover, in the (unlikely)

event that neutrinos are dominant even near here-now, the model above with ρ interpreted as the energy density of neutrinos would be more important than the Einstein-de Sitter model (cf. Sections 6.1.10 and 6.6.4).

EXERCISE 6.2.12

Assumptions as in Proposition 6.2.2, show: (a) $\hat{T}$ obeys the timelike convergence condition (cf. Section 4.3.7a). (b) Let $\hat{E}$ be the stress-energy tensor of F. Then $\hat{T} + \hat{E}$ is a normal stress-energy tensor which obeys the timelike convergence condition. (c) If $\hat{T} = \rho Z \otimes Z$ for some reference frame Z—that is, if $\hat{T}$ has the form of a dust stress-energy tensor, then: no component of $\mathcal{M}$ is a photon gas F unless F $\equiv 0$; if any component is a perfect fluid, it is a dust with reference frame Z; and if any component is a particle flow (P, η, m, e) with η somewhere nonzero, then $m \neq 0$ and $P = mZ$ wherever $\eta \neq 0$.

EXERCISE 6.2.13

Let M be a simple cosmological spacetime with Einstein tensor G. Suppose $G = T$, where $\hat{T}$ is a stress-energy tensor which is normal and obeys the timelike convergence condition. Suppose div ∂_4 is somewhere nonnegative. Show the following generalizations of results in Section 6.0. (a) ∂_4 is again canonically determined—for example, as the eigenvector reference frame of $\bar{T}$. The results in Section 6.0.2 on the expansion, synchronizability, and so on, of ∂_4 remain valid verbatim. R satisfies $\dot{R} > 0$. (b) $\mathcal{F}$ again has a finite lower bound, which can be taken as 0 without loss of generality. Then u^4 is again the maximum proper time since the beginning, in the sense of Exercise 6.0.14d. (c) The isometry group $\mathcal{G}M$ is again isomorphic to the isometry group of Euclidean 3-space (contrast Exercise 6.0.17b.)

EXERCISE 6.2.14. ROBERTSON–WALKER SPACETIMES

Though we will not use them in an essential way, we give some very famous examples of models with extra adjustable parameters. Let $(\mathbb{R}^4, g_0)$ be Minkowski spacetime, $H \subset \mathbb{R}^4$ be the subset defined by $(u^4)^2 = 1 + \sum_{\rho=1}^{3}(u^\rho)^2$ and $u^4 > 0$. (a) Show that H is a three-dimensional submanifold homeomorphic to $\mathbb{R}^3$ and that g_0 induces a Riemannian metric h on H. Show that the Riemannian 3-manifold (H, h) satisfies $\text{Ric}(X, X) = -2$ $\forall$ unit vector X. (b) Let $M = H \times \mathcal{F}$, where $\mathcal{F}$ is an open interval in $\mathbb{R}$. Denote the projections by $\pi: M \to H$ and $u^4: M \to \mathcal{F}$, and let $R: \mathcal{F} \to (0, \infty)$. Then $g = -du^4 \otimes du^4 + (R \circ u^4)^2 \pi^* h$ defines a Lorentzian metric on M. (M, g), oriented and time-oriented in the obvious way, is called a *Robertson–Walker spacetime of negative spatial curvature*; here "negative" refers to the minus sign in $\text{Ric}(X, X) = -2$. *Robertson–Walker spacetimes of positive spatial curvature* are defined correspondingly, with (H, h) replaced by $(\mathcal{S}^3, h)$ where h is the metric induced on the unit 3-sphere by the Euclidean metric of $\mathbb{R}^4$ (cf. Section 0.0.9). Those of *zero spatial curvature* are merely simple cosmological spacetimes.

6 Cosmology

For a 3-dimensional Riemannian manifold, constancy of the Ricci curvature is equivalent to constancy of the sectional curvature. $(H, \mathbf{h})$ above is then a Riemannian manifold of sectional curvature -1. The Robertson–Walker spacetimes may therefore be characterized as those of the form $M \times \mathscr{F}$, where $\mathscr{F}$ is an open interval, M is a simply connected space form (cf. J. A. Wolf [1]), and the Lorentzian metric is $-du^4 \otimes du^4 + (R \circ u^4)^2 \pi^* \mathbf{h}$ in the above notation except $\mathbf{h}$ is now the Riemannian metric of M.

Now replace assumption (b) of Proposition 6.2.7 by the requirement that M be a Robertson–Walker spacetime of negative spatial curvature. Show from the Einstein field equation that without loss of generality one may take $\mathscr{F} = (0, \infty)$ and take R to be a function defined by: $R(u) = A(\cosh t - 1)$, where t is implicitly defined by $u = A(\sinh t - t)$, and $A \in (0, \infty)$. A is the extra adjustable parameter mentioned. (c) Derive the analogous result for the positive curvature case. (d) For a Robertson–Walker spacetime with $\dot{R} \neq 0$, define the *deceleration parameter* $q = -\ddot{R}R/\dot{R}^2$. Show that in Einstein–de Sitter spacetime, $q = \frac{1}{2}$ and can be interpreted as a measure of deceleration in the sense that $q = |a|/|v|^2$, where a and v are the relative 3-acceleration and relative 3-velocity, respectively, of Proposition 2.3.7. Show also that in case (b) above $0 < q < \frac{1}{2}$, and in case (c), $\frac{1}{2} < q$.

Robertson–Walker spacetimes are characterized as the spatially homogeneous isotropic spacetimes in the following sense: if a spacetime locally admits a 6-dimensional group of local isometries whose orbits are spacelike, then it is locally isometric to a Robertson–Walker spacetime; the converse is obvious. In view of the strong empirical evidence for spatial homogeneity and isotropy (Section 6.1), these spacetimes are clearly the most reasonable zeroth order approximations to try.

Despite the differences in geometry and topology, all the Robertson–Walker spacetimes whose Einstein tensor equals a dust stress-energy tensor as above make similar predictions for all observable effects (cf. Section 6.2.1e). The causal past of here-now is topologically trivial in case (c), even though the spacetime itself is not. ("We have not yet had time to see even half-way around the universe.") In a realistic treatment, one should regard all these spacetimes as a single one-parameter set, with Einstein–de Sitter spacetime as a typical representative, characterized by the fact that the deceleration parameter is 1/2.

6.3 The Einstein–de Sitter model

6.3.1 Plan

Mathematically, the natural approach might now be to prove some general theorems about cosmological models. Instead we continue the zig-zag procedure of model building outlined in Section 6.2.1a.

More specifically, let $(M, \mathscr{M}, z)$ be the Einstein–de Sitter model (Section 6.2.9) throughout the rest of Section 6.3. We will compare $(M, \mathscr{M}, z)$ to the data and outline its virtues; then, in subsequent sections, discuss its inadequacy for handling events near the big bang and go on to present some less idealized models. Overall, we shall demonstrate that, while not all of M is an exact model for all of physical spacetime, at least an open submanifold of M is a useful zeroth order approximation for part of the history of the universe.

In view of $Z = \partial_4$ in Section 6.2.5, our model predicts that the cosmological frequency ratio $\imath$ is $\imath = (u^4z/u^4x)^{2/3} > 1$ (Proposition 6.0.8), in qualitative agreement with $\imath > 1$ for the observed $\imath$ (Section 6.1.5). Also, rather trivially, note that our comment in Section 6.2.8 about the match between empirical and predicted symmetries applies. For recall that each isometry ψ of M onto itself obeys $\psi_*\partial_4 = \partial_4$ and $u^4 \circ \psi = u^4$, and hence such an isometry is an automorphism of the whole model $(M, \mathcal{M}, z)$—that is, $\mathcal{M} = (Z, \rho) = (\psi_*Z, \rho \circ \psi)$, and $u^4\psi z = u^4 z$.

6.3.2 Energy density

In view of Proposition 6.2.7, the model predicts the observed energy density should be $T(Z, Z) = \rho[du^4(\partial_4)]^2(z) = \rho z = (4/3)(u^4z)^{-2} = (4/3)(10^{10} \text{ years})^{-2}$. The observed energy density has a very uncertain value which is roughly comparable (Section 6.1.10). This counts as a significant achievement of the model.

> Not as a major triumph, since current estimates favor a rather lower value of the observed energy density. Among models with extra adjustable parameters, those in Exercise 6.2.1b, using Robertson–Walker spacetimes of negative spatial curvature, can fit any observed lower value by adjusting the extra parameter (Exercise 6.4.12b).

6.3.3 The Hubble law for neighbors

By using the interpretation Section 6.2.5a that the history of a galaxy is modelled by an integral curve of ∂_4, we can get a preliminary comparison of the Einstein–de Sitter model's predictions with the empirical Hubble Law. Proceeding naively for the moment, we regard a moderately nearby galaxy as a neighbor W of our own galaxy (cf. Sections 2.0.3 and 2.1.2 and Definition 2.3.2); we than have from Proposition 2.3.7 that the 3-velocity of the galaxy relative to us, now, is $v = [2/(3u^4z)]W$.

Suppose $|W| < 10^8$ years. Then we may make the following identifications (Section 6.1.6), (a) Our local rest-space $Z^\perp$ is $\mathbb{R}^3$ and the figure in Section 6.1.7 applies (cf. Section 2.1.2). (b) $|W| = |\vec{W}| = d_L = d_A$, where $\vec{W}$ is the Euclidean vector of Section 6.1.7, while d_L and d_A are luminosity and area distance, respectively; and $|v| = |\vec{v}| = v$ in the notation of Section 6.1.5 and Exercise 6.1.7. Thus we can compare the prediction $|W| = (3u^4z/2)|v|$ of our model with the empirical result $d_L = t_H v$ when $|W| \ll 10^{10}$ years. We have $3u^4z/2 = 15$ billion years and $t_H \cong 18$ billion years. In view of the uncertainties in measuring t_H (Section 6.1.7) and in guessing $u^4z = 10^{10}$ years (Section 6.2.9), the approximate agreement is quite encouraging for this model (and

6 Cosmology

for similar ones). We henceforth write t for the *predicted Hubble time* of the Einstein–de Sitter model—that is, $t = 3u^4z/2 = 15$ billion years.

The naive calculation just given is somewhat unsatisfying: it works with spacelike vectors in a tangent space while the actual physics involved is modelled by lightlike curves in the manifold itself.

We now begin an analysis that is more realistic, because it works with photons; it leads to a sharper confrontation between theory and observations, because it applies also to galaxies so distant-early that they cannot be regarded as neighbors—that is, $|W| < 10^8$ need not be assumed.

Suppose a galaxy at $x \in M$ can send a photon to z (Figure 6.3.4a). Let d be the distance, within the level surface $u^4 = u^4z$, between z and the point y where the inextendible ∂_4 integral curve through x intersects the level surface. Of course d is not directly observable; for example, if the galaxy blows up at x' and thus fails to reach y altogether, we shall not learn of the catastrophe for many millions of years. But the next proposition shows that in our model the cosmological frequency ratio $z(x, z)$ of Proposition 6.0.8 can replace d. The proposition also shows that z can replace the cosmological time difference $u^4z - u^4x$ and that for $\varkappa = z - 1$ much less than 1, $\varkappa$ is directly proportional to both d and $u^4z - u^4x$.

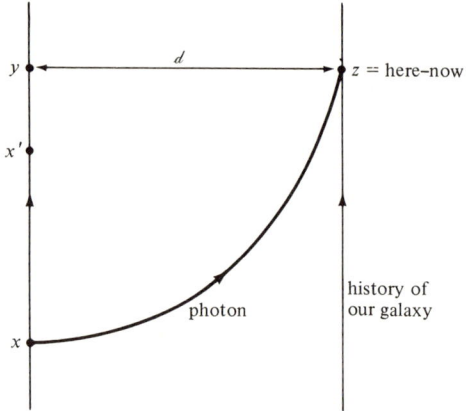

Figure 6.3.4a

In contrast to d, $u^4z - u^4x$ is in principle rather directly measurable. For example, by analyzing the spectrum (color; cf. Section 5.5.3) of a galaxy, it should be possible to infer how old the stars in that galaxy are. One would then usually guess that u^4x should be about the same number or perhaps a little larger. In practice, this kind of measurement is as yet almost wholly useless (cf. Gunn and Oke [1]).

Proposition 6.3.4. *Let t be the predicted Hubble time, and d be as above. Then:* (a) $d = d(z)$, *where* $d: (1, \infty) \to (0, 2t)$ *is the increasing function* $d(u) = 2t(1 - u^{-1/2})$. (b) $u^4z - u^4x = (2t/3)(1 - z^{-3/2})$. (c) *For* $\varkappa \ll 1$, $d(z) \cong t\varkappa \cong u^4z - u^4x$.

6.3 The Einstein–de Sitter model

PROOF. By Section 6.0.3, we have $d = (u^4z)^{2/3}\delta(x, z)$, where we use the notation of Section 6.0.12. Now for any isometry ψ we know that $u^4 \circ \psi = u^4$, $\delta(\psi x, \psi z) = \delta(x, z)$, and $\imath(\psi x, \psi z) = \imath(x, z)$ (Section 6.0). It thus suffices to prove the proposition for the case where x and z lie on the standard light signal of Example 5.2.2. In that case we can determine u^4z, u^4x, $\delta(x, z)$, and finally d in terms of t and $\imath$ via the following chain of results: (d) $u^4z = 2t/3$ (Section 6.3.3). (e) $u^4x = \imath^{-3/2}u^4z$ (Proposition 6.0.8). (f) $\delta(x, z) = 3[(u^4z)^{1/3} - (u^4x)^{1/3}]$ (Exercise 5.4.5a).

Thus

$$d = (u^4z)^{2/3}\delta(x, z) = (2t/3)^{2/3}3[(2t/3)^{1/3} - \imath^{-1/2}(2t/3)^{1/3}]$$
$$= 2t(1 - \imath^{-1/2})$$

as claimed in (a). The image of $\imath$ is $(1, \infty)$ by Lemma 6.0.10. Trivially, $d(u) = 2t(1 - u^{-1/2})$ is increasing and has range $(0, 2t)$. Thus (a) holds.

(d) and (e) together imply (b). Finally, (c) follows from the fact that the functions of $\imath$ in (a) and (b) are analytic for $\imath \in (0, \infty)$ and can therefore be approximated by using Taylor series. For example, $1 - \imath^{-1/2} = 1 - (1 + \varkappa)^{-1/2} = \frac{1}{2}\varkappa + O(\varkappa^2)$ so $2t(1 - \imath^{-1/2}) \cong t\varkappa$ for $\varkappa \ll 1$. □

Now observing the frequency ratio for various distant galaxies does not by itself check a model or the Hubble law. One needs some independent measurements. In the next three subsections, we shall describe a particular set of measurements that can bring about a closer confrontation between the Einstein–de Sitter model and data. These involve measuring the luminosity distance d_L (Sections 5.5.5 and 6.1.6a) of a galaxy and the solid angle the galaxy subtends on our celestial sphere (Section 5.1.3). To analyze what the Einstein–de Sitter model predicts, we shall first discuss a relationship between the intrinsic cross-sectional area of a galaxy and that of its image on the celestial sphere.

6.3.5 An auxiliary 2-sphere

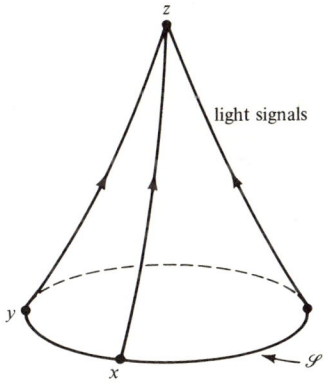

6 Cosmology

Suppose $x \in M$ and there exists a light signal from x to $z =$ here-now (figure above). Fixing x and z, we define $\mathscr{S} \subset M$ by: $\mathscr{S} = \{y \mid \imath(y, z) = \imath(x, z)$ and there is a light signal from y to $z\}$, where $\imath$ is the cosmological frequency ratio. Thus, roughly speaking, the astronomer identifies $\mathscr{S}$ by looking for all galaxies whose observed red shift is the same as that of the galaxy at x.

Proposition 6.3.5a. $y \in \mathscr{S}$ *iff there is a spatial rotation around* z, $\psi: M \to M$, *such that* $\psi x = y$.

> *Proof.* Suppose such a ψ exists. Then ψ preserves the time orientation (Section 6.0.5c) as well as geodesics. By the definition of a light signal (Section 5.2), ψ preserves light signals. Thus there is a light signal from y to z and, by Lemma 6.0.9, $\imath(x, z) = \imath(\psi x, \psi z) = \imath(y, z)$. Hence $y \in \mathscr{S}$.
> Conversely suppose $y \in \mathscr{S}$. Let ψ_y be an isometry that carries the light signal from y to z into the standard light signal (Exercise 5.2.5). Define ψ_x similarly. Then $u^4 \psi_x z = u^4 z = u^4 \psi_y z$ (Section 6.0.5); since u^4 increases monotonically along the standard photon, this implies $\psi_x z = \psi_y z$. Similarly the fact that $\imath$ is monotonic along the standard photon (Lemma 6.0.10) implies $\psi_x x = \psi_y y$. Thus the isometry $\psi = \psi_y^{-1} \circ \psi_x$ obeys $\psi z = z$ and $\psi x = y$. ψ is by definition a spatial rotation around z. □

Corollary 6.3.5b. $y \in \mathscr{S}$ *iff* $u^4 y = u^4 x$ *and* $\delta(x, z) = \delta(y, z)$. *Moreover,* $\mathscr{S}$ *is a spacelike 2-manifold diffeomorphic to* $\mathscr{S}^2$. *The metric induced on* $\mathscr{S}$ *by* $\mathbf{g}$ *is that of a 2-sphere in* $\mathbb{R}^3$ *with radius* $(u^4 x)^{2/3} \delta(x, z)$.

The proof is immediate from Exercise 6.0.16.

Define a map χ from our celestial sphere $\mathscr{S}_z$ into $\mathscr{S}$ as follows. For each $W \in \mathscr{S}_z$, $Z - W$ is a future-pointing lightlike vector and thus determines a unique inextendible light signal through z. By the definition of $\mathscr{S}$ this light signal intersects $\mathscr{S}$ at the unique point, say y, where $\imath(x, z) = \imath(y, z)$; then $y = \chi W$, by definition. χ is one-one since $Z - W = a(Z - W')$ with $W, W' \in Z^\perp$ implies $W = W'$. Denote the metric of $\mathscr{S}_z$ by $\mathbf{h}_z$ (cf. Section 5.1.3). Then a short argument, left as Exercise 6.3.19 following, gives:

(c) $\chi: \mathscr{S}_z \to \mathscr{S}$ is a diffeomorphism and $\chi^* \mathbf{h} = [(u^4 x)^{2/3} \delta(x, z)]^2 \mathbf{h}_z$.

6.3.6 Idealizations

Now consider a distant galaxy. Suppose as before that for some point x on the history of its center, there is a light signal from x to z. We now make some idealizations and approximations.

(a) Assume that the galaxy history intersects $\mathscr{S}$ of Section 6.3.5 in an integration region $\Delta\mathscr{S} \subset \mathscr{S}$ (cf. Section 5.5.1) and assume the area A of $\Delta\mathscr{S}$ is known from observations (cf. Section 6.1.2 and 6.1.6, and Exercise 6.1.15).
(b) Assume we get a light signal from the distant galaxy iff that light signal contains here-now and intersects $\Delta\mathscr{S}$.

6.3 The Einstein-de Sitter model

By our assumptions, the set $\Delta \mathscr{S}_z = \chi^{-1}(\Delta \mathscr{S})$ of those points on our celestial sphere at which we see the galaxy has a well-defined area $\Omega \in (0, 4\pi)$. By Section 6.3.5c, $\Omega/A = 4\pi/A$, where A is the intrinsic area of $\mathscr{S}$—that is, $A = 4\pi(u^4 x)^{4/3} \delta^2(x, z)$. Now as the next proposition shows more explicitly, A can be expressed in terms of the cosmological frequency ratio $\imath$ for (x, z) and the predicted Hubble time t; moreover, A was assumed known. Thus the model makes a prediction, which can in principle be checked directly, for Ω at a given cosmological red shift $z = \imath - 1$, as follows; the notation is as in Proposition 6.3.4a, with $d = d(\imath)$.

Proposition 6.3.7. $\Omega/A = (\imath/d)^2$—that is, the area-distance is $d_A = d/\imath$.

Proof. $A = 4\pi(u^4 x)^{4/3} \delta^2(x, z)$. Using the results in the proof of Proposition 6.3.4, we have:

$$A = 4\pi \imath^{-2} \left(\frac{2t}{3}\right)^{4/3} \cdot 9 \left[\left(\frac{2t}{3}\right)^{1/3} - \imath^{-1/2} \left(\frac{2t}{3}\right)^{1/3} \right]^2$$

$$= 16\pi \imath^{-2} [t(1 - \imath^{-1/2})]^2 = 4\pi \imath^{-2} d^2(r).$$

Thus $\Omega/A = 4\pi/A$ gives $\Omega/A = (\imath/d)^2$. By our operational definition (Section 6.1.6b), $d_A = (A/\Omega)^{1/2} = d/\imath$. □

Proposition 6.3.8. (a) *The luminosity distance is* $d_L = \imath d$. (b) *Regarded as a function of* $\imath$, d_L *is an increasing onto function:* $(1, \infty) \to (0, \infty)$.

Remarks. We shall postpone the (long winded) proof of (a) to the Appendix of this chapter (Theorem 6.7.5). However, note that the much more general result $d_L = \imath^2 d_A$ quoted (without proof) in Section 6.1.6 gives (a) directly from Proposition 6.3.7. Given (a), (b) is immediate from Proposition 6.3.4.

6.3.9 Focusing

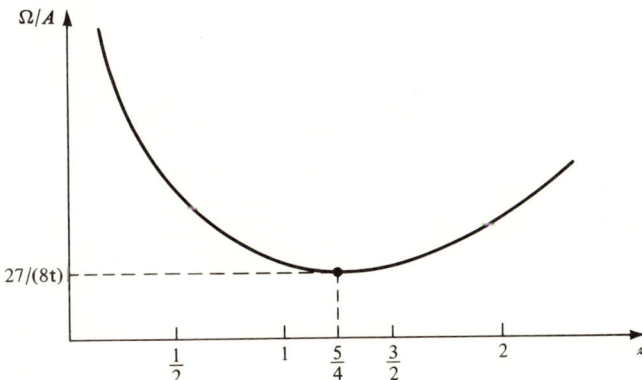

Before commenting on the relation of these two results to observations, we make a few comments. Recall $z = \imath - 1$. (a) For $z \to 0$, Proposition 6.3.7 with A fixed formally gives $\Omega \to \infty$. This merely means that if a galaxy were

6 Cosmology

right on top of us, our assumption $\Omega \ll 4\pi$ would fail and our definition of d_A would be inapplicable (see Exercise 6.1.15). (b) Using Propositions 6.3.4 and 6.3.7, we find for Ω/A the preceding graph, with a minimum at $z = 5/4$. Thus, if we imagine observing galaxies of the same intrinsic area A, and could do the observation for red shifts greater than $5/4$, we should find that Ω is an increasing function of $\imath$ for $\imath = z + 1 > 9/4$.

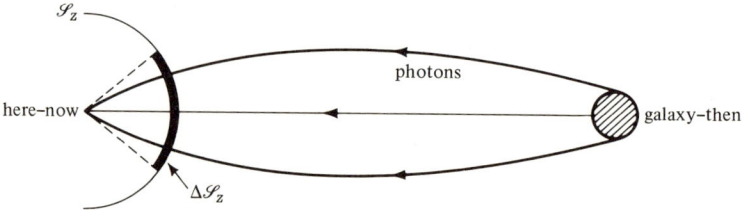

Intuitively speaking, the reason Ω is big is focusing. We give an intuitive picture in "space," and also suppress one "spatial" dimension.

The dotted lines on the left are tangent to the world lines of the incoming photons as shown, and the heavy arc marked off on $\mathscr{S}_z$ represents the apparent size of the galaxy according to the actual observer at here-now. We draw the photon lines curved to indicate intuitively that spacetime curvature tends to focus lightlike geodesics in Einstein–de Sitter spacetime (that the curvature of a space slice vanishes is irrelevant). Since $\mathscr{S}_z$ is in a tangent space, it is the tangents at here-now that determine $\Delta \mathscr{S}_z$. As the picture perhaps suggests, the net effect is to make the area of $\Delta \mathscr{S}_z$ "bigger than it should be." The more spacetime there is between us and the galaxy, the more focusing.

6.3.10 Comparison with observations

For almost half a century, an enormous amount of work has been done in trying to compare such predictions as Propositions 6.3.8 and 6.3.7 to the observations (cf. Tamman [1]). At present the situation is this. (a) For galaxies, considerable data are available in the range $0 < z < 0.465$. (b) For $z \ll 1$, Propositions 6.3.8 and 6.3.7 give $d_L \cong d_A \cong tz$ in view of Proposition 6.3.4c. The model therefore makes a prediction for the value of t once $\{d_A, z\}$ or $\{d_L, z\}$ are determined. Here, in contrast to Section 6.3.3, the relevant quantities (d_A, d_L and z) can be regarded as directly measurable. Thus we have now justified more carefully our earlier comments in Section 6.3.3 on the satisfactory agreement between the observed value $t_H \cong 18$ billion years and the model's prediction $t = 15$ billion years. (c) The graph below shows the red shift z at luminosity distance d_L for more than 50 galaxies with comparatively large red shifts (cf. Section 6.1.5). The data are from Gunn and Oke [1]. The highest dot is the radio galaxy 3C 295 mentioned in Section 6.1.5.

The upper curve is the prediction of the Einstein–de Sitter model. The lower curve was obtained using a Robertson–Walker spacetime of negative spatial curvature, with the extra adjustable parameter chosen so that the deceleration parameter here-now is $1/4$ (cf. Exercise 6.2.14). As the figure suggests, current data on galaxy red shifts and luminosities are compatible with various models, including the Einstein–de Sitter model (Sandage and Tamman [1], Gunn and Oke [1]). The same statement holds for the further data on d_A (cf. the graph on p. 173 in Peebles [1]). (d) The quasar data goes to much higher redshifts, where the predictions of different models differ considerably (cf. Lang *et al.* [1]). However, for quasars, d_L is harder to find because of uncertainties about the absolute luminosity L; in many cases d_A cannot even be estimated. (e) One of the most serious problems is trying to correct for evolutionary effects. We are looking so far back in time that the galaxies or quasars may not yet be sufficiently similar to more familiar ones closer to here-now for our estimates of d_L or d_A to be reasonably accurate. (f) On balance, the Einstein–de Sitter model fits the galaxy and quasar observations almost as well as any other model, and better than most, even though the others have extra adjustable parameters.

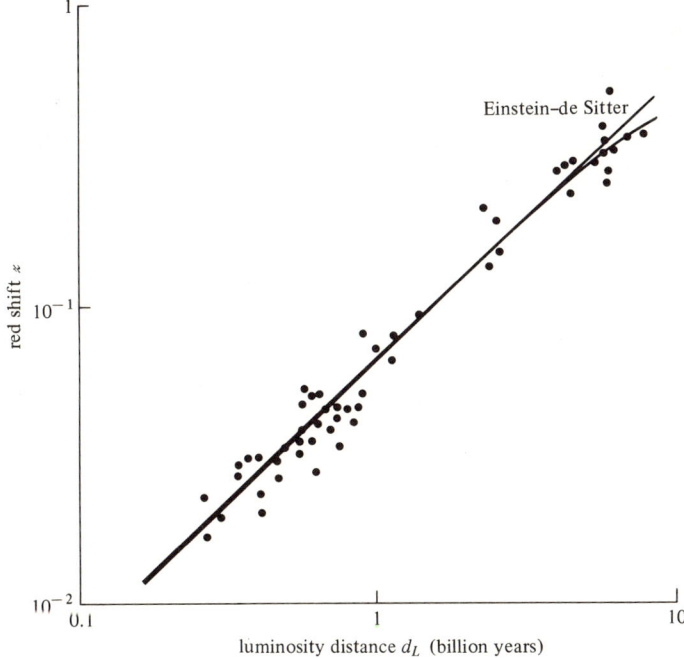

Among models with such parameters, the current data seem to favor the negative spatial curvature Robertson–Walker models over the positive spatial curvature ones. A few years ago the situation was the opposite.

We now consider another kind of observation. Suppose one restricts attention to galaxies of a particular kind—for example, radio galaxies whose absolute luminosity is in some given range and whose spectrum obeys some given restrictions. Suppose one counts all those whose luminosity distance d_L is less than some given value d_0. The answer according to the Einstein–de Sitter model is given by Proposition 6.3.13 below; it replaces the naive guess $N = (4\pi/3)nd_0^3$ in Section 6.1.11. We begin with some preparatory material.

6.3.11 Number counts: the matter model

(a) Let us assume that the galaxies of interest can be modelled by a particle flow (P, η) of rest-mass m, where m is the average rest-mass of one such galaxy and $P = m\partial_4$. The idea is that our dust matter model $\mathcal{M}$ of Section 6.2.9 is itself a superposition of many such particle flows, all having P proportional to ∂_4 (cf. Exercise 3.15.8b); the precise value of m will be irrelevant here. (b) Let us further assume η is "spatially homogeneous" in the sense that $\eta \circ \psi = \eta$ for all isometries ψ of M into itself. Thus by Section 6.0.5h, η "depends only on time"—that is, there exists a function $n: (0, \infty) \to [0, \infty)$ such that $\eta = (1/m)(n \circ u^4)$. The factor $(1/m)$ has been inserted so that $n(u)$ may have the intuitive interpretation of "number of galaxies per unit 3-volume at cosmological time u"; compare Example 3.2.2. (c) Finally, let us assume that galaxies of the chosen type are conserved—that is, div $(\eta P) = 0$. This assumption is unrealistic if galaxies are subject to strong evolutionary effects. For then a galaxy can change its parameters (e.g., its total luminosity or its spectrum) so much in billions of years that it "drops out" of the given particle flow (P, η) and joins another. The relevant time period is that between x of Figure 6.3.4a, when the galaxy sends us light, and y in that figure, when the galaxy is supposed to make its due contribution to $n(u^4 z)$ above. Similarly, newcomers joining our particle flow would violate div $(\eta P) = 0$. Our comparison to observations below thus involves not only the Einstein–de Sitter model but also the homogeneity and conservation assumptions just made.

6.3.12 Number counts; geometry

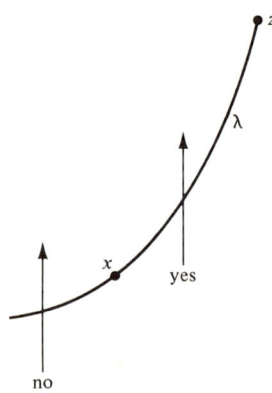

6.3 The Einstein-de Sitter model

Suppose z = here-now lies on the standard light signal $[\lambda]$. Given our maximum luminosity distance $d_L = d_0$, there is a unique frequency ratio $\imath_0 \in (1, \infty)$ such that $d_0 = \imath_0 d(\imath_0)$ by Proposition 6.3.8, and thus by Proposition 6.0.10 a unique point x on $[\lambda]$ such that the luminosity distance $d_L(z, Z)$ measures for a galaxy at x is d_0. Then by the same two propositions a galaxy whose history intersects the standard light signal contributes to our counted number iff the intersection point is between x and z (figure). Thus the actual situation of interest is that depicted pictorially below, once with two dimensions suppressed and then with less detail but only one dimension suppressed.

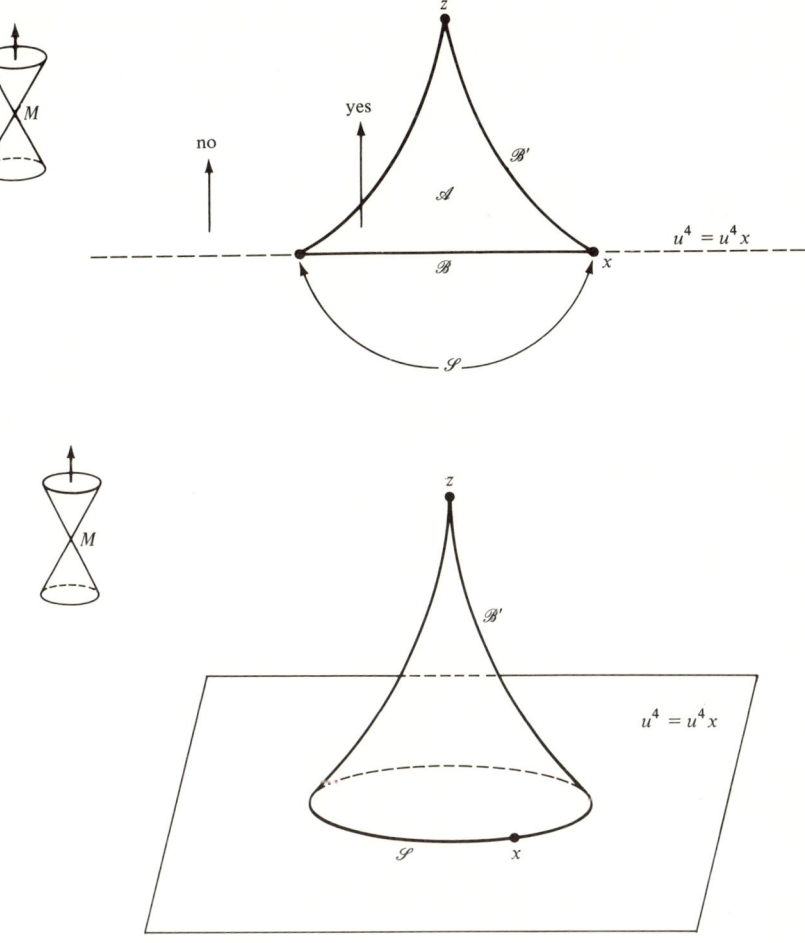

Here $\mathscr{A}$ is that compact subset of our causal past (Proposition 6.0.13) for which $u^4 \geq u^4 x$; formally, $y \in \mathscr{A}$ iff $u^4 z \geq u^4 y \geq u^4 x$, and $\delta(z, y) \leq 3[(u^4 z)^{1/3} - (u^4 x)^{1/3}]$. The boundary of $\mathscr{A}$ consists of the following submanifolds: a submanifold $\mathscr{B}$, diffeomorphic to the open unit 3-ball, given explicitly

6 Cosmology

by $u^4y = u^4x$ together with $\delta(y, z) < \delta(x, z)$; the 2-sphere $\mathscr{S}$ of Section 6.3.5; a (lightlike) 3-submanifold $\mathscr{B}'$, diffeomorphic to $\mathscr{S}^2 \times \mathbb{R}$, given explicitly by $u^4z > u^4y > u^4x$ together with $\delta(z, y) = 3[(u^4z)^{1/3} - (u^4y)^{1/3}]$; and z itself. As usual we assign $\mathscr{B}$ and $\mathscr{B}'$ the orientation induced from that of $\mathscr{A}$.

By the definition of total number in Section 3.2 the number N of galaxies whose luminosity distance is less than or equal to d_0 is $N = \int_{\mathscr{B}'} \eta i(P)\Omega$.

Proposition 6.3.13. *Suppose $d_0 \in (0, \infty)$ and let $\imath_0$ be the unique cosmological frequency ratio such that $\imath_0 d(\imath_0) = d_0$. (Notation as in Proposition 6.3.4). Then $N = (4\pi/3)n(u^4z)d_0^3 = [4\pi n(u^4z)/3]d_0^3/\imath_0^3$, where $d_0 = d(\imath_0)$.*

Proof. Since div $(\eta P) = 0$, $N = -\int_{\mathscr{B}} \eta i(P)\Omega$ by Stokes' theorem. By Exercise 3.2.5, $\eta = (1/m)n(u^4z)(u^4z)^2(u^4)^{-2}$. Now $i(P)\Omega = mi(\partial_4)\Omega = -m(u^4)^2 du^1 \wedge du^2 \wedge du^3$, where we have used Lemma 6.2.6 for Ω. Together we get:

$$N = n(u^4z)(u^4z)^2 \int_{\mathscr{B}} du^1 \wedge du^2 \wedge du^3.$$

Since $\mathscr{B}$ can be regarded as the ball around the origin in $\mathbb{R}^3$ with radius $3[(u^4z)^{1/3} - (u^4x)^{1/3}]$, the last integral over $\mathscr{B}$ equals

$$(4\pi/3)\{3[(u^4z)^{1/3} - (u^4x)^{1/3}]\}^3.$$

Replacing u^4z by $2t/3$ and u^4x by $\imath_0^{-3/2} \cdot 2t/3$ (cf. the proof of Proposition 6.3.4), we obtain:

$$N = \frac{4\pi n(u^4z)}{3} [2t(1 - \imath_0^{-1/2})]^3 = \frac{4\pi}{3} n(u^4z)d^3(\imath_0).$$

Proposition 6.3.8 finishes the proof. $\square$

6.3.14 Remarks

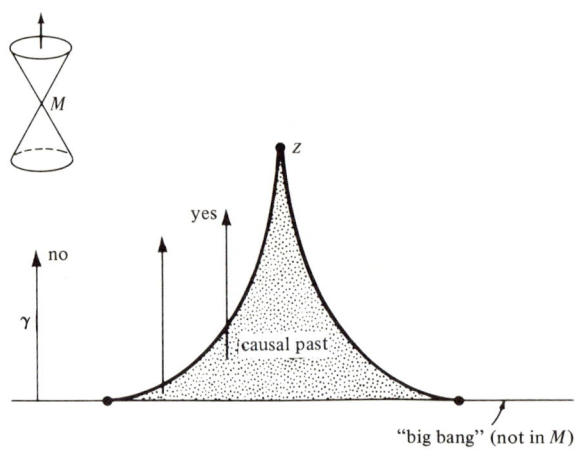

(a) For $\imath \ll 1$, we can regard d as the Newtonian distance d_N (cf. Section 6.1.6) and then the result given coincides with the naive Newtonian result $N = (4\pi/3)nd_N^3$. Of course then we also have $d_0 \simeq d_0$, since $\imath \simeq 1$. However, for $z \gg 1$ the results of the Einstein–de Sitter model are different. In particular, for $z \to \infty$, $N \to (4\pi/3)n(u^4z)(2t)^3$ by Proposition 6.3.4, so we can in principle count at most a certain finite number of galaxies (cf. the figure and Exercise 6.3.19). (b) Note in the figure that the observer γ in the reference frame ∂_4 has not yet had a chance to communicate with us since the big bang.

6.3.15 Comparison with observations

Proposition 6.3.13 relates N both to the maximum luminosity distance d_0 and to the red shift corresponding to d_0. Both kinds of predictions have been intensively investigated. Looking at quasars and other radio sources with radio telescopes is the key measurement. At least for the quasars, red shifts as large as $z \simeq 3.5$ are involved. One suspects that some of the sources with unknown red shift may have even larger values of z. Despite heated controversies it seems reasonably clear, and we shall assume, that these large red shifts are cosmological (Section 6.0.7), rather than being due to weird physics in objects within 10^8 years of here-now (cf. Longair [1]). Then the radio observations penetrate very far back in time indeed (see Figure 6.5.2b following).

The data seem to contradict every reasonable known model which assumes, as we did in Section 6.3.11b,c above, the conservation of sources and spatial homogeneity (Longair [1], Petrosian [1]). In particular there seems to be very strong evidence that there were more bright radio sources billions of years ago than there are near here-now. Thus the results suggest the universe was different in the past, as one might indeed expect if something like a big bang occurred.

Although this is not certain, the nonconservation of sources in the early universe, rather than spatial nonhomogeneity, is believed to be accountable for the contradiction of the data with Proposition 6.3.13. This would mean it is Assumption 6.3.11c, and not necessarily the Einstein–de Sitter model itself, that is at fault. For instance, one can get a roughly consistent picture using the Einstein–de Sitter model if one assumes, instead of div$(\eta \boldsymbol{P}) = 0$, a suitable modification that corresponds to the gradual fizzling out of radio galaxies and quasars since the universe was about one-tenth of its present age. However, the uncertainties about the kind of evolution that occurred preclude any attempt to single out just one model using the number count data (Rees [1]). Thus the uncertainties about matter models are crucial, here as elsewhere.

> The famous steady-state cosmology requires both spatial homogeneity and, on the average, lack of source evolution. The number count data seems definitely inconsistent with the steady-state cosmology. Nowadays, the microwave data (Section 6.1.9) are considered even stronger evidence against the steady-state cosmology by most authors. At least for the present, this unusually beautiful model, so enticing because of its absolute minimum of adjustable parameters, is dead as a doornail.

6 Cosmology

6.3.16 Less is more

Before shooting down the Einstein–de Sitter model in Section 6.4, we owe it a tribute. As we have tried to indicate by a few examples, it does actually provide "a common meeting ground for an enormous variety of observational data" (Section 6.2.1g). Basic, very intuitive, mathematically instructive, perhaps as accurate physically near here-now as any other extensively analyzed cosmological model, and mellowed by 55 years of vigorous give-and-take, it focuses on essentials and is free of all fancy frills. Regarding it as an exact model of nature would be sheer nonsense; but it is a truly elegant zeroth-order approximation.

EXERCISE 6.3.17

In Exercise 6.2.11 on the "pure radiation universe," assume $u^4z = 10^{10}$ years. (a) Using an argument similar to that of Section 6.3.3, show the predicted Hubble time is 2×10^{10} years. (b) Find expressions for the area-distance as a function of cosmological frequency ratio (i.e., modify Proposition 6.3.7) and for the number count N as a function of area-distance (cf. Proposition 6.3.13). (c) Generalizing (a), consider a simple cosmological spacetime M such that $\dot{R} > 0$ and model galaxy centers by integral curves of ∂_4. With $z =$ here-now, show that the predicted value of the Hubble time is $(R/\dot{R})(u^4z)$. (d) Generalizing further, show that (c) remains valid if M is replaced by any Robertson–Walker spacetime for which $\dot{R} > 0$.

EXERCISE 6.3.18

Prove Corollary 6.3.5c, either by brute force or by using the rotation group $\mathcal{O}^3$ suitably.

EXERCISE 6.3.19

Let $(M, \mathcal{M}, z)$ be the Einstein–de Sitter model and suppose in Section 6.3.11 $(m\partial_4, m^{-1}n \circ u^4)$ is a particle flow that models all galaxies simultaneously (not just one kind as in Section 6.3.11; thus $m > 0$ should now be the average rest-mass for all kinds of galaxies). (a) Assume that within the level surface $u^4 = 10^{10}$ years, the average distance between galaxy centers is 10^7 years. Using the integrals in Section 3.2, argue that $n(10^{10}) = 10^{-21}$(years)$^{-3}$. (b) Assume galaxies are conserved (Section 6.3.11c) and assume we can see all the way back to the big bang. Show that the observable universe then contains about 10^{11} (100 billion) galaxies. (c) For what value of m does the stress energy tensor $\hat{T}$ of $(m\partial_4, m^{-1}n \circ u^4)$ obey $T(Z, Z) = (10^{10}$ years$)^{-2}$ (cf. Section 6.1.10)?

6.4 Simple cosmological models

Following our plan of Section 6.3.1, we now discuss why the Einstein–de Sitter model is inadequate near the big bang and discuss an improved model (Definition 6.4.4). The discussion centers on the microwave radiation (Section 6.1.9) and leads up to some rather exciting current speculations on the early universe.

6.4 Simple cosmological models

Let $(M, \mathcal{M}, z)$ be the Einstein–de Sitter model. $\mathcal{M}$ is a dust, so we cannot regard $\mathcal{M}$ as including the microwave photons (Section 6.2.8; Exercise 6.2.12). Nevertheless, we can still hope that minor *ad hoc* modifications of the model would suffice to account for the microwave data (Section 6.1.9). In making these modifications, one could go so far as ignoring, if necessary, one or more of the original assumptions (Section 6.2.3) of a cosmological model; compare the discussion of Section 6.2.1a concerning model-building at this point. We give a typical modified model below (Section 6.4.1) and then show that, unfortunately, it completely breaks down at early times (Proposition 6.4.2), thus indicating that the whole Einstein–de Sitter model is beyond salvage near the big bang.

6.4.1 *The microwave photons*

Let $(M, \mathcal{M}, z)$ be the Einstein–de Sitter model and let $\hat{T}$ be the stress-energy tensor of $\mathcal{M}$. Now define a matter model $\mathcal{M}_0$ on M to be the superposition (Example 3.13.3) of the dust $\mathcal{M}$ and a rest-mass zero perfect fluid $\mathcal{M}_p = (\rho_p, \frac{1}{3}\rho_p, \partial_4)$, with ρ_p yet to be determined. Let $\hat{T}_p$ be the stress-energy tensor of $\mathcal{M}_p$, so that $\hat{T}_p = (\rho_p/3)[4\partial_4 \otimes \partial_4 + \hat{g}]$ (cf. the notation and discussion in Section 6.1.10). $\mathcal{M}_p$ models the microwave photons; the motivation is given by the spatial isotropy of M itself and the observed isotropy of the microwave radiation on the one hand, and the discussions in Sections 3.15 and 5.7 on the other. To specify ρ_p, we assume: (a) ρ_p is spatially homogeneous in the sense that $\partial_\mu \rho_p = 0 \,\forall \mu = 1, 2, 3$. This again matches the isometries of M and is also suggested by the (rough) spatial homogeneity of various observed quantities (Section 6.1). There is strong, though not quite decisive, empirical evidence that near here-now photon collisions have a negligible effect on T_p (Peebles [1]; Weinberg [1]). We therefore make assumption (b): div $\hat{T}_p = 0$ (cf. Proposition 3.9.2 with $E = 0 = F$). Then $(M, \mathcal{M}_0, z)$, together with the matter equations (a) and (b) above, is our modified Einstein–de Sitter model.

To arrive at our estimate that near here-now div $\hat{T}_p = 0$ holds to good approximation, one must first try to think through all the possible collisions a photon can undergo, using the laws of microphysics. For example, one has available well known microphysical models for collisions in which electrons and photons are incoming. Similarly, the microphysics governing photon-photon collisions is well enough understood that one can predict such collisions are exceedingly rare near here-now. For each possible kind of collision allowed by microphysics one must then estimate, using the available knowledge of what matter is present near here-now, whether T_p is significantly affected by such collisions. Having done this, one finally arrives at the idealization mentioned (cf. Sciama [1]). The entire argument is tricky, because we are not really sure of our matter models even near here-now, and leaves a few cosmologists skeptical.

When extrapolating to early times, as we plan to do, the assumption div $\hat{T}_p = 0$ also has limitations to be discussed in the next two sections. Very generally speaking, in building models of the early universe, one usually begins with some very simple matter equations [assumption (b) in the present case; Section 6.3.11c is another example]. These give a preliminary idea how the relevant quantities (ρ_p in the present case) behave. Then one can gradually work out more realistic matter models and equations. Section 6.5 illustrates this procedure. If one tries instead to consider all possible forms of matter and all possible matter-matter interactions *ab initio*, one gets a hopeless mess. Compare our comments in Section 6.2.1.

> In the present context, the details of our assumptions are not so critical since all we are trying to do is to shoot down the Einstein–de Sitter model. It turns out that modifying, for example, assumption (b) merely alters the way in which the model gets shot down; no reasonable set of assumptions can save the model.

$(M, \mathcal{M}_0, z)$ differs from the Einstein–de Sitter model in having one more component in the matter model $\mathcal{M}_0$ and in having two extra matter equations. Moreover, the stress-energy tensor of $\mathcal{M}_0$ is $\hat{T} + \hat{T}_p = \rho \partial_4 \otimes \partial_4 + (\rho_p/3)[4\partial_4 \otimes \partial_4 + \hat{g}]$ (see Example 3.13.3), and therefore the Einstein field equation for $(M, \mathcal{M}_0, z)$ does not hold (Section 6.2.9 and the proof of Proposition 6.2.7). Assumption 6.2.3a not being valid in it, $(M, \mathcal{M}_0, z)$ is not a cosmological model according to the definition in Section 6.2.5. However, the Einstein field equation would hold for $(M, \mathcal{M}_0, z)$ if one agreed to neglect the contribution of the rest-mass zero perfect fluid to the stress-energy tensor (proof of Proposition 6.2.7)—that is, if one neglects the effects of the microwaves on the spacetime M (cf. Sections 3.9 and 3.12). Thus if we insist on the validity of the Einstein field equation, we may regard $(M, \mathcal{M}_0, z)$ as a "cosmological model" in which the rest-mass zero perfect fluid is "test matter" (Section 6.1.10d). Near here-now, this idealization is reasonable (Exercise 6.1.14); in early epochs it is not, as we shall now show.

Proposition 6.4.2 shows precisely in what sense the modified Einstein–de Sitter model $(M, \mathcal{M}_0, z)$ is self-defeating: after neglecting the contribution of T_p to the Einstein equation, it goes on to predict that for early times T_p actually dominates the stress-energy tensor of its own matter model. In the proposition the relevant value of a is about $a = 10^{-4}$ (Section 6.1.10). Lemma 6.4.3, which is used in proving the proposition, adopts the notation of Lemma 6.2.6 and says, roughly speaking, that as we go backwards in time toward a big bang everything gets denser at least as fast as R^{-3}. It can be applied both to "test matter" and to matter that significantly influences spacetime, so it will be rather useful when we introduce the generalizations the proposition suggests. A preliminary motivation for the assumption $\dot{R} > 0$ in Lemma 6.4.3 is given by Exercise 6.2.13a.

Proposition 6.4.2. *$(M, \mathcal{M}_0, z)$ as above, let $\rho = T(\partial_4, \partial_4)$ be the energy density of the dust $\mathcal{M}$. Define $a \in (0, \infty)$ by $a = \rho_p z/\rho z$. Then $\rho_p = a(u^4 z/u^4)^{2/3} \rho$; in particular $T(\partial_4, \partial_4) \ll T_p(\partial_4, \partial_4)$—that is, $\rho \ll \rho_p$—whenever $u^4 \ll a^{3/2} \times 10^{10}$ years.*

Lemma 6.4.3. *Suppose M is a simple cosmological spacetime with $\dot{R} > 0$. Let $\mathcal{M} = (\rho, p, \partial_4)$ be a perfect fluid on M such that $\partial_\mu \rho = 0 = \partial_\mu p$ for $\mu = 1, 2, 3$, and that the stress-energy tensor $\hat{T}$ of $\mathcal{M}$ obeys $\operatorname{div} \hat{T} = 0$. Then:*
(a) $\forall y \in M$, *the inequalities*

$$\left(\frac{R(u^4 y)}{R \circ u^4}\right)^3 \leq \frac{\rho}{\rho y} \leq \left(\frac{R(u^4 y)}{R \circ u^4}\right)^4$$

hold in the region determined by $u^4 \leq u^4 y$. Moreover: (b) *The first (respectively, second) inequality is an equality valid on all of M iff we have the dust case $p = 0$ (respectively, rest-mass zero case $p = \rho/3$).*

Proof of Lemma 6.4.3. $\operatorname{div} \hat{T} = 0$ implies $\operatorname{div}(\rho \partial_4) = -p \operatorname{div} \partial_4$ (Proposition 3.15.3), which is equivalent to $\partial_4 \rho = -(\rho + p) \operatorname{div} \partial_4$. Now $\operatorname{div} \partial_4 = 3(\dot{R}/R) \circ u^4$ (Lemma 6.2.6c) and $0 < \rho \leq \rho + p \leq (4/3)\rho$ (Definition 3.15.2). We thus get $-3(\dot{R}/R) \circ u^4 \geq (\partial_4 \rho)/\rho \geq -4(\dot{R}/R) \circ u^4$. Since $\partial_\mu \rho = 0 \, \forall \mu = 1, 2, 3$ and $\dot{R} > 0$, an elementary integration gives (a). (b) follows from the fact that the first (respectively, second) inequality is an equality in

$$\rho \leq \rho + p \leq \frac{4}{3}\rho,$$

and thus in the other conditional inequalities above, iff $p = 0$ (respectively, $p = \rho/3$). □

Proof of Proposition 6.4.2. By (b) of the lemma, we obtain:

$$\rho = (\rho z)[R(u^4 z)]^3 (R \circ u^4)^{-3},$$
$$\rho_p = (\rho_p z)[R(u^4 z)]^4 (R \circ u^4)^{-4}.$$

Dividing, we get $\rho/\rho_p = a^{-1} \dfrac{R \circ u^4}{R(u^4 z)}$. □

In view of $\rho \ll \rho_p$ for $u^4 \ll 10^{10}$ in Proposition 6.4.2, we cannot systematically analyze the microwave photons or the early universe until we generalize the Einstein–de Sitter model's assumption that the matter model is a dust. The following generalization has been very intensively investigated during the last 20 years; attempts have been made to apply it to times as early as 10^{-42} seconds after a big bang.

> The number 10^{-42} is that suggested by the value of Planck's quantum constant; compare the fine-print comment below Section 5.1.3.
> Of course the attempts are speculative. If the reader takes the view that any attempt to extrapolate to times earlier than 2000 B.C. are extremely dubious, he will have the enthusiastic support of many observational astronomers and the (possibly grudging) respect of thoughtful theorists. However, it does seem clear that to explain the universe near here-now one must investigate the early universe and there seems to be no natural stopping place along the route of this strange journey into the past. There is something awesome, even frightening about the conjecture that our current theories may conceivably apply to such early times.

Definition 6.4.4. A *simple cosmological model* is a cosmological model $(M, \mathcal{M}, z)$ such that: (a) (M, g) is a simple cosmological spacetime, (b) $M = \mathbb{R}^3 \times (0, \infty)$ and $R \to 0$ as $u^4 \to 0$; (c) $u^4 z = 10^{10}$ years.

Here the motivations for, and limitations of, assumption (a) have already been outlined (Sections 6.2.3 and 6.2.8). In (b), taking $\mathcal{F} = (0, a), a \in (0, \infty]$, involves no further loss of generality given (a) and the assumptions (Section 6.2.3) of a cosmological model (Exercise 6.2.13b). The remaining parts of (b) are that $a = \infty$ and that $R \to 0$ as $u^4 \to 0$. Both of these can usually be proved once a sufficiently detailed matter model is chosen (cf. Proposition 6.2.7 and Exercise 6.2.11 for two examples). Thus (b) in effect acts as an extremely general assumption on $\mathcal{M}$, which replaces the very specific requirement that $\mathcal{M}$ be a dust.

For a perfect fluid that obeys the requirements of Lemma 6.4.3, our assumption (b) implies that the energy density $\rho \to \infty$ as $u^4 \to 0$. For this (and other) reasons we can again regard $u^4 \to 0$ as approach to a big bang and expect that, qualitatively speaking, matter is becoming ever denser in this limit. Then, in view of Exercise 6.2.13, u^4 can again be interpreted as maximum proper time since the big bang. Thus the motivations for, and limitations of, the final assumption $u^4 z = 10^{10}$ years are just as before (Section 6.2.9).

Let $(M, \mathcal{M}, z)$ be a simple cosmological model. We again designate u^4 as the *cosmological time*. Exercise 6.2.13 shows that the comoving reference frame is ∂_4, that ∂_4 is expanding, and that $\dot{R} > 0$.

We can now use a simple cosmological model to cure the particular disease indicated by Proposition 6.4.2. The resulting model, detailed in Proposition 6.4.5, is superior to the Einstein–de Sitter model or the modified Einstein–de Sitter model of Section 6.4.1 in that the microwave radiation is built into this model from the beginning in the form of a rest-mass zero perfect fluid. In particular, as a consequence of the influence of this rest-mass zero perfect fluid on spacetime (via the Einstein field equation), the resulting spacetime is no longer the Einstein–de Sitter spacetime. Of course, even this model has its own limitations (Sections 6.2.1a,b and 6.6).

Proposition 6.4.5. *Let $(M, \mathcal{M}, z)$ be a simple cosmological model such that:* (a) $\mathcal{M}$ *is a superposition of a dust* $\mathcal{M}_1 = (\rho_1, \partial_4)$ *and a rest-mass zero perfect fluid* $\mathcal{M}_2 = (\rho_2, \rho_2/3, \partial_4)$; (b) *the stress-energy tensor* $\hat{T}_2$ *of* $\mathcal{M}_2$ *obeys* div $\hat{T}_2 = 0$; (c) $\partial_\mu \rho_2 = 0$, *for* $\mu = 1, 2, 3$. *Let a be the positive number defined by* $\rho_2 z = a\rho_1 z$. *Then, up to equivalence (Exercise 6.0.18) the following hold:* (d) R *is the function determined implicitly by* $u = [R(u) + b]^{1/2} \times [R(u) - 2b] + 2b^{3/2}$, *where* $b = a\{(u^4 z)/[(1 - 2a)(a + 1)^{1/2} + 2a^{3/2}]\}^{2/3}$; (e) $\rho_1 = (4R^{-3}/3) \circ u^4$; *and* (f) $\rho_2 = (4bR^{-4}/3) \circ u^4$.

6.4.6 Remarks

Before giving the proof we make a few comments.

(a) The interpretation of interest here is that $\mathcal{M}_1$ models the matter in galaxies and $\mathcal{M}_2$, as mentioned above, models the microwave photons. In this context, a is about 10^{-4} and all three assumptions in Proposition 6.4.5. have been motivated earlier in this section. However, the proposition has other applications (cf. Section 6.6.4). (b) Note that conditions 6.4.5d–f determine $(M, \mathcal{M}, z)$ uniquely if it exists. We are thus presenting our model in the form of a uniqueness theorem; Exercise 6.4.10 states the corresponding existence result and gives hints on its proof. (c) Note that as $a \to 0$, one has $b \to 0$, $\rho_2 \to 0$, and $R(u) \to u^{2/3}$; thus the whole model approaches the Einstein–de Sitter model, as one would expect.

Proof of Proposition 6.4.5. Let (M, z) be a simple cosmological model such that Assumptions 6.4.5a–c hold. The Einstein field equation $G = T_1 + T_2$ implies div $\hat{T}_1 = 0$, because of (b) and div $\hat{G} = 0$; it also implies $\partial_\mu \rho_1 = 0 \, \forall \mu \in (1, 2, 3)$, because of (c) and Lemma 6.2.6b. Thus Lemma 6.4.3 implies both $\rho_1 = (4b_1/3)(R \circ u^4)^{-3}$ and $\rho_2 = (4/3)b(R \circ u^4)^{-4}$ for some $b_1, b \in (0, \infty)$. Using the diffeomorphism determined by $u^4 \to u^4$, $u^\mu \to (b_1)^{-1/3} u^\mu$ for $\mu = 1, 2, 3$, we can and shall assume $b_1 = 1$ without loss of generality (cf. Exercise 6.0.18a). Thus (e) and (f) hold. Using Lemma 6.2.6b again we now get $3(\dot{R}/R)^2 = (4/3)(R^{-3} + bR^{-4})$. Since $R \to 0$ as $u^4 \to 0$ we can integrate to obtain $u^4 = 3/2 \int_0^R v \, dv/(v + b)^{1/2} = (R + b)^{1/2}(R - 2b) + 2b^{3/2}$. To complete the proof of (d), we note from (e) and (f) that $\rho_2 z = a\rho_1 z$ iff $b = aR(u^4 z)$ iff $b = a\{10^{10}/[(1 + a)^{1/2}(1 - 2a) + 2a^{3/2}]\}^{2/3}$. □

Now that we have a model (Proposition 6.4.5) that takes into account the influence of the microwave photons on spacetime, we must next make sure that, if we abandon the Einstein–de Sitter model we are not throwing out the baby with the bath water—that for the observations analyzed in Section 6.3 the model (Proposition 6.4.5) is no worse than the Einstein–de Sitter model.

In fact such is the case. The observations mentioned in Section 6.3 concern only red shifts less than nine, in most cases very much less. Thus we are concerned with at most the u^4 range determined by $(1/10)R(u^4 z) \le R(u^4) \le R(u^4 z)$ (Exercise 6.0.17); here R is given by Proposition 6.4.5d with $a \simeq 10^{-4}$ and $u^4 z = 10^{10}$ years. Numerical estimates (using Taylor series) show that throughout this entire u^4 range $R(u^4)$ differs from $(u^4)^{2/3}$ (the Einstein–de Sitter behavior) by less than 1%. The same then holds for all the predicted effects in Section 6.3 and such small changes are negligible compared to the empirical uncertainties.

As an example, note from Proposition 6.4.5d that $a = 10^{-4}$ gives $b \simeq 10^{-4} \times 10^{20/3}$ and thus $R(10^{10}) \simeq (1 + 10^{-4})10^{20/3}$. The predicted value of the Hubble time is now $t = (R/\dot{R})(10^{10})$ (Exercise 6.3.17c). The model gives $R/\dot{R} = R(du^4/dR) = (3/2)R^2(R + b)^{-1/2}$ (Proposition 6.4.5d). Thus $t \simeq (3/2)(1 + \tfrac{3}{2} \cdot 10^{-4}) \times 10^{10}$—that is, the predicted Hubble time differs from that predicted by the Einstein–de Sitter model by a bit more than 0.01%. Compared to the 15% observational inaccuracy of the Hubble constant (Section 6.1.7), and the still larger uncertainty in choosing $u^4 z = 10^{10}$ years (Section 6.2.9), a discrepancy of $(3/2) \times 10^{-4}$ is grotesquely small.

Thus the model (Proposition 6.4.5) shares the crude but important virtues of the Einstein–de Sitter model without having the disease diagnosed at the start of this section. Why didn't we use it *ab initio*? Because it is much clumsier than the Einstein–de Sitter model, is no better near here-now and, as we shall see shortly, has diseases of its own.

EXERCISE 6.4.7. THE BIG BANG

Let $(M, \mathcal{M}, z)$ be a simple cosmological model such that $\mathcal{M}$ is a superposition of perfect fluids $\{(\rho_A, p_A, \partial_A) \mid A = 1, \ldots, N\}$. (a) Generalize Lemma 6.4.3 by showing $\sum_{A=1}^{N} \rho_A \to \infty$ as $u^4 \to 0$. (Intuitively: "*overall*, matter gets denser as we approach the big bang".) (b) Use Example 3.12.1 to construct a case where $N = 2$ and $\rho_2 \to 0$ as $u^4 \to 0$.

> (b) Shows that not every matter component need be present at the big bang "$u^4 = 0$": some may be made later. In fact probably helium and deuterium *are* made later (Section 6.5). There is some indication that when quantum models are used for very early times in some general models (a) can fail: conceivably all forms of matter were made, by quantum process, after the big bang. See Misner [1].

EXERCISE 6.4.8. PHOTONS

Let $(M, \mathcal{M}, z)$ be a simple cosmological model. (a) Show that the causality and chronology relations for M can again be found from those on Minkowski spacetime by using conformal invariance. (b) Let $\lambda: \mathcal{E} \to M$ be an inextendible, freely falling photon. Show: $\mathcal{E} = (a, \infty)$, $a \in \mathbb{R}$; and $u \to u^4 \lambda u$ determines an increasing function from (a, ∞) onto $(0, \infty)$.

EXERCISE 6.4.9. MAXIMALITY

In Exercise 6.2.13, suppose $\mathcal{F} = (0, \infty)$ and $Ru \to 0$ as $u \to 0$. Show M is maximal by the following steps: $(\ln R)(u) \to -\infty$ as $u \to 0$; $(\dot{R}/R)(u) \to \infty$ as $u \to 0$; $G(\partial_4, \partial_4) \to \infty$ as $u \to 0$; and each inextendible future-pointing lightlike geodesic has the form $\lambda: (a, \infty) \to M$ where $u^4 \lambda u \to 0$ for $u \to a$ (cf. Proposition 1.3.2 and Exercise 5.2.7).

EXERCISE 6.4.10. EXISTENCE

To show the existence of a model that obeys all the conditions in Proposition 6.4.5, proceed as follows. $\forall b \in (0, \infty)$ define a C^∞ onto function $u: (0, \infty) \to (0, \infty)$ by $u(R) = (3/2) \int_0^R v \, dv/(v + b)^{1/2}$. Since $u' > 0$, its inverse function $R: (0, \infty) \to (0, \infty)$ also satisfies $R' > 0$. Now let M be the simple cosmological spacetime characterized by: $\mathcal{F} = (0, \infty)$ and R is as above. Then with $\rho_1 = (4R^{-3}/3) \circ u^4$ and $\rho_2 = (4R^{-4}/3) \circ u^4$, $\mathcal{M}_1 = (\rho_1, 0, \partial_4)$ and $\mathcal{M}_2 = (\rho_2, \rho_2/3, \partial_4)$ are perfect fluids on M. Let $\mathcal{M}$ be the superposition of $\mathcal{M}_1$ and $\mathcal{M}_2$ and choose $z \in M$ so that $u^4 z = 10^{10}$ years. (a) Show $(M, \mathcal{M}, z)$ is a simple cosmological model. (Hint: To check the Einstein field equation use Lemma 6.2.6; to check maximality use Exercise 6.4.9.) (b) Show that assumptions 6.4.5a–c hold. (Hint: For div $\hat{T}_2 = 0$, use Proposition 3.15.3.) (c) Finally show by algebra that if b is chosen as in Proposition 6.4.5d, then $\rho_2 z = a \rho_1 z$.

EXERCISE 6.4.11

Show with a pocket calculator (and/or Taylor series) that, in Proposition 6.4.5d with $a = 10^{-4}$, $R(10^{-6} \text{ years})/R(10^{10} \text{ years}) \cong 10^{-10}$.

EXERCISE 6.4.12

(a) Let $(M, \mathcal{M}, z)$ be a simple cosmological model and let $\hat{T}$ be the stress-energy tensor of M. Show that $T(Z, Z) = 3t^{-2}$, where $t = (R/\dot{R})(u^4 z)$ is the predicted Hubble time (Exercise 6.3.17c). (b) In Definition 6.4.4, replace the assumption that M be a simple cosmological spacetime by the assumption that M be a Robertson–Walker spacetime of negative spatial curvature. Show that then $T(Z, Z) = 3at^{-2}$, where $t = (R/\dot{R})(u^4 z)$ is the predicted Hubble time (Exercise 6.3.17c) and a can be adjusted to have any value in the range $(0, 1)$ even if $\mathcal{M}$ is assumed to be a dust.

6.5 The early universe

We outline the most nearly standard current model for the history of the universe from the early epoch where, as one now believes, helium was formed by nucleosynthesis until the present. $(M, \mathcal{M}, z)$ is a simple cosmological spacetime throughout. What we shall actually use is the subset of M determined by 10^{-6} years $\leq u^4 \leq 10^{10}$ years.

> 10^{-6} years $\cong 30$ seconds. What happened still earlier—a big bang or some completely different behavior governed by as yet unknown physics—is irrelevant provided $\mathcal{M}$ obeys certain assumptions on our subset. Moreover, replacing M by some other appropriate Robertson–Walker spacetime leads to a similar model.

One main stimulus for current investigations of the early universe comes from the fact that near here-now there seems to be no mechanism that can produce the observed Planck spectrum of the microwave radiation (Section 6.1.9). We thus start with a famous result that indicates that this Planck spectrum may be a relic from a much earlier epoch.

Let us model the microwave radiation by a photon distribution function F on the spacetime M of a simple cosmological model $(M, \mathcal{M}, z)$. Assume F is conserved on $\mathbb{R}^3 \times \mathscr{E} \subset M$, where $\mathscr{E}$ is an interval in $(0, \infty)$. Discussions in Sections 5.7.6 and 6.4.1 indicate the motivations for, and limitations of, this assumption.

Proposition 6.5.1. *Suppose there exist constants $a \in \mathscr{E}$ and $\mathsf{T}_a \in (0, \infty)$ such that F is Planck with temperature T_a for $(x, \partial_4 x)$ whenever $x \in M$ and $u^4 x = a$. Then F is Planck with temperature $\mathsf{T} = [R(a)/R(u^4 y)]\mathsf{T}_a$ for $(y, \partial_4 y)$ whenever $y \in \mathbb{R}^3 \times \mathscr{E}$.*

Proof. Suppose $y \in \mathbb{R}^3 \times \mathscr{E}$ and $Y \in \mathscr{L}_y^+$. Let $\lambda : \mathscr{F} \to M$ be the inextendible freely falling photon for which $\lambda 0 = y$ and $\lambda_* 0 = Y$. Then there is a unique $b \in \mathbb{R}$ such that $u^4 \lambda b = a$, and $\lambda[0, b] \subset \mathbb{R}^3 \times \mathscr{E}$ (Exercise 6.4.8b). $\mathsf{F}(y, Y) = \mathsf{F}(\lambda b, \lambda_* b)$ since F is conserved in $\mathbb{R}^3 \times \mathscr{E}$. Now since F is Planck for $(\lambda b, \partial_4 \lambda b)$, $\mathsf{F}(\lambda b, \lambda_* b) = P(e'/k\mathsf{T}_a)$, where P is the Planck function (Example 5.5.4), k is the Boltzmann constant and e' is the energy $(\lambda b, \partial_4 \lambda b)$ measures for λ. Hence $\mathsf{F}(y, Y) = P(e'/k\mathsf{T}_a)$. But $g(\lambda_* b, \partial_4 \lambda b) = [R(u^4 y)/R(a)]g(\partial_4 y, Y)$ (Exercise 6.0.17d), so $e' = -g(\lambda_* b, \partial_4 \lambda b) = [R(u^4 y)/R(a)]e$, where e is the energy $(y, \partial_4 y)$ measures for λ. Thus with $\mathsf{T} = [R(a)/R(u^4 y)]\mathsf{T}_a$, $\mathsf{F}(y, Y) = P(e/k\mathsf{T})$. This holds $\forall Y \in \mathscr{L}_y^+$, so F is Planck for $(y, \partial_4 y)$ with temperature T. $\square$

Thus, if we can somehow explain why F should be Planck at some early cosmological time $u^4 = a$ and collision-free for $u^4 \geq a$, the proposition yields an explanation for the observed Planck spectrum. Now assuming F Planck for early times is plausible. At early times matter was denser (Lemma 6.4.3, Exercise 6.4.7); high densities favor collisions, which in turn tend to establish thermal equilibrium (Example 5.5.4c). But more details on the matter model are needed to make this explanation stick.

To get more details on the matter in the universe at early times, one needs at least a crude, preliminary estimate of the temperature in order to estimate what kinds of matter were present (cf. Section 6.5.2c). Then one can, in particular, see if our argument above for the Planck spectrum holds water.

To make such a crude estimate of the temperature we can, for example, use Propositions 6.4.5 and 6.5.1. Choosing the early time as $a = 10^{-6}$ years for definiteness, Proposition 6.5.1 gives the estimate $\mathsf{T}(10^{-6} \text{ years}) = [R(10^{10} \text{ years})/R(10^{-6} \text{ years})]\mathsf{T}(10^{10} \text{ years})$. Thus by the data in Section 6.1.9 and by Exercise 6.4.11, $\mathsf{T}(10^{-6} \text{ years}) \simeq 10^{10} \times 2.7$ Kelvins is our estimate. Considering how simple-minded the assumptions of Propositions 6.4.5 and 6.5.1 actually are, the estimate is tolerably accurate. The currently accepted value, obtained only after a long process of model-building and using a fair amount of microphysics (Weinberg [1]), is roughly a factor of 10 less.

> One intuitive reason the temperature is lower at early times in the more accurate estimate is that the latter takes into account the photons that have been emitted in collisions meanwhile. This gives a higher temperature now for a given early photon temperature or, equivalently, a lower photon temperature earlier for a given photon temperature now. By neglecting collisions above we have thus overestimated the early photon temperature.

6.5 The early universe

6.5.2 The primordial fireball

The reader will have noticed that trying to understand the early universe by building models bears an unfortunate resemblance to trying to read a book backwards. We now outline the main results, proceeding forward in time for vividness.

To do this concisely we shall have to postpone doubts and scruples till Section 6.6. For the moment, we assume that a simple cosmological model $(M, \mathscr{M}, z)$ applies and that the simplest version of the current party line on the early universe is substantially correct. Proofs will be omitted and, where convenient, terms such as "temperature," "anti-neutrinos," and so on, will be used without further explanation.

(a) To specify the simple cosmological spacetime $M \in (M, \mathscr{M}, z)$ completely, one finds the function R by integrating the Einstein field equation, using the matter model $\mathscr{M}$ much as in Proposition 6.2.7, Exercise 6.2.11, and Proposition 6.4.5. In practice, $\mathscr{M}$ is so complicated that computer integrations or the kind of rough estimates indicated by Exercise 6.5.3 following are used. For our purposes, it will suffice to recall that R is increasing. We now turn to $\mathscr{M}$.

(b) The following spacetime diagram summarizes the results. It shows M and the causal past of here-now, much as in Sections 6.0.12 and 6.3.12.

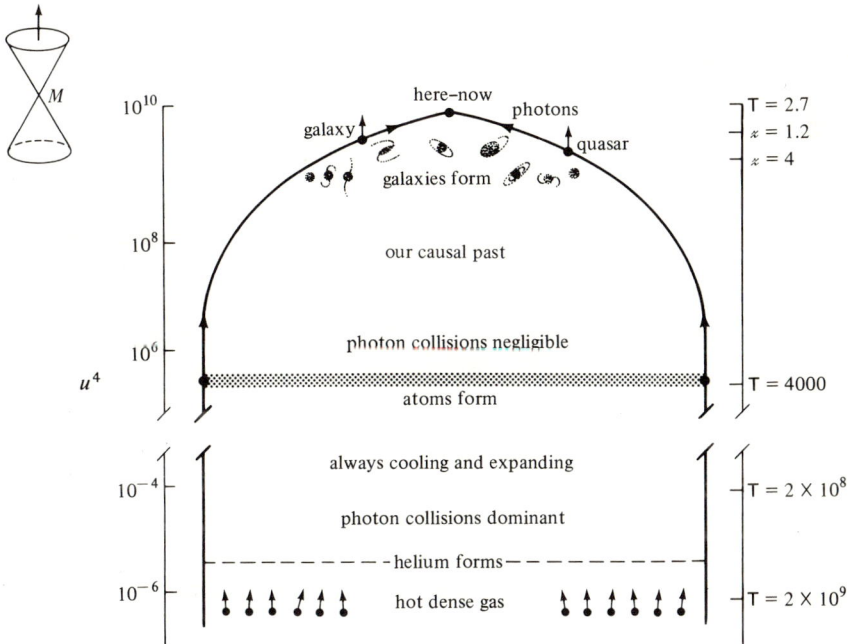

Figure 6.5.2b. The observable universe and the primordial fireball.

However, the vertical scale is taken to be proportional to $\log u^4$ rather than u^4 as otherwise all features of interest would be squeezed into an illegibly thin layer next to the big bang; the interval $10^{-4} < u^4 < 10^5$ has been shortened. The horizontal axis is u^3; two spatial dimensions are suppressed as before. For four temperatures, measured in Kelvins as always, and two red shifts, we have indicated the corresponding space slice.

(c) We begin our discussion of the diagram at the bottom, $u^4 = 10^{-6}$ years $\simeq \frac{1}{2}$ minute after the big bang. The matter is a hot, dense gas, the primordial fireball. Protons, neutrons, electrons, and photons are present. These are colliding very frequently and are all in thermal equilibrium. The temperature is about 2 billion Kelvins. Other particles are also present, notably neutrinos and antineutrinos.

In principle, the particles can combine. A proton and an electron can come together to form an ordinary hydrogen atom, with the proton as nucleus; a proton and neutron can combine to form a deuterium ("heavy hydrogen") nucleus: two protons and two neutrons can combine (in several steps) to form an ordinary helium nucleus; adding two electrons would then give an ordinary helium atom; and so on.

But at 2 billion Kelvins these compound bodies do not last long. For example, almost the instant a proton and neutron combine to form a deuterium nucleus they are blown apart again by a collision with, say, a high-energy photon. Thus the total amount of hydrogen atoms, helium nuclei, and so on, is negligible.

> It is in trying to settle such points that one needs crude preliminary estimates of the temperature before trying to choose a matter model.

(d) As u^4 increases, the temperature and the densities decrease, roughly in the way given by Proposition 6.5.1 and by Lemma 6.4.3 or Exercise 6.4.7. Now nucleosynthesis becomes possible. For at lower temperatures and densities, collisions are less violent and less frequent. Deuterium and helium nuclei have a higher life expectancy. In fact, at about $u^4 = 3$ minutes many helium nuclei are formed and contrive to survive until further cooling and expansion end the danger of their wholesale destruction. This is the helium creation epoch, sketched on the diagram.

Calculations show, remarkably enough, that shortly after this epoch, say at $u^4 = 10^{-5}$ years $\simeq 5$ minutes, the proportion of helium nuclei to protons is about that actually observed near here-now (Weinberg [1], Wagoner [1]). The gas now also contains electrons, photons, small amounts of deuterium and other nuclei, and other particles. Cosmological nucleosynthesis stops at this point. Much later, stars form, and in their interiors one again has the extreme conditions required for element cooking; but that is another story.

If one allows an extra parameter, one can of course fit the nuclear abundance data more closely. Once again, the data give preference to the Robertson–Walker spacetimes of negative spatial curvature over those of positive spatial curvature (Wagoner [1]).

(e) Somewhat later, at $u^4 = 10^{-4}$ years, the temperature has dropped to a few hundred million Kelvins. Atoms cannot withstand such temperatures, though nuclei can. Thus the electrons are still wandering around freely, rather than being captured by nuclei to make atoms. Collisions between these free electrons and protons are still so frequent that the photons must have a Planck spectrum (Sunyaev [1]; cf. Example 5.5.4c).

(f) However, the mixture continues to expand and cool. After a comparatively long time, when u^4 is about three hundred thousand years, as shown on the diagram, hydrogen and helium atoms form. Thereafter, say at $u^4 = 10^6$ years, the scarcity of free electrons means that collisions involving photons become and remain negligible as far as their effect on the photon distribution function is concerned. Thus, in view of Proposition 6.5.1, we can explain the observed Planck spectrum (Section 6.1.9) if we can explain why the photon distribution function should be Planck at $u^4 = 10^6$ years. In view of our comments in (e), it only remains to discuss the period between $u^4 = 10^{-4}$ years and $u^4 = 10^6$ years, during which collisions involving photons are neither completely dominant nor wholly negligible. That requires a separate, rather detailed, argument (Sunyaev [1]) whose outcome is favorable for the proposed explanation. Thus our big bang model does furnish a reason for the observed Planck spectrum. The observed numerical value of slightly more than 2.7 degrees is consistent with the nuclear abundance argument (d).

The whole argument is by no means airtight. But, as often happens in astrophysics, the absence of plausible alternatives lends it force.

(g) One can ask, on this model, how far back in time and out in space we are looking when we measure the microwaves. The answer is indicated by the two heavy dots on the space slice labelled T = 4000 in the diagram. These dots represent a 2-sphere. In effect, we are looking from inside at the surface of the primeval fireball, behind which (in time) lurks the big bang.

(h) In our picture, galaxies and quasars form from the hydrogen and helium gas at a cosmological time of about 1 billion years (cf. Section 6.3.15 for the evidence on this point). We have shown a quasar about as distant-early as any whose red shift has been observed as well as the galaxy 3C295 mentioned in Section 6.1.5.

In all this awesome panorama, only a tiny region near here-now is known to contain beings capable of compassion, reason, and the shock of recognition. Let us hope that, despite our insanities, the region extends further into the future than current estimates suggest.

6 Cosmology

Exercise 6.5.3

Let $(M, \mathcal{M}, z)$ be a simple cosmological model. Lemma 6.4.3 gave qualitative results on matter by using the limiting cases $p = 0$ and $p = \frac{1}{3}\rho$. A similar result holds for spacetime, as follows. Let $\mathcal{M}$ be a perfect fluid. Denote the subset $u^4 \leq u^4 z$ of M by N. Recall that $t = (R/\dot{R})(u^4 z)$ denotes the predicted Hubble time (Exercise 6.3.17c). Show that on N,

$$\left(\frac{3u^4}{2t}\right)^{2/3} \leq \frac{R \circ u^4}{Ru^4 z} \leq \left(\frac{2u^4}{t}\right)^{1/2},$$

where the first (respectively, second) equality holds on N iff $\mathcal{M}$ is a dust (respectively, rest-mass zero perfect fluid) on N.

6.6 Other models

We briefly discuss those further models that can be motivated as attempts to cure some specific disease of the simple cosmological models. In principle, this will carry our plan (Section 6.3.1) of gradually getting a more and more accurate overall picture of our universe to the limits set by the gaps in our current theoretical and empirical knowledge of microphysics and macrophysics. However, for brevity we will be quite vague at this stage.

6.6.1 Robertson–Walker models

When more data come in, it will eventually be necessary to use models that have extra adjustable parameters. For example, in all simple cosmological models the predicted Hubble time t is related to the predicted value ρ_0 of the total energy density an actual observer measures by $\rho_0 = 3t^{-2}$ (Exercise 6.4.12a). Sooner or later ρ_0 and t will both be measured to such high accuracy that the idealization $\rho_0 t^2 = 3$ will become inappropriate (no one in his right mind expects exact equality). Then one will need at least one extra parameter. Similar comments apply to the other data. Models using Robertson–Walker spacetimes are the natural generalization if one wants exactly one further adjustable parameter (cf. Exercise 6.2.14b).

> The negative spatial curvature case is at the moment favored over the positive spatial curvature case by a number of independent results; see Sections 6.3.2, 6.3.10, and 6.5.2d. As yet, this is very far from decisive; compare the comments in Rees [1].

6.6.2 Electromagnetic fields

Galaxies and quasars have magnetic fields that may be cosmological, rather than local, in origin. Thus models that drop our assumption (Section 6.2.4) that $F = 0$ continue to receive a little attention.

6.6.3 Inhomogeneities

The actual universe has no exact symmetries; for example, there are fairly big clumps of galaxies. But in typical cosmological models, the spacetime has a nontrivial isometry group. Various attempts have been made to cure this serious disease. Elegant global theorems, applicable to more nearly generic spacetimes, have been found (Hawking and Ellis [1]). And linearized perturbations without symmetries have been studied in great detail (cf. Weinberg [1], Fischer and Marsden [1]). The linearized perturbation theory has proved valuable—for example, in the discussion of the origin of galaxies (see Longair [1], especially the surveys by Silk and by Doroshkevich *et al.*, and see Jones [1]).

In addition, many cosmological spacetimes whose isometry group, though still nontrivial, is at least smaller than the isometry group of Proposition 6.0.5 have been analyzed (MacCallum [1]; Ryan and Shepley [1]). Most of these are of interest primarily as mathematical examples.

6.6.4 Other matter models

As emphasized earlier, our empirical information on the matter content of the universe has serious gaps, so alternate matter models must be analyzed. For example, detecting neutrinos is very difficult (cf. Section 3.8). Even if the energy density due to neutrinos is larger here-now than all the other contributions mentioned in Section 6.1.10c, we would not necessarily be able to detect the neutrinos directly via collision experiments. If neutrinos are dominant near here-now, we should be using the pure radiation universe of Exercise 6.2.11 and Section 6.3.17 or the model of Proposition 6.4.5 with $a > 1$ or some similar model.

> However, models of the very early universe, obtained for example by considering $u^4 \simeq 10^{-12}$ years in a model similar to that discussed in Section 6.5.2, suggest that the neutrino contribution to the energy density $T'(Z, Z)$ in Section 6.1.10 is even smaller than the photon contribution $T_p(Z, Z)$ in Section 6.1.10 (cf. Weinberg [1]). Thus if standard extrapolations backward in time are even roughly accurate, neglecting the neutrino contribution near here-now, as done in the text, is a highly accurate approximation.

Many similar uncertainties have been discussed, particularly as regards the matter content of the early universe, and corresponding matter models diffcrent from that outlined in Section 6.5.2 have been investigated.

> Some of these models assume that in a sufficiently large region the number of anti-protons equals the number of protons, though the latter predominate near here-now; others postulate that at early times much of the matter was in the form of "superbaryons"—particles whose baryon number is much larger in magnitude than one (see Section 3.8 for a brief description of baryon number); and so on (see Weinberg [1] and Wagoner [1] for brief comments and further references). One must also consider models in which black holes form the dominant matter component at early times though there are some indications to the contrary (cf. Carr [1]).

6.6.5 Quantum effects

For hot, dense matter quantum effects are dominant. Moreover, recall that Planck's quantum constant has the value $h = (6 \times 10^{-43} \text{ seconds})^2$. One believes that if the scalar curvature S (or some other curvature measure such as $G(\partial_4, \partial_4)$ in a simple cosmological spacetime) becomes as large as h^{-1}, quantum gravitational effects should become important. On both counts the nonquantum assumptions used in defining simple cosmological models become physically unrealistic near the big bang these models predict. For this reason, even if for no others, the models are inapplicable for u^4 sufficiently small.

The use of quantum matter models (DeWitt [1]) and/or quantum gravitational models (cf. Misner [1]) is the indicated cure for this disease; as yet no definitive results have emerged.

Finally, we mention that there are many cosmological models that are motivated by other considerations and many more that are not motivated by any physically plausible arguments.

6.7 Appendix: luminosity distance in the Einstein–de Sitter model

For completeness, we now outline how the relation $d_L = \imath d$ of Proposition 6.3.8 for luminosity distance d_L, which we used in comparing the Einstein–de Sitter model to observations, can be proved. $(M, \mathcal{M}, z)$ is the Einstein–de Sitter model throughout this section.

We shall need a model for a galaxy emitting light, and choose the simplest detailed one available. Intuitively speaking, our model will consist of a point source sending out photons of a given color equally in all directions.

> Using a galaxy model corresponding to an extended body is possible, though much less convenient, and gives equivalent results.

6.7.1 An emitting galaxy; assumptions

The reader may find the example of Section 6.7.2 helpful in reading the present section.

(a) We regard the history of the galaxy as an integral curve γ of ∂_4 (cf. Sections 6.1.3a and 6.2.5). To take into account the fact that the galaxy was presumably formed at some time, rather than having been around ever since the big bang, we assume γ has a past endpoint $x \in M$ (as defined in Section 3.8). For convenience we do not take the image of γ to include x (though the closure must) and we assume γ has no future endpoint. To model the photons emitted we shall use a photon beam—that is, a particle flow (P, η) of rest-mass zero.

(b) P and η will be defined only on the region $\mathcal{U}$ obtained by deleting the image of γ from the chronological future of x (shown dotted in the figure).

6.7 Appendix: luminosity distance in the Einstein–de Sitter model

Indeed, this open region consists of points to which γ can actually send freely falling photons (cf. Exercise 6.0.15). The excision of γ is also needed mathematically, to insure that P of Proposition 6.7.3 is a well-defined vector field (also compare the figure).

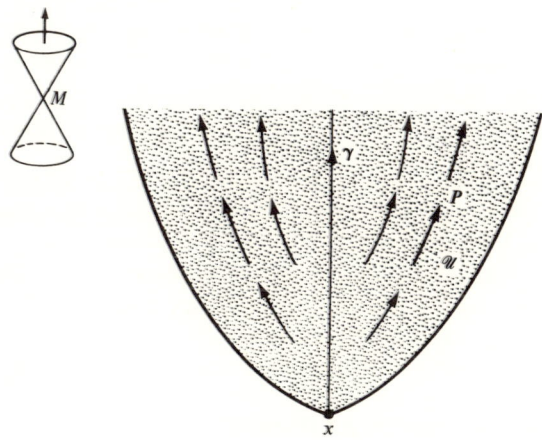

(c) γ models the emitter, so we insist that each inextendible integral curve of P have a past endpoint on γ (this is consistent with (b); cf. Section 3.8).
(d) Since we are dealing with photons, we demand $D_P P = 0$ (motivation in Section 5.1.1).
(e) We demand that photons be conserved after they are emitted—that is, $\operatorname{div}(\eta P) = 0$. This assumption is known to be slightly unrealistic, since in practice, for example, clouds within our own galaxy will absorb some of the photons before they reach here-now. However, a half century of hard observational work indicates that to take absorption into account one need simply make certain standard corrections to the photon-conserved model; thus we can afford to ignore this point.
(f) The requirement that the emission be the same in all directions can be formalized by requiring that for each spatial rotation ψ around x (as defined in Section 6.0.5g) $\psi_* P = P$ and $\eta \circ \psi = \eta$; note here that $\psi \mathcal{U} = \mathcal{U}$.
(g) Finally, the condition that each emitted photon have the same color can be imposed by requiring that, for some $e \in (0, \infty)$, $\lim_{n \to \infty} -g(P x_n, \partial_4 x_n) = e$ whenever $x_n \in M$ and $\lim_{n \to \infty} x_n$ exists and is a point on (the image of) γ. Here e is interpreted physically as the energy γ measures at emission since $\partial_4 x_n$ approaches the tangent to γ at $\lim_{n \to \infty} x_n$; that e is the same at each past endpoint in (c) means that at emission each photon has the same color (cf. Section 5.1).

Given x and e, these conditions fix γ and P uniquely (Exercise 6.7.6). However, they do not fully determine η. Intuitively, the reason is that we have not yet specified how many photons γ emits during any one given second of γ's proper time: have not yet said how bright γ is intrinsically at any instant. Now suppose we count all the photons present at cosmological time b—that is, in the level surface $u^4 = b$, and also count these present at a slightly later cosmological time $b + \varepsilon$. Since photons are conserved, it is intuitively plausible that the difference is the number of photons γ emitted during the interval ε and the product of this difference with the energy e of one photon is the total energy γ sent out during that interval, a measure of γ's absolute luminosity.

More formally, for each $u \in (a, \infty)$, let $\mathscr{B}_u$ be the intersection of the level surface $u^4 = u$ with the region $\mathscr{U}$ defined in part (b) above. Then (Section 3.2) the number $N(u)$ of photons in $\mathscr{B}_u$ is $N(u) = -\int_{\mathscr{B}_u} i(\eta P)\Omega$. We require that this (improper) integral exist (cf. Section 5.6) and be a smooth function of u. We define the *absolute luminosity of γ at proper time* $u \in (a, \infty)$ to be $L(u) = e dN/du$. By the heuristic discussion above, $L(u)$ indeed coincides with the absolute luminosity as described verbally in 5.5.5. Exercise 6.7.7 discusses the identification more systematically.

We now give an example of our model for a galaxy emitting light.

6.7.2 An emitting galaxy: an example

Suppose $\gamma: (a, \infty) \to M$ is given by $\gamma u = (0, 0, 0, u)$. Thus our requirement on γ in Section 6.7.1a holds with $x = (0, 0, 0, a)$. In Section 6.7.1b, $\mathscr{U}$ is determined by: $y \in \mathscr{U}$ iff $u^4 y > u^4 x$ and $0 < \delta(x, y) < 3[(u^4 y)^{1/3} - (u^4 x)^{1/3}]$ (Exercise 6.0.15). One obtains an example of a photon beam (P, η) obeying the conditions in Section 6.7.1 by integrating the equations $D_P P = 0 = \text{div}(\eta P)$ subject to suitable auxiliary conditions, much as in Proposition 3.2.3, Theorem 3.8.3, and Theorem 3.11.2. Rather than go through the whole process once more, we pull the final result out of a hat and check that it works—that is, give an existence proof. The integration process, disguised as a uniqueness proof, will be left as Exercise 6.7.6. To state the final result, we will use two auxiliary functions on $\mathscr{U}$: $r = [\sum_{\rho=1}^{3} (u^\rho)^2]^{1/2}$ and $T = (u^4)^{1/3} - r/3$. Note that on $\mathscr{U}$, both functions are C^∞, that the image of $\mathscr{U}$ under r is $(0, \infty)$ and that the image of $\mathscr{U}$ under T is $(a^{1/3}, \infty)$.

> From a physical point of view, T^3 will act as a kind of "retarded time" or "phase" (somewhat analogous to the phase ϕ of an electromagnetic plane wave Section 3.7.3) in the sense that an instantaneous observer (x, X) who measures the particle flow gets information on the behavior of the source γ, not at time $u = u^4 x$, but at γ's earlier proper time $u = T^3 x$.
>
> r is merely δ of Section 6.0.12 with one argument held fixed, and the comments there on δ also apply to r.

Proposition 6.7.3. *Let x, γ, $\mathscr{U}$, r and T be as above, let $L = (a, \infty) \to (0, \infty)$ be a C^∞ function, and suppose $e \in (0, \infty)$. On $\mathscr{U}$, define the function $\eta = (1/4\pi e^2)(L \circ T^3)(u^4)^{-4/3} r^{-2}$ and define $\boldsymbol{P}$ as the vector field physically equiv-*

6.7 Appendix: luminosity distance in the Einstein–de Sitter model

alent to $-d(eT^3)$. Then (P, η) is a photon beam on $\mathscr{U}$ which obeys the requirements Section 6.7.1c–g; moreover, $L(u)$ is the absolute luminosity of γ at proper time u.

Proof. From the definition of r, $dr = r^{-1} \sum_{\rho=1}^{3} u^\rho du^\rho$. Using the form of g we find that the vector field physically equivalent to dr is $(u^4)^{-2/3}\partial$, where $r\partial = (u^4)^{-2/3} \sum_{\rho=1}^{3} u^\rho \partial_\rho$ and thus $g(\partial, \partial) = 1$. The definition of P now gives $P = eT^2(u^4)^{-2/3}(\partial_4 + \partial)$. Thus $g(P, P) = 0$. Thus P is a lightlike future-pointing vector field on $\mathscr{U}$; since in addition η is a smooth, nonnegative function $\mathscr{U}$, (P, η) is a photon beam on $\mathscr{U}$ as asserted. We now check the conditions of Section 6.7.1.

Exercise 1.0.5 shows directly that the geodesic condition (d)—that is, $D_P P = 0$, holds. To prove the spatial isotropy condition (f) note that for each spatial rotation ψ about x, $u^4 \circ \psi = u^4$ (Section 6.0.5a) and $r \circ \psi = r$ (Section 6.0.12a). Thus we also have $T \circ \psi = T$ and $\eta \circ \psi = \eta$ as required; moreover $T \circ \psi = T$, $\psi^* dT = dT$ and $\psi^* g = g$ give $\psi_* P = P$ in view of the definition of P as physically equivalent to $d(eT^3)$. Thus Section 6.7.1f holds. To verify that the "energies equal" condition 6.7.1g holds, note that $-g(\partial_4, P) = eT^2(u^4)^{-2/3}$ on $\mathscr{U}$ and that whenever we approach the image of γ from within $\mathscr{U}$, $r \to 0$ and thus $T \to (u^4)^{1/3}$. Hence $-g(\partial_4, P) \to e$ as required. Checking the remaining two conditions (Section 6.7.1c and e) is a bit harder.

To prove the conservation condition $\operatorname{div}(\eta P) = 0$, we first note the auxiliary result that $(\partial_4 + \partial)T = dT(\partial_4 + \partial) = [(u^4)^{2/3}/e^2 T^4]g(P, P) = 0$. Applying Exercise 3.6.4d several times we now have

$4\pi e \operatorname{div}(\eta P)$
$= \operatorname{div}[(L \circ T^3)T^2(u^4 r)^{-2}(\partial_4 + \partial)]$
$= (u^4 r)^{-2}(\partial_4 + \partial)[(L \circ T^3)T^2] + (L \circ T^3)T^2 \operatorname{div}[(u^4 r)^{-2}(\partial_4 + \partial)]$
$= 0 + (L \circ T^3)T^2\{(u^4)^{-2}\partial_4[r^{-2}] + r^{-2}\operatorname{div}[(u^4)^{-2}\partial_4]$
$\qquad + r^{-2}\partial[(u^4)^{-2}] + (u^4)^{-2}\operatorname{div}(r^{-2}\partial)\}$
$= T^2(L \circ T^3)r^{-2}\operatorname{div}[(u^4)^{-2}\partial_4] + T^2(L \circ T^3)(u^4)^{-2}\operatorname{div}(r^{-2}\partial).$

We claim both of these last two divergences vanish. In fact $\operatorname{div}[(u^4)^{-2}\partial_4]$ $= \partial_4[(u^4)^{-2}] + (u^4)^{-2}\operatorname{div}\partial_4 = -2(u^4)^{-3} + 2(u^4)^{-3} = 0$ where we have used Lemma 6.2.6c. Similarly

$d[i(r^{-2}\partial)\Omega]$
$= d[r^{-3}(u^1 du^2 \wedge du^3 - u^2 du^1 \wedge du^3 + u^3 du^1 \wedge du^2) \wedge du^4] = 0$

by our earlier expression for dr, so $\operatorname{div}(r^{-2}\partial) = 0$. Combining, we have $\operatorname{div}(\eta P) = 0$ as required.

To check the condition in Section 6.7.1c, let λ be an inextendible integral curve of P in $\mathscr{U}$ and we must show λ has a past endpoint. Because P is spatially isotropic, we may use a rotation around x to reduce considerations to the case: $u^1(\lambda t) = u^2(\lambda t) = 0$ and $u^3(\lambda t) > 0$ for some $t \in \mathbb{R}$. An explicit integration of P then shows: there exists $b \in (a^{5/3}, \infty)$ such that $\lambda: (b, \infty) \to M$ and $\lambda s = (5eb^{2/5}/3)(0, 0, 3s^{1/5} - 3b^{1/5}, s^{3/5})$. Thus λ has the past endpoint $(0, 0, 0, b^{3/5})$ on γ, as required.

Thus our example obeys all the conditions in Section 6.7.1. It remains to identify L as the absolute luminosity. Notation as in Section 6.7.1, fix $u \in (a, \infty)$ and let $\mathscr{B}_u$ be the intersection of the level hypersurface $u^4 = u$ with $\mathscr{U}$. Denoting the inclusion $\mathscr{B}_u \to M$ by τ, we have $\tau^* du^4 = 0$. Thus

$$\tau^*[i(\eta P)\Omega] = (-4\pi e)^{-1}(L \circ T^3) \cdot T^2 r^{-2} du^1 \wedge du^2 \wedge du^3$$

6 Cosmology

by Lemma 6.2.6c. Using the natural area form on $\mathcal{S}^2$ (Section 0.0.9) we now have

$$N(u) = -\int_{\mathcal{B}_u} i(\eta P)\Omega$$

$$= (4\pi e)^{-1} \int_{\mathcal{B}_u} (L \circ T^3) T^2 r^{-2} du^1 \wedge du^2 \wedge du^3$$

$$= (4\pi e)^{-1} \int_{\mathcal{S}^2} \zeta \int_0^{3u^{1/3} - 3a^{1/3}} (u^{1/3} - \tfrac{1}{3}r)^2 L([u^{1/3} - \tfrac{1}{3}r]^3) dr$$

$$= (3/e) \int_{a^{1/3}}^{u^{1/3}} v^2 L(v) dv.$$

Hence $edN/du = (3/e)(e/3)u^{-2/3}u^{2/3}L(u) = L(u)$. □

6.7.4 The Newtonian limit

In our example, the stress-energy tensor of the particle flow (P, η) is $\hat{T} = \eta P \otimes P = (4\pi)^{-1}(L \circ T^3)T^4(u^4)^{-4/3}\mathscr{d} \cdot [(\partial_4 + \partial) \otimes (\partial_4 + \partial)]$, where $\mathscr{d} = (u^4)^{2/3}r$ is the distance analyzed in Proposition 6.3.4. Hence the energy density $\ell = T(\partial_4 x, \partial_4 x)$—that is, the apparent luminosity, measured by any comoving instantaneous observer $(x, \partial_4 x)$ on $\mathcal{U}$—is $\ell = (4\pi)^{-1}(L \circ T^3) \times [(u^4)^{-1/3}T]^4 \mathscr{d}^{-2}$. Note that as we approach γ—that is, for an instantaneous comoving observer very near the source—the following computational and conceptual simplifications occur: $T \to (u^4)^{1/3}$; $(u^4)^{-1/3}T \to 1$; $(x, \partial_4 x)$ is nearly at rest with respect to γ; and $\mathscr{d}$ can be conceptually identified with Newtonian distance d_N. Thus our general relativistic model duly recaptures the Newtonian result $\ell = L/4\pi d_N^2$ in the appropriate limit—that is, when relative velocity and spacetime curvature can be neglected.

We are now finally in a position to discuss the law with which the Einstein–de Sitter model replaces this Newtonian one. Suppose the actual observer (z, Z) measures the apparent luminosity ℓ of a distant-early giant elliptical galaxy, modelled as we have just indicated. Without essential loss of generality, we may use the model of Proposition 6.7.3 directly (Exercise 6.7.6), taking $z \in \mathcal{U}$. Suppose the absolute luminosity L evaluated at $(Tz)^3$ is known (Sections 6.1.2 and 6.1.6). Writing L_0 for $L([Tz]^3)$, we now get the following result, equivalent to Proposition 6.3.8a via the definition $\ell = L/(4\pi d_L^2)$ in Section 6.1.6a.

Theorem 6.7.5. $\ell = (L_0/4\pi)[\imath d(\imath)]^{-2}$.

PROOF. Let y be the point on γ such that there is a light signal from y to z. Then $\imath = (u^4 z/u^4 y)^{2/3}$ and the function $r(z)$ of Proposition 6.7.3 $= \delta(y, z) = 3[(u^4 z)^{1/3} - (u^4 y)^{1/3}]$—that is, $T(z) = (u^4 y)^{1/3}$. We now have from Section 6.7.4 and Proposition 6.3.4 that

$$\ell = \{(L_0/4\pi)[(u^4)^{-1/3}T]^4 \mathscr{d}^{-2}\}(z) = (L_0/4\pi)(u^4 x/u^4 z)^{4/3} \mathbf{d}^{-2}(\imath)$$
$$= (L_0/4\pi)[\imath d(\imath)]^{-2}. \quad \square$$

6.7 Appendix: luminosity distance in the Einstein–de Sitter model

EXERCISE 6.7.6

Show that the photon beam of Proposition 6.7.3 is actually the general case in the following sense. If (γ', P', η', e) is a quadruple that obeys the assumptions in Section 6.7.1, then there is an isometry ψ such that (γ, P, η, e), defined to be $(\psi \circ \gamma, \psi_* P', \eta' \circ \psi, e)$, is given by the explicit equations of Proposition 6.7.3.

EXERCISE 6.7.7

The method we used at the end of Section 6.7.1 to identify the absolute luminosity physically depends in a crucial way on the assumption (Section 6.7.1e) that photons are conserved after emission. We give an alternate identification method, which is easier to generalize to the case that photon collisions are allowed. Specifically, show that $L(u) = 4\pi \lim [d^2 T(X, X)]$ where: (a) the limit is taken for any sequence of instantaneous observers (x, X) in $\mathcal{U}$, with x approaching the point y on γ for which $u^4 y = u$ and X approaching the tangent to γ at y; (b) d is the "spatial distance" from x to y—in the limit it will not matter which concept of spatial distance in Section 6.1.6 is used.

7 Further applications

7.0 Review and notation

Throughout this chapter, $(\mathscr{S}^2, h, \zeta)$ is the unit sphere in $\mathbb{R}^3$ (Section 0.0.9) and (M, g, D) is a spacetime.

Suppose $x \in \mathscr{S}^2$. Then there is precisely one Killing vector field K_x on $\mathscr{S}^2$ such that: (a) $K_x x = 0$. (b) Let η be the 1-form physically equivalent to K_x with respect to h—that is, $\eta(X) = h(K_x, X) \,\forall X$ tangent to $\mathscr{S}^2$. Then $d\eta$ is consistent with the orientation of $\mathscr{S}^2$—that is, $d\eta = f\zeta$ for some nonnegative function f on $\mathscr{S}^2$. (c) max $h(K_x, K_x) = 1$, where the maximum is taken over all points of $\mathscr{S}^2$.

This K_x corresponds, up to sign, to rotations around the axis determined by x and the origin of $\mathbb{R}^3$. For example, let $I: \mathscr{S}^2 \to \mathbb{R}^3$ be the inclusion and suppose $Ix = (1, 0, 0)$. Then $I_* K_x = (u^2 \partial_3 - u^3 \partial_2) \circ I$. We call K_x the *generator of rotations around the x-axis*.

Exercise 7.0.1

Suppose $x \in \mathscr{S}^2$ and η is as above. Show that there exist $y, z \in \mathscr{S}^2$ such that $h = \eta \otimes \eta + \zeta \otimes \zeta + \theta \otimes \theta$, where ζ, θ are, respectively, the 1-forms physically equivalent to K_y and K_z.

Exercise 7.0.2

Let (N, g, D) be a normal Schwarzschild spacetime (Section 1.4.2) and let $P: N \to \mathscr{S}^2$ be the projection. Let $\gamma: \mathscr{E} \to N$ be a curve such that, for every generator of rotations K as above, $h(P_* \gamma_*, K \circ P \circ \gamma) = $ constant. Show that the curve $P \circ \gamma: \mathscr{E} \to \mathscr{S}^2$ lies in a great circle.

7.1 Preview

In this chapter we briefly discuss applications of general relativity to the solar system, to black holes, and to gravitational waves. As in Chapter 6, we analyze simple models intensively, rather than try to survey all models of current interest.

Each topic we shall discuss is related, directly or indirectly, to stars. Sections 7.2 to 7.4 are applicable mainly to a normal star, such as our sun. Section 7.5 concerns "collapsed" stars. Possible sources for the gravitational waves analyzed in Section 7.6 are very dense stars in violent motion.

To get reasonably realistic models—Newtonian or relativistic—for the history of the inside of a star requires a more sophisticated analysis of matter than we have given in Chapter 3. Such models are discussed by Weinberg [1] and by Misner, Thorne, and Wheeler [1]. We shall content ourselves with models for the history of the region outside a star. To place our models in perspective, we now informally outline current ideas on stellar life cycles and end-states.

7.1.1 *The history of a star*

A star is born when a large cloud of gas pulls itself together gravitationally. As the cloud gets denser, it heats up. Eventually, it gets so hot and dense that nuclear burning begins. Usually the most important nuclear reaction is the conversion of hydrogen to helium, mentioned in Chapter 6.

The star now settles down in a state of near equilibrium. It is held together by gravity. It is held apart by pressure effects, which arise because the core is very hot and dense. Like most observed stars, our sun is in such a near-equilibrium state. It will remain so for another 6×10^9 years approximately. In some stars one can directly see the above conflict between gravity and pressure: the star throbs periodically as one side or the other temporarily gains the upper hand.

Now a star in such a near-equilibrium state is living on borrowed time. Photons and other particles escape from its surface. This tends to cool the star and thus decrease the pressure effects mentioned. Nuclear burning is required to prevent cooling and maintain pressure. Sooner or later the normal nuclear fuels run out, while the gravity never gives up. Then what?

Sometimes, in fact probably most of the time, the result is very violent. The star begins to collapse rapidly and its temperature rises. New kinds of nuclear fuel, hitherto too cool or dilute to burn, catch fire. Flinging these last resources into the fight, the star explodes. In some cases the explosion remains as bright as a galaxy of 10^{11} normal stars for several days.

But the gravity is still there. Even if the star has fragmented, each fragment will collapse under its own gravity unless it somehow contrives to hold itself apart. Thus we are led to discuss stellar end-states: states that are either in true equilibrium or at least do not change significantly in a time of 10^{10} years.

7 Further applications

7.1.2 Stellar end-states

The end-state of a star or fragment is determined primarily by three kinds of effects: gravitational effects, electromagnetic interactions between individual particles in the star, and a quantum effect known as the Pauli exclusion principle. Roughly, the gravity tends to collapse the object while the other two effects tend to keep it apart. If the object is very dense, nuclear forces (cf. Section 0.1) also play an important role whose details are currently not wholly understood. In a detailed analysis one must also take into account the rotation of the object and the net electric charge on it. However, throughout this chapter, we confine attention to the case of negligible rotation and negligible net electric charge.

The parameter that determines the end-state is then the active mass of the object. An object more massive than the moon but not much more massive than Jupiter simply becomes a "planet" with a density comparable to the density of the earth. For active masses considerably larger than that of Jupiter and less than 3/2 that of the sun the typical end-state is a white dwarf star, with a density perhaps 10^6 the density of water. The gravity just outside a white dwarf star is sufficiently intense that discrepancies of about one part in 10^4 between Newtonian models and general relativistic models arise.

For active masses somewhat larger than that of the sun the end state may be a neutron star, such as the pulsar at the center of the Crab nebula. The density is about 10^{15} the density of water, comparable to the density of an atomic nucleus; the radius is perhaps 10 miles ($\simeq 5 \times 10^{-5}$ seconds). As far as is known, the best available model for the exterior of an irrotational neutron star with no net electric charge is an appropriate submanifold of a normal Schwarzschild spacetime. Newtonian models are not very useful.

Suppose the active mass is twice that of the sun, or more. Current nonquantum models predict that a black hole must be formed. Roughly speaking, the gravity wins; the star collapses to densities beyond the scope of current physics, perhaps to infinite density. No signals can escape from it. In the models of Section 7.5, the star's only influence on the rest of the universe is its gravitational field, left behind like the grin of the Cheshire cat. As we write, there is a vigorous controversy on whether or not black holes have been detected (cf. the article by Thorne and Zeldovich in DeWitt and DeWitt [1]).

We now turn to the general relativistic models.

7.2 Stationary spacetimes

Many of the gravitational fields important in physics are, intuitively speaking, time-independent. For example, the gravitational field of the earth is time-independent to the extent that we neglect vibrations of the earth and other small effects. Now, even an observer near the earth can arrange to measure a changing local geometry merely by, say, moving from a point very near the earth to a more distant point. Thus one cannot define time independence of

7.2 Stationary spacetimes

a gravitational field in terms of arbitrary observers. Instead, one characterizes time independence by postulating the existence of a reference frame whose observers do not experience any changes in the local geometry. The precise definition follows.

Let Z be a reference frame on M (Section 2.3.1). Z is defined as *stationary* iff there is a positive function f on M such that fZ is a Killing vector field (i.e., the Lie derivative $L_{fZ}g$ vanishes; cf. Section 3.6.3). Z is defined as *static* iff Z is stationary and irrotational. M is *stationary* (respectively, *static*) iff there exists on M a stationary (respectively, static) reference frame Z. An observer $\gamma: \mathscr{E} \to M$ is *stationary* iff he is in some stationary reference frame.

> Stationary spacetimes were originally defined as spacetimes such that $\partial_4 g_{ij} = 0$ in at least one coordinate system. Some coordinate versions of the above definitions are outlined in Optional exercise 8.4.9.
>
> While a stationary spacetime corresponds to the gravitational field generated by a time-independent source, "static" usually means that in addition the source does not rotate. The existence of a static reference frame is thus an extreme idealization (cf. Exercise 7.2.2a).

Let Z be a stationary reference frame and $\gamma: \mathscr{E} \to M$ be an observer in Z. Then, with f as above, $X = fZ$ is Killing, and $\gamma_* = Z \circ \gamma = [1/f \circ \gamma](X \circ \gamma)$. Suppose $u, \hat{u} \in \mathscr{E}$ and $\mathscr{U} \subset M$ is a sufficiently small open neighborhood of γu. Then there exists an element ψ of the flow of X whose domain contains $\mathscr{U}$ and such that $\psi \gamma u = \gamma \hat{u}$. Let $\phi = \psi|_{\mathscr{U}}$. Then $\phi: \mathscr{U} \to \phi \mathscr{U}$ is an isometry and $\phi_*(\gamma_* u) = \gamma_* \hat{u}$. In this sense γ always observes the same local geometry as his proper time increases and, intuitively, feels he is "not moving with respect to the gravitational field." We define a stationary reference frame on M as *absolute* iff there are no other stationary reference frames on M. We define an observer in M to be *at rest* iff γ is in an absolute, stationary reference frame. Unless M is quite special, no observer in M is at rest; but the special spacetimes are important in practice.

> You, the reader, probably do have the intuitive feeling you are at rest even though you are not (we hope) freely falling. In simple models of the earth, a person sitting in a room is in fact at rest in the technical sense defined above.

If there is an absolute, stationary reference frame on M, one can make precise the concept of "gravitational redshift," which is often used in physics. Let Z be a stationary reference frame, and let $\lambda: [a, b] \to M$ be a freely falling photon. Then the frequency ratio $\imath$ for $([\lambda], Z\lambda a, Z\lambda b)$ is $\imath = g(\lambda_* a, Z\lambda a)/g(\lambda_* b, Z\lambda b)$ (see Section 5.4). Now, in general, there may be many freely falling photons going from λa to λb with distinct worldlines (cf. Exercise 7.2.3). The following proposition shows that $\imath$ is in fact independent of these photons in question, and depends only on λa and λb.

7 Further applications

Proposition 7.2.1. $z = |X\lambda b|/|X\lambda a|$.

PROOF.

$z = g(\lambda_* a, Z\lambda a)/g(\lambda_* b, Z\lambda b) = (|X\lambda b|/|X\lambda a|) \cdot [g(\lambda_* a, X\lambda a)/g(\lambda_* b, X\lambda b)]$.
Since X is Killing and λ is geodesic, $(g \circ \lambda)(\lambda_*, X \circ \lambda) = $ constant (see Section 3.6.3). □

In particular, suppose Z above is absolute, $(x, y) \in M \times M$, and there exists at least one freely falling photon λ such that $\lambda a = x$ and $\lambda b = y$. Then the frequency ratio z above is uniquely determined by (x, y). $z - 1$ is then defined as the *gravitational redshift for x observed at y*.

> *Newtonian analogue.* In the Newtonian limit $z - 1$ is the gravitational potential difference between x and y (Optional exercise 9.3.3). If x is "lower down" than y, $z - 1$ is, roughly, the fractional energy loss the photon suffers as it climbs upward against gravity.

EXERCISE 7.2.2

Show: (a) Einstein–de Sitter spacetime is not stationary; (b) $(1 - 2\mu/r)^{-1/2} \partial/\partial t$ is a static reference frame on a normal Schwarzschild spacetime (Section 1.4.2).

EXERCISE 7.2.3

Let N be a normal Schwarzschild spacetime, and let $q \in \mathscr{S}^2$ and $\hat{q} \in \mathscr{S}^2$ be antipodal, and $x \in N$ be the point determined by $(tx, rx, Px) = (0, b, q)$. Show that there is a b such that for some $y \in N$: $ry = b$, $Py = \hat{q}$ and more than one photon worldline goes from x to y.

EXERCISE 7.2.4

Let Z be a static reference frame on a simply connected spacetime M. Show that there exists a function $h: M \to \mathbb{R}$ such that $Zh = 0$ and $g(D_Y Z, \cdot) = g(Y, Z)dh \; \forall$ vector field Y on M.

7.3 The geometry of Schwarzschild spacetimes

Throughout this section, the notation will be as in Section 1.4.2 except that we shall let U denote the disconnected manifold $N \cup B$. Thus $8\pi\mu$ is an active mass $\in (0, \infty)$; N is a normal Schwarzschild spacetime; B is a Schwarzschild black hole; $dr|_N$ is spacelike, while $-dr|_B$ is timelike future pointing.

In later sections we shall use Schwarzschild spacetimes to analyze the solar system and to discuss collapsed objects. The way in which they will be used is roughly indicated by Figure 7.3.1.

7.3 The geometry of Schwarzschild spacetimes

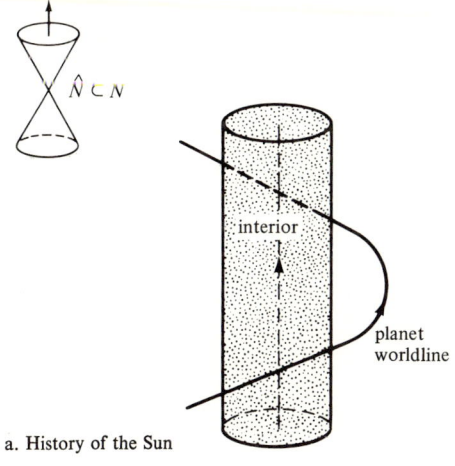

a. History of the Sun

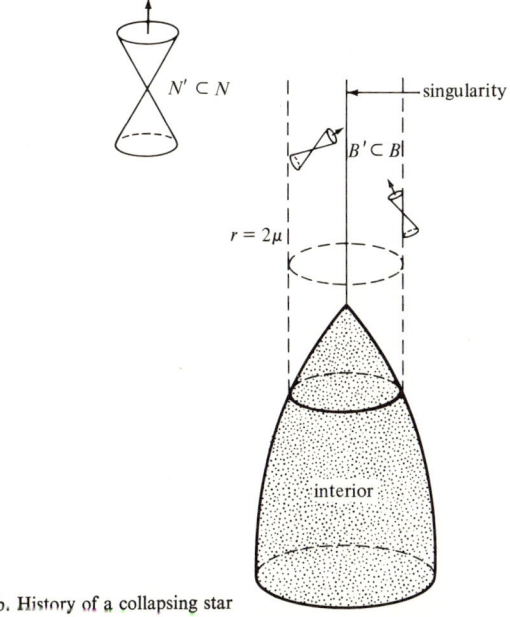

b. History of a collapsing star

Figure 7.3.1. Applications of Schwarzschild spacetimes.

In Figure 7.3.1a "Interior" indicates the history of the sun's interior, with the worldline of the center shown as a dotted line. The "cross sections" are diffeomorphic to the unit ball in $\mathbb{R}^3$; they are shown as discs. We shall use an open submanifold $\hat{N}$ of N to model the history of the vacuum region outside "Interior." Figure 7.3.1b indicates the history of a collapsing object. Our model for the unshaded region will be an open submanifold N' of N "glued

221

7 Further applications

onto" an open submanifold B' of B. In neither case will a model for "interior" be given.

This section analyzes the geometric properties of $U = N \cup B$. We need geometric interpretations of t and r to make precise our comment in Section 1.4.2 that, on N, t is a kind of time and r a kind of radius. We also need information on freely falling particles. As in Proposition 1.4.4, we start by computing curvatures.

Define a tensor field $j: U \to T_2{}^0 U$ by $j = g - r^2 P^* h$. Define a tensor field $T: U \to T_4{}^0 U$ as follows. $\forall x \in U$ and $\forall X_1, X_2, X_3, X_4 \in U_x$: $T(X_1, X_2, X_3, X_4) = (1/4)\{[j(X_1, X_3)(P^*h)(X_2, X_4) - (1 \Leftrightarrow 2)] - [3 \Leftrightarrow 4]\}$, where $(1 \Leftrightarrow 2)$ and $[3 \Leftrightarrow 4]$ denote interchanges—for example, if $F \in T_2{}^0(U_x)$, $F(X_1, X_2) - (1 \Leftrightarrow 2) \equiv F(X_1, X_2) - F(X_2, X_1)$. Then T has all the symmetry properties stated in Section 1.0.2a–c. Let $\hat{R}: U \to T_4{}^0 U$ be physically equivalent to the curvature tensor of U.

Proposition 7.3.2. $\hat{R} = (4\mu/r^3)[2(dr \wedge dt) \otimes (dr \wedge dt) - r^2 T + 2r^4(P^*\zeta) \otimes (P^*\zeta)]$.

> *Proof.* Since the reader has worked through each step of Proposition 1.4.4 an outline will suffice. Let (χ^1, χ^2) be an orthonormal basis of 1-forms on a nonempty open subset of $\mathscr{S}^2$. Then locally on N, $\omega^A = rP^*\chi^A$, $\omega^3 = (1 - 2\mu/r)^{-1/2} dr$, $\omega^4 = (1 - 2\mu/r)^{1/2} dt$, $A \in (1, 2)$, is an orthonormal basis of 1-forms. Section 1.0.3d,e imply that the nonzero connection forms are:
>
> $\omega_3{}^4 = \omega_4{}^3 = [(1 - 2\mu/r)^{1/2}]' \omega^4$; $\omega_A{}^3 = -\omega_3{}^A = -(1 - 2\mu/r)^{1/2} P^* \chi^A$;
>
> and $\omega_B{}^A = P^* \chi_B{}^A$, where $'$ denotes d/dr and $\{\chi_B{}^A\}$ are the connection forms for (χ^1, χ^2). Section 1.0.3b,c give, locally on N:
>
> $$\hat{R} = (4\mu/r^3)\Big\{ 2(\omega^4 \wedge \omega^3) \otimes (\omega^4 \wedge \omega^3) + \sum_{A,B=1}^{2} 2(\omega^A \wedge \omega^B) \otimes (\omega^A \wedge \omega^B)$$
> $$- \sum_{A=1}^{2} (\omega^4 \wedge \omega^A) \otimes (\omega^4 \wedge \omega^A) + (\omega^3 \wedge \omega^A) \otimes (\omega^3 \wedge \omega^A) \Big\}.$$
>
> The Proposition for N now follows. The proof for B is identical apart from appropriate sign changes. □

Corollary 7.3.3. *Both N and B are vacuum but not flat.*

PROOF. $\hat{R}(\partial/\partial t, \partial/\partial r, \partial/\partial t, \partial/\partial r) = (\mu/r^3)[dr(\partial/\partial r)dt(\partial/\partial t)]^2 = \mu/r^3 \neq 0$ by Proposition 7.3.2. Thus $\hat{R}|_N \neq 0 \neq \hat{R}|_B$, and neither N nor B is flat. Computing **Ric** by algebraic manipulations similar to those in Proposition 1.4.4 gives **Ric** $= 0$. By Exercise 1.4.7b, **Ric** $= 0$ implies that the Einstein tensor **G** vanishes. □

It seems intuitively clear that $r \to \infty$ corresponds to approaching spatial infinity on N. But Lorentzian manifolds are tricky and such intuitive arguments can be misleading, so we give a formal definition. In the notation of Section 3.6.1, define $f: U \to \mathbb{R}$ by $f = R^{ijkl} R_{ijkl}$. Roughly speaking, f

measures the overall magnitude of the curvature; even observers in an opaque box can determine f locally by relative acceleration measurements (Proposition 2.3.3). A particle or observer $\gamma: (a, b) \to M$ is said to *escape to infinity* iff $\lim_{u \to b} f\gamma u = 0$. Roughly, $z \in M$ is "near infinity" iff fz is "very small." The following corollary of Proposition 7.3.2 characterizes $r/\mu^{1/3}$ and shows, among other things, that γ above escapes to infinity iff $\lim_{u \to b} r\gamma u = \infty$.

Corollary 7.3.4. $f = 144\mu^2/r^6$.

PROOF. Let $\tilde{T}$ be the (4, 0)-tensor field physically equivalent on U to T in Proposition 7.3.2; let A be the (2, 0)-tensor field physically equivalent to $P^*\zeta$. Using the Lorentzian metric g we find that the (4, 0) tensor field $\tilde{R}$ physically equivalent to $\hat{R}$ on U is given by

$$\tilde{R} = (4\mu/r^3)[2(\partial/\partial r \wedge \partial/\partial t) \otimes (\partial/\partial r \wedge \partial/\partial t) - r^{-2}\tilde{T} + 2r^{-4}A \otimes A].$$

Taking the appropriate four traces of $\hat{R} \otimes \tilde{R}: M \to T_4^4 M$ gives the desired result. □

We now introduce a canonically determined static reference frame on N, characterize $dt|_N$, and characterize $(\partial/\partial t)|_N$. We emphasize that the following geometric and physical interpretations of $t|_N$ are not valid for $t|_B$. Let X be a vector field on N.

Corollary 7.3.5. *X is Killing and future pointing iff there exists an $a \in (0, \infty)$ such that $a\partial/\partial t = X$.*

PROOF. Suppose $X = a\partial/\partial t$, $a \in (0, \infty)$. Then X is future pointing. Moreover, each element in the flow of X is an isometry $N \to N$. Thus X is Killing (cf. Section 3.6.3).

To prove the converse, assume henceforth that X is Killing. Then each element $\psi: \mathcal{U} \to N$ in the flow of X is an isometry $\psi: \mathcal{U} \to \psi\mathcal{U}$. Thus if $\tilde{R}$, $\hat{R}$, and f are as in Corollary 7.3.4, we have: $(\psi^*\hat{R})x = \hat{R}x \; \forall x \in \mathcal{U}$; $(\psi_*\tilde{R})x = \tilde{R}x \; \forall x \in \psi\mathcal{U}$; and thus $f \circ \psi = f|_{\psi\mathcal{U}}$. Thus $Xf = 0$, which implies $Xr = 0$ and thus $L_X r = 0$, where L_X is the Lie derivative. Since $L_X d = dL_X$, we now get $L_X[(1 - 2\mu/r)^{-1} dr] = 0$. Since $\partial/\partial r$ is physically equivalent to $[(1 - 2\mu/r)]^{-1} dr$ and $L_X g = 0$, we get $L_X(\partial/\partial r) = 0$. We now exploit the two conditions $L_X r = 0 = L_X(\partial/\partial r)$.

$Xr = 0$ implies there exists a C^∞ function $\ell: N \to \mathbb{R}$ and a vector field $Y: N \to TN$ such that $Q_* Y = 0$ and $X = \ell(\partial/\partial t) + Y$. We now have $L_X(\partial/\partial r) = 0 \Rightarrow L_{\partial/\partial r} X = 0 \Rightarrow (\partial/\partial r)\ell = 0$ and $L_{\partial/\partial r} Y = 0 \Rightarrow D_{\partial/\partial r} Y = D_Y(\partial/\partial r)$. Using the explicit expressions of the connection forms in the proof of Proposition 7.3.2, we obtain $D_Y \partial/\partial r = rY$. This implies $D_{\partial/\partial r} Y = rY$, $\Rightarrow (\partial/\partial r)\{r^{-2} g(Y, Y)\} = 0$. Thus $g(X, X)$, which equals $-\ell^2(1 - 2\mu/r) + r^2[r^{-2} g(Y, Y)]$, becomes positive for r sufficiently large unless $g(Y, Y) = 0$. Assume now X is future pointing. Then $g(Y, Y) = 0$ and, since Y is spacelike, $Y = 0$. Thus $X = \ell(\partial/\partial t)$. The following lemma then concludes the proof.

7 Further applications

Lemma. *Let W be a nowhere zero Killing vector field and let ℓ be a C^∞ function. If ℓW is also a Killing field, then ℓ is a constant.*

PROOF. Suppose $x \in M$. By Section 3.6.1c, $g(D_X(\ell W), Y) + g(X, D_Y(\ell W)) = 0 = (X\ell)g(W, Y) + (Y\ell)g(X, W) \, \forall X, Y \in M_x$. Suppose $g(X, W) = 0$. Since $Wx \neq 0$ we can choose Y such that $g(Y, W) \neq 0$ and conclude $X\ell = 0$. Suppose $g(X, W) \neq 0$. Choosing $Y = X$ gives $X\ell = 0$ again. Thus $X\ell = 0 \, \forall X \in M_x$. Since M is connected and x was arbitrary, $\ell =$ constant. □

This lemma remains valid so long as W is not identically zero. See Exercise 7.3.14.

Note that for $a = 1$, $X = \partial/\partial t$ has the following additional property: $\forall$ observer $\gamma: (b, \infty) \to N$ who escapes to infinity, $g(X \circ \gamma, X \circ \gamma) \to -1$ as $u \in (b, \infty)$ approaches infinity. Define $Z = (1 - 2\mu/r)^{-1/2}(\partial/\partial t)$. We now characterize $dt|_N$.

Corollary 7.3.6. *Z is a static, absolute reference frame on N; $\hat{t}: N \to \mathbb{R}$ is a time function for Z such that $Z\hat{t} \to 1$ as $r \to \infty$ iff $d\hat{t} = dt|_N$; an observer γ in N is at rest iff $r \circ \gamma =$ constant $\in (2\mu, \infty)$ and $P \circ \gamma =$ constant $\in \mathscr{S}^2$.*

PROOF. Chase down the definitions. □

Note that Z is synchronizable but not proper time synchronizable. Thus Z is not geodesic, and an observer in N who is at rest is not freely falling. An observer γ in Z can compare the compromise, radar synchronized time t (cf. Section 5.3.1) with his own proper time u. For r very large, the explicit form given above shows $dt(Z) \simeq 1$; then $u \simeq t \circ \gamma + b$, $b \in \mathbb{R}$. But for observers at rest at small values of r, and for observers not at rest, t and the proper time in general disagree.

Suppose $a \in (2\mu, \infty)$ and $b \in \mathbb{R}$. Define $\mathscr{S} = (r^{-1}a) \cap (t^{-1}b) \subset N$, where r^{-1} and t^{-1} denote complete inverse images. The reader should verify the following properties.

(a) $\forall a, b$ as above, $\mathscr{S}$ is an imbedded, spacelike 2-submanifold diffeomorphic to $\mathscr{S}^2$; the induced metric on $\mathscr{S}$ is $a^2 h$ so the intrinsic area of $\mathscr{S}$ is $4\pi a^2$.
(b) By Corollaries 7.3.4 and 7.3.6 the collection of all such $\mathscr{S}$ is canonically defined in any representative of the gravitational field represented by N, even if r and t are not given *ab initio*.
(c) $\forall p \in N$, there exists precisely one such $\mathscr{S}$ that contains p. Thus we have characterized $r|_N$ ($rp = \sqrt{A/4\pi} \, \forall p \in N$, where A is the area of the sphere $\mathscr{S}$ which contains p). $r|_B$ has a wholly similar characterization.

As long as the above discussion is kept in mind it becomes permissible to think of $t|_N$ as *the* time, canonically determined up to an additive constant, and to think of r as *the* (area-) radius. To measure

7.3 The geometry of Schwarzschild spacetimes

$r(p)$ one can, in principle, measure the area of $\mathscr{S}$ above by ordinary surveying techniques.

To discuss observations in the solar system, we shall need information on freely falling particles in Schwarzschild spacetimes. Let $\gamma\colon \mathscr{E} \to U$ be a future-pointing causal geodesic. Then $\delta = P \circ \gamma$ is a curve in $\mathscr{S}^2$ and we can regard $r \circ \gamma\colon \mathscr{E} \to (0, \infty)$ as a function—say, ρ—over δ. $\forall x \in \mathscr{S}^2$, let K_x be the generator of rotations about the x axis (Section 7.0). Define $\partial = (\partial/\partial t) \circ \gamma\colon \mathscr{E} \to TU$. Proposition 7.3.7 gives "first integrals" ("constants of the motion") for γ.

Proposition 7.3.7. *There exist unique numbers* $\mathsf{E} \in \mathbb{R}$, $J \in [0, \infty)$, $J_x \in [-J, J]$ *such that:*

(a) $g(\gamma_*, \partial) = -\mathsf{E}$;
(b) $\rho^4 h(\delta_*, \delta_*) = J^2$;
(c) $\rho^2 h(\delta_*, K_x) = J_x$.

PROOF. Since $\partial/\partial t$ is Killing, (a) follows from Section 3.6.3e. Similarly, the vector field $V\colon U \to TU$ determined by $P_* V = K_x$ and $Q_* V = 0$ is Killing and $\rho^2 h(\delta_*, K_x) = g(V, \gamma_*)$, so (c) follows for $J_x \in \mathbb{R}$. (b) and $J_x \in [-J, J]$ now follow from Exercise 7.0.1. □

Newtonian analogues. In Newtonian physics, a freely falling particle in a spherically symmetric gravitational field has conserved angular momentum $\vec{J}$. If the gravitational field is time-independent, the particle also has conserved energy E, consisting of its Newtonian kinetic energy and its Newtonian gravitational potential energy. For the case of a particle $\gamma\colon \mathscr{E} \to N$ with nonzero rest mass m, E above is analogous to $E + m$ (cf. Optional exercise 9.3.3). Now $h(\delta_*, \delta_*)$, from its definition, can be interpreted as the velocity of the curve γ projected on the unit sphere; the reader may consult the fine-print paragraph in Section 7.4.4 for further details of this interpretation. Thus the Newtonian analogue of $h(\delta_*, \delta_*)$ is ω^2, where ω is the Newtonian angular velocity. The Newtonian analogue of ρ is r_N, which is the radial function of the Newtonian particle in terms of angle. Since Newtonian physics gives $|\vec{J}| = r_N^2 \omega$ (Alonzo–Finn [1]), the Newtonian analogue of the quantity $\rho^4 h(\delta_*, \delta_*)$ is $|\vec{J}|^2$. By (b) of Proposition 7.3.7, the Newtonian analogue of J is then $|\vec{J}|$. J_x is analogous to the first component of $\vec{J}$.

7.3.8 Auxiliary terms

Many Newtonian concepts that cannot be applied to general spacetimes can be applied to Schwarzschild spacetimes. To avoid ambiguities we list some definitions. Let $\gamma\colon \mathscr{E} \to U$ be a particle or observer in N or B, and let $\alpha\colon \mathscr{E} \to N$ be a particle or observer. Let K_x be as in Section 7.0. Define $J\colon \mathscr{E} \to [0, \infty)$ by $J^2 = (r \circ \gamma)^4 h(P_* \gamma_*, P_* \gamma_*)$. Ju is called the *total angular momentum* of γ at $u \in \mathscr{E}$. The function $J_x = (r \circ \gamma)^2 h(P_* \gamma_*, K_x)\colon \mathscr{E} \to \mathbb{R}$ defines the *angular*

7 Further applications

momentum in ("*around*") *the x direction*; note that $-J \leq J_x \leq J$ (Exercise 7.0.1). Define E: $\mathscr{E} \to (0, \infty)$ by E $= -g(\alpha_*, \partial/\partial t)$; E$u$ is called the *total energy of* α *at* $u \in \mathscr{E}$. Note that E$u > 0$ since $(\partial/\partial t)|_N$ is future pointing. If γ (respectively α) is freely falling, then J and J_x (respectively, E) are constants, by Proposition 7.3.7.

γ is said to *maintain direction* iff $J = 0$. α is said to: (a) *go in, maintain height, go out*, respectively, iff $dr(\alpha_*) < 0$, $dr(\alpha_*) = 0$, $dr(\alpha_*) > 0$, respectively; (b) *go directly in, hover, go directly out*, respectively, iff α maintains direction and goes in, maintains height, goes out, respectively; (c) *circle* iff α maintains height and $(P \circ \alpha)\mathscr{E} \subset \mathscr{S}^2$ is an arc of a great circle.

7.3.9 Freely falling photons in normal Schwarzschild spacetime

A complete classification of freely falling particles in N or B using Proposition 7.3.7 is clumsy. In this subsection, we give a complete description of the freely falling photons in normal Schwarzschild spacetime that maintain direction. Exercise 7.3.10 discusses one other special case that admits a simple description.

Thus let $q \in \mathscr{S}^2$ be fixed throughout this discussion. We will show that any inextendible freely falling photon $\lambda: \mathscr{E} \to N$ such that $P \circ \lambda = q$ and $dr(\lambda_* e) > 0$ for some $e \in \mathscr{E}$ is equivalent (in the sense of Section 5.0.2) to the curve $u \to (P\lambda u, r\lambda u, t\lambda u) = (q, u, u + 2\mu \ln u) \forall u \in (2\mu, \infty)$. The two conditions of $P \circ \lambda = q$ and $dr(\lambda_*) > 0$ somewhere therefore determine a light signal $[\lambda]$ that goes directly out (Section 7.3.8). To see this, note that since $\partial/\partial t$ is Killing by Corollary 7.3.5, we have by virtue of Section 5.0.2 that, for $\lambda: \mathscr{E} \to N$ as above: $g(\lambda_*, \lambda_*) = 0$ and $g(\lambda_*, \partial/\partial t) = a \in \mathbb{R}$. Now $a < 0$ since λ_* and $\partial/\partial t$ are both future pointing and $\partial/\partial t$ is timelike (Exercise 1.1.9c). These two equations give:

$$\frac{d}{du}(r\lambda u) = \pm \left(1 - \frac{2\mu}{r\lambda u}\right)\left[\frac{d}{du}(t\lambda u)\right]$$

$$\left(1 - \frac{2\mu}{r\lambda u}\right)\left[\frac{d}{du}(t\lambda u)\right] = -a.$$

Integrating, using $dr(\lambda_* e) > 0$ and using a positive affine reparametrization gives the representative stated above. It may be observed that the light signal observes a "coordinate singularity" in a finite parameter interval—that is, the above representative of $[\lambda]$ obeys $r \circ \lambda \to 2\mu$ and $\hat{g}(dt, dt) \circ \lambda \to \infty$ as $u \to 2\mu$. By Proposition 1.3.2, this suggests that (N, g) may not be maximal. In fact, (N, g) is not maximal, as we shall see in Section 7.5.

Suppose in the above, instead of requiring $dr(\lambda_* e) > 0$ for some $e \in \mathscr{E}$, we require $dr(\lambda_* e) < 0$ for some $e \in \mathscr{E}$. Then we would obtain a unique light signal that goes directly in and is represented by $u \to (q, -u, u - 2\mu \ln(-u)) \forall u \in (-\infty, -2\mu)$. These two light signals are called, respectively, the *directly outgoing* and *directly incoming standard light signals in N*.

7.3 The geometry of Schwarzschild spacetimes

EXERCISE 7.3.10

Verify the details in the following table which summarizes the freely falling particles $\gamma\colon \mathscr{E} \to N$ in normal Schwarzschild spacetime with rest-mass m, total angular momentum J, total energy E and the indicated properties. For the notation, $\beta\colon (-\infty, \infty) \to \mathscr{S}^2$ is a unit speed great circle, while $H \in [3\mu, \infty)$ and $\Delta = \mu/H \in (0, \tfrac{1}{3}]$ are constants.

$\mathscr{E}$	$(r \circ \gamma)s$	$(P \circ \gamma)s$	$(t \circ \gamma)s$	m	J	E	Property
$(2\mu, \infty)$	s	$q \in \mathscr{S}^2$	$s + 2\mu \ln s$	0	0	1	Directly in
$(-\infty, -2\mu)$	$-s$	$q \in \mathscr{S}^2$	$s - 2\mu \ln(-s)$	0	0	1	Directly out
$(-\infty, \infty)$	H	$\beta(Js/H^2)$	s	$\lvert(1-3\Delta)^{1/2}\rvert$	$(\mu H)^{1/2}$	$1-2\Delta$	Circles

Newtonian analogue. For a particle with inertial mass unity in a circular orbit at height $\lvert \vec{x} \rvert$ above the center of the sun (cf. the preceding fine-print section), the third row of the table gives very similar results when $\lvert \vec{x} \rvert = H \gg 3\mu$. But when $H = 3\mu$, we have $m = 0$, corresponding to a photon circling; Newtonian physics cannot handle such situations.

EXERCISE 7.3.11

Let $\gamma\colon \mathscr{E} \to N$ be an observer, let $A\colon \mathscr{E} \to TN$ be γ's world acceleration, and let $\alpha\colon (c, d) \to N$ be a freely falling particle with rest-mass m. Show the following.

(a) γ is at rest iff γ hovers iff γ maintains height and direction, which implies $A = [(\mu/r^2)(\partial/\partial \mathbf{r})] \circ \gamma$, which implies $\lvert A \rvert = \mu/[r^2(1 - 2\mu/r)] \circ \gamma$, which implies γ is not freely falling. Thus to stay at rest the observer must world accelerate ("upward").
(b) γ maintains direction iff there exists $q \in \mathscr{S}^2$ such that $P \circ \gamma = q$.
(c) α maintains height iff α circles and $P \circ \alpha$ is constant speed.
(d) The total energy E of α is constant; moreover, α escapes to infinity iff $d = \infty$ and $E > m$.

Newtonian analogues. Consider a rocket of unit inertial mass that hovers above the North Pole of the earth at distance $\lvert \vec{x} \rvert$ from the center of the earth. To balance the earth's gravity the rocket motor must exert a constant upward force $\vec{F}$ of magnitude $\lvert \vec{F} \rvert = \mu/r^2$, where $8\pi\mu$ is the Newtonian active-mass of the earth. In general relativity only $\vec{F}$, not the earth's gravity, would count as a force. $\lvert A \rvert$ in part (a) may be regarded as the corresponding upward acceleration, with $2\mu/r$ a general relativistic correction term. In (d) the condition $E > m$ corresponds to the Newtonian statement that a particle has escape velocity iff its total Newtonian energy E, analogous to $E - m$, is greater than zero.

EXERCISE 7.3.12

Let $\hat{M}$ be a spacetime that represents the gravitational field represented by N. Show there exists a projection $\hat{P}\colon \hat{M} \to \mathscr{S}^2$ with properties corresponding to those of $P: N \to \mathscr{S}^2$. Show that if $x, y \in \hat{M}$, $\hat{P}x$, and $\hat{P}y$ are given and $\hat{P}x \neq \hat{P}y$, then $\hat{P}$ is uniquely determined.

EXERCISE 7.3.13

Let $\hat{M} \subset N$ be defined by $\hat{M} = r^{-1}(a, \infty)$ with $a \in (2\mu, \infty)$. Show $\hat{M}$ is a spacetime and, with Z as in Corollary 7.3.6, $Z|_{\hat{M}}$ is an absolute, static reference frame on $\hat{M}$.

EXERCISE 7.3.14

(a) A submanifold N of M is called *totally geodesic* iff any geodesic of M tangent to N at one point must lie completely in N. Show that the zero set of a Killing vector field is a totally geodesic submanifold (possibly disconnected). (b) Use (a) and Exercise 3.6.6 to show that the lemma of Corollary 7.3.5 remains valid if W is only required to be nonzero at one point.

7.4 The solar system

7.4.1 Idealizations

In this section we use the normal Schwarzschild spacetime N to analyze certain observable effects in the solar system. Let's regard the history of the solar system outside the sun as a spacetime together with various observers and various photons. The observers idealize planets, moons, rockets, distant stars, and so on, as well as actual observers on earth. Some of the photons are emitted by the sun (or a terrestrial radar set), bounce off a planet, and hit a photographic plate on earth; other photons come to us from a distant star through the sun's gravitational field, and so on. To get a theory of the solar system, astronomers take a large number of photon observations and some comparatively simple model with a few adjustable parameters. They get a set of best-fit values for the parameters (or a clear-cut inconsistency). In the rest of this section we will sometimes talk as if some specific observation determines one specific parameter, some other observation indicates that a general relativistic model is better than a Newtonian one, and so on. These are oversimplifications; really the game is to look for overall consistency.

Let $a = 2.32$ seconds, $\mu = 4.92 \times 10^{-6}$ seconds. Then $4\pi a^2$ is approximately the surface area of the sun and $8\pi\mu$ is approximately the active mass of the sun. Let N be the normal Schwarzschild spacetime of active mass $\bar{m} = 8\pi\mu$. Note that $2\mu \ll a$. Throughout the rest of this section (M, g, D) will be the spacetime determined by setting $M = r^{-1}(a, \infty)$, an open submanifold of N. Using M to model the history of the solar system outside the sun involves neglecting the gravity of the planets and many other small effects (cf. Section 2.1.2). We will talk as if M were a physically exact model; moreover, we will analyze measurements by using the absolute, static reference frame Z on M (cf. Exercise 7.3.13).

7.4 The solar system

Newtonian analogues. Newtonian descriptions of the solar system usually use an oversimplified model of the sun and use "the inertial reference frame which is at rest with respect to the center of the sun." Often the detailed physical definition of that reference frame is treated, overoptimistically, as intuitively obvious.

7.4.2 Operational definitions and relativistic effects

It is in principle possible to determine the following structures on M by measuring techniques already discussed.

(a) The absolute, static reference frame Z (Section 7.2, Exercise 7.3.13).
(b) $t: M \to \mathbb{R}$, given one choice of time origin by one observer in Z (Proposition 2.3.5, Sections 5.3 and 7.2, and Corollary 7.3.6).
(c) $r: M \to (a, \infty)$, for example as an area-radius (Section 7.3).
(d) μ above, for example by applying Proposition 2.3.3 and Corollary 7.3.4.
(e) $P: M \to \mathscr{S}^2$, given two reference directions (Exercise 7.3.12).

Other predictions of the model, such as the accelerations (Exercise 7.3.11a) frequency ratios (Section 7.2), and planetary or photon worldlines (Section 7.3.9) can then in principle be tested by observations. In the rest of this section we describe some tests particularly important for relativity. It will be convenient to define $\Delta: M \to (0, \mu/a)$ by $\Delta = \mu/r$. Thus $\Delta(x) < 10^{-6} \forall x \in M$. Often Newtonian results and relativistic ones differ by terms of order Δ. In particular, unless one has observations accurate to one part in a million, no harm is done if r and/or t are measured by Newtonian techniques different from the ones outlined above. But when small relativistic corrections are of interest, the exact meaning of r and t becomes crucial.

7.4.3 Gravitational redshifts

Suppose there is a photon λ from $x \in M$ to $y \in M$. By Proposition 7.2.1 and Corollary 7.3.5, the frequency ratio $\imath$ for $([\lambda], Zx, Zy)$ is

$$\imath = [1 - 2\Delta(y)]^{1/2}/[1 - 2\Delta(x)]^{1/2}.$$

For x on the surface of the sun we have, by Section 7.4.1, $\Delta(x) \simeq 2 \times 10^{-6}$; for y = here-now using the distance of the earth from the sun gives $\Delta(y) \simeq 10^{-8}$. Thus $\imath \simeq 1 + 2 \times 10^{-6}$ in this case. This prediction is consistent with the observations. However, the surface of the sun is quite messy, and the interpretation of the observations is controversial. More convincing measurements of the redshift effect under discussion use the gravitational field of a white dwarf star or of the earth rather than that of the sun. The latter gives results consistent with a corresponding model of the earth, the accuracy being better than 1% (Weinberg [1]). Unfortunately, no such frequency ratio observation can be counted as a sharp test of general relativity. Almost all competing theories give the same prediction to first order in Δ (Misner-Thorne-Wheeler [1]).

7 Further applications

The fact that $z > 1$ in the above is usually not regarded as a recession of Zx from Zy; compare the fine-print paragraph in Example 5.4.3. Rather, people say that the sun's gravity fields robs the climbing photon of some kinetic energy, thereby decreasing its frequency by the time it gets up to the earth. Compare the fine-print paragraph after Proposition 7.2.1.

7.4.4 Radar

Radar measurements have recently become quite useful tests. The simplest case is the following. Consider observers $\gamma, \hat{\gamma}$ in the absolute reference frame Z. Suppose $P \circ \gamma = P \circ \hat{\gamma}$, and $r \circ \gamma < r \circ \hat{\gamma}$. You may imagine both observers hovering above the same point on the sun with $\hat{\gamma}$ higher. Suppose γ emits a photon λ_1 which bounces off $\hat{\gamma}$ and returns as a photon λ_2. In the terminology of Section 7.3.9, $[\lambda_1]$ and $[\lambda_2]$ are, respectively, directly outgoing and directly incoming standard light signals. γ can measure the proper time difference $u_2 - u_1$.

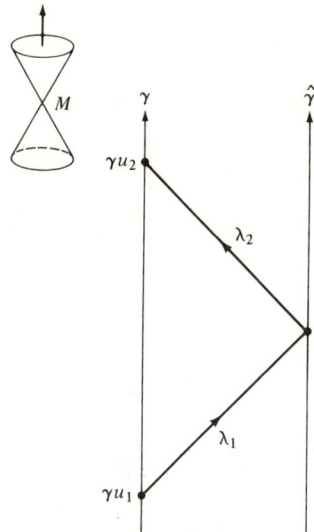

Let $b = r \circ \gamma$, $\hat{b} = r \circ \hat{\gamma}$. Section 7.3.9 and some algebra show that $(t \circ \gamma)u_2 - (t \circ \gamma)u_1 = 2[\hat{b} - b + 2\mu \ln (\hat{b}/b)]$. Now

$$u_2 - u_1 = [1 - 2\mu/b]^{1/2}[(t \circ \gamma)u_2 - (t \circ \gamma)u_1],$$

since $Z = (1 - 2\Delta)^{-1/2}(\partial/\partial t)$. Thus if μ, b, and $\hat{b}$ are known to γ he can check his observed value of $u_2 - u_1$ against the predicted value

$$u_2 - u_1 = 2[1 - 2(\mu/b)]^{1/2}\left[\hat{b} - b + 2\mu \ln \frac{\hat{b}}{b}\right].$$

7.4 The solar system

Newtonian theory would give simply the limiting behavior for b and $\hat{b}$ very large—namely:

time-lag $= u_2 - u_1 = 2(\hat{b} - b) = 2 \cdot$ distance $= 2\cdot$(distance/speed of light).

Measurements of the effect in more complicated analogues of the situation here discussed indicate that the predicted deviations from the Newtonian result are correct; the accuracy is about 5% (Misner–Thorne–Wheeler [1]).

Of course the simplest use of radar measurements is actually not in checking general relativity, but in determining, say, $\hat{b}$, given b and μ. But when one has enough interrelated radar measurements, some of them serve as a check on general relativity.

Some other measurements are interpreted by analyzing "particle orbits," which are easier to observe than particle worldlines. Roughly, a particle orbit is determined by giving the radius as a function of angle. Specifically, let (γ, m) be a freely falling particle of rest-mass m with inextendible geodesic $\gamma: \mathscr{E} \to M$, total energy $\mathsf{E} \in (0, \infty)$ and total angular momentum $J \in [0, \infty)$ (cf. Section 7.3.8). The character of $\mathscr{E} \subset \mathbb{R}$ will depend on the details of the particle motion. For example, if the particle is ejected from the sun's surface and eventually escapes to infinity (Exercise 7.3.11d), then $\mathscr{E} = (d, \infty)$, with $d \in (-\infty, \infty)$ and $(r \circ \gamma)e \to a = 2.32$ seconds as e approaches d. In the following discussion we shall, however, restrict attention to the case $\mathscr{E} = \mathbb{R}$. As will be seen more explicitly in an example below, assuming $\mathscr{E} = \mathbb{R}$ is consistent only if J is sufficiently large; in particular, we henceforth assume $J \neq 0$.

By Proposition 7.3.7c and Exercise 7.0.2 $(P \circ \gamma)\mathscr{E} \subset \mathscr{S}^2$ is contained in a great circle; moreover, $J \neq 0$ implies $(P \circ \gamma)_*$ is nowhere zero. Thus there is a diffeomorphism $\kappa: (-c, c) \to \mathscr{E}(c \leq \infty)$ such that $P \circ \gamma \circ \kappa: (-c, c) \to \mathscr{S}^2$ is an arc of a unit speed great circle in $(\mathscr{S}^2, h)$. If κ is any such diffeomorphism, the function $\tilde{\gamma}$ defined by $\tilde{\gamma} = r \circ \gamma \circ \kappa: (-c, c) \to (a, \infty)$ is called an *orbit function* for (γ, m).

> Without essential loss of generality, we may assume $(P\gamma\kappa)0$ is the North Pole of the unit sphere. Then, since the curve $(P \circ \gamma \circ \kappa): (-c, c) \to \mathscr{S}^2$ is unit speed and lies in a great circle of the unit sphere, we may think of each $\phi \in (-c, c)$ as the (oriented) angle that $\tilde{\gamma}\phi$ makes with the North Pole (see figure).
>
> Moreover, in view of the fact that Δ in Proposition 6.4.2 is very small everywhere on M we may regard $\tilde{\gamma}\phi - a$ as, to good approximation, the height of the particle above the surface of the sun when the particle has angle ϕ with respect to the North Pole.

Newtonian analogue. In Newtonian physics, consider a spherically symmetric body centered at the origin of $\mathbb{R}^3$. A particle moving in the body's gravitational field will remain within some 2-plane $\mathbb{R}^2 \subset \mathbb{R}^3$ (Alonzo and Finn [1]). Let (r_N, ϕ) be standard polar coordinates for $\mathbb{R}^2$. $\tilde{\gamma}$ is like $r_N(\phi)$, which specifies the particle path without direct reference to time.

7 Further applications

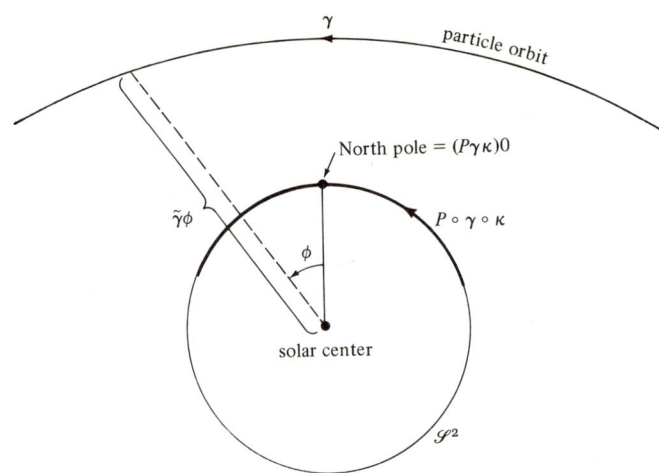

Let Δ, γ, $\tilde{\gamma}$, κ, E, and J be as above. For convenience define $f = \Delta \circ \gamma \circ \kappa$; $\mathbf{a} = (\mu m/J)^2 \in (0, \infty)$ and $\mathbf{b} = [(\mu E/J)^2 - \mathbf{a}] \in (-\mathbf{a}, \infty)$. Note that $f = \mu/\tilde{\gamma}$ and that $f \ll 1$. The following proposition is basic to the study of particle orbits in the solar system.

Proposition 7.4.5. *f obeys the ordinary differential equation $(f')^2 = \mathbf{b} + 2\mathbf{a}f - f^2 + 2f^3$.*

PROOF. If we write $\gamma_* = a_1 \partial + a_2(\partial/\partial r) \circ \gamma + \delta_*$, where each $a_i \in \mathbb{R}$ and $\partial = (\partial/\partial t) \circ \gamma$ as in Proposition 7.3.7 and $Q_*\delta_* = 0$, then $a_2 = dr(\gamma_*) = (r \circ \gamma)'$ and Proposition 7.3.7a implies that $a_1 = E(1 - 2\Delta \circ \gamma)^{-1}$. Hence,

$$-m^2 = g(\gamma_*, \gamma_*)$$
$$= -a_1^2(1 - 2\Delta \circ \gamma) + a_2^2(1 - 2\Delta \circ \gamma)^{-1} + \frac{J^2}{(r \circ \gamma)^2},$$

where the last term is a consequence of Proposition 7.3.7c. Substituting the values of a_1 and a_2 into this equation and using some algebra, we get

$$-m^2 = (1 - 2\Delta \circ \gamma)^{-1}(-E^2 + [(r \circ \gamma)']^2) + \frac{J^2}{(r \circ \gamma)^2}.$$

Composing both sides on the right with κ and using the definition of f, we obtain

$$-m^2 = (1 - 2f)^{-1}(-E^2 + [(r \circ \gamma)' \circ \kappa]^2) + \frac{J^2 f^2}{\mu^2}.$$

Multiplying both sides by $(1 - 2f)(\mu/J)^2$ and substituting the values of **a** and **b** yield:

$$-\mathbf{b} + 2\mathbf{a}f - f^2 + 2f^3 = [(r \circ \gamma)' \circ \kappa]^2 \left(\frac{\mu}{J}\right)^2.$$

7.4 The solar system

It remains to identify the right side with $(f')^2$. By the definition of κ, $P \circ \gamma \circ \kappa$ is a unit speed curve in $(\mathcal{S}^2, h)$. This means $1 = [h(\delta_*, \delta_*) \circ \kappa] \cdot (\kappa')^2 = (J\kappa')^2/\tilde{\gamma}^4$. Moreover, $\tilde{\gamma}' = [(r \circ \gamma)' \circ \kappa]\kappa'$. Hence

$$[(r \circ \gamma)' \circ \kappa]^2 (\mu/J)^2 = (\mu\tilde{\gamma}'/\tilde{\gamma}^2)^2 = (f')^2. \qquad \square$$

Since $f \ll 1$, the $(2f^3)$ term may be dropped from Proposition 7.4.5 in a first order approximation. Thus the differential equation of Proposition 7.4.5 reads:

$$(f')^2 = b + 2af - f^2.$$

Newtonian analogue. Notation and set-up as in the preceding fine-print section, Newtonian physics shows that $r_N(\phi)$ satisfies the following equation:

$$\left[\frac{d}{d\phi}\left(\frac{1}{r_N}\right)\right]^2 = b + a\left(\frac{1}{r_N}\right) - \left(\frac{1}{r_N}\right)^2,$$

where a, b are constants. Since $f = \mu/\tilde{\gamma}$ and $\tilde{\gamma}$ is like r_N, one may regard this Newtonian equation as a first-order approximation to Proposition 7.4.5.

7.4.6 Bending of light

One application of Proposition 7.4.5 is the computation of how much the sun's gravity "bends" light rays.

Let $\gamma: \mathscr{E} \to N$ be a freely falling photon, f, b, and $a = 0$ be as above. Thus $(f')^2 = b - f^2 + 2f^3$. Since we are only considering the case $\mathscr{E} = \mathbb{R}$, we may interpret γ as a photon "coming from infinity and going off to infinity" without running into the sun (cf. Section 7.3.11). Since $f \ll 1$, one may simplify and consider first the approximate equation $(f')^2 = b - f^2$, with maximal solution $f(\phi) = (b)^{1/2} \cos \phi$. Thus the orbit function

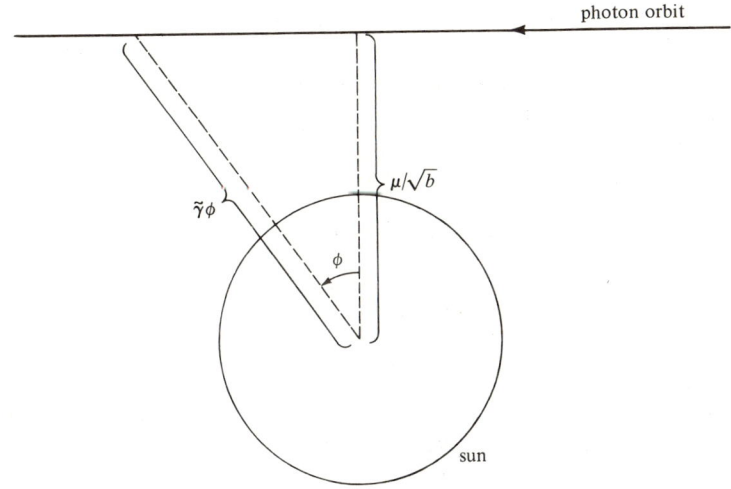

7 Further applications

$\tilde{\gamma}: (-c, c) \to (a, \infty)$ equals $\tilde{\gamma}(\phi) = \mu((\mathbf{b})^{1/2} \cos \phi)^{-1}$. Since $\tilde{\gamma}$ is a C^∞ function, we see that $(-c, c) = (-\pi/2, \pi/2)$. From $\tilde{\gamma}(\phi) \cos \phi =$ constant, one recognizes that $\tilde{\gamma}$ describes a straight line (see figure). Note that the photon does not hit the sun iff $\mu/(\mathbf{b})^{1/2} > a$, or equivalently, iff $J^2 > \mathsf{E}^2 a$.

Intuitively, if a light signal has a straight-line orbit unbent by the sun, then its complete orbit subtends a total angle of π at the solar center. Just as intuitively, if f (and thus $\tilde{\gamma}$) is defined on $(-c, c)$ with $c > \pi/2$, then the angle subtended by the light signal is $2c > \pi$. One then interprets this as the bending of the light signal by the gravity of the sun, the total bending angle being $2c - \pi$ (see Figure 7.4.6a).

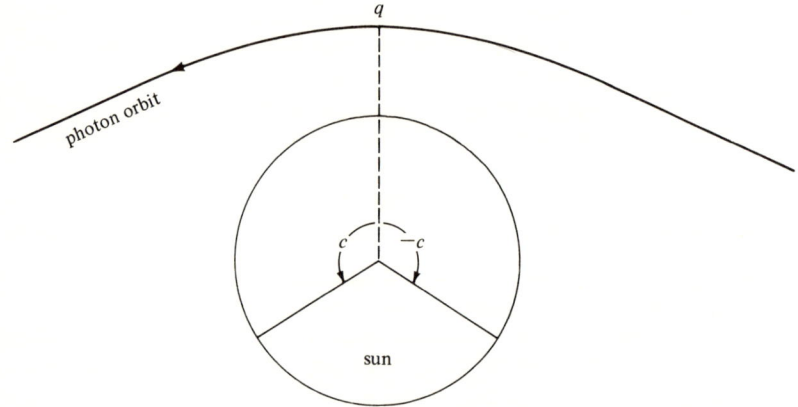

Figure 7.4.6a

We now prove, completely within the mathematical framework, that if $f: (-c, c) \to (0, \mu/a)$ is the maximal solution of the differential equation

(*) $$f' = (\mathbf{b} - f^2 + 2f^3)^{1/2}$$

of Proposition 7.4.5, then $c > \pi/2$. It will be necessary to first make explicit those assumptions on f that are dictated by the underlying physics: (i) $f < 1/4$, (ii) f is twice continuously differentiable, and (iii) $\mathbf{b} > 0$. Indeed, for the solar system, (i) is implied by $f = \mu/\tilde{\gamma} < \mu/a \cong 2 \times 10^{-6}$, (ii) is implied by $\tilde{\gamma}$ being C^∞ and nowhere zero, and (iii) is implied by $\mathbf{b} = (\mu\mathsf{E}/J)^2 > 0$ since we have already assumed $\mathsf{E} > 0$ throughout.

> Note that since we shall deal with f in the following proof only as a solution of (*), it is necessary to have assumptions like (i) to (iii) before physically meaningful conclusions about f can be drawn. For instance, the function $(x, y) \to (\mathbf{b} - y^2 + 2y^3)^{1/2}$ not being Lipschitzian at $(0, \rho)$, where ρ is a real root of the polynomial $(\mathbf{b} - y^2 + 2y^3)$, a priori (*) will have several distinct solutions; one solution is furnished by the constant function $f(\phi) = \rho$.
>
> Note also that under assumptions (i) to (iii), (*) defines a unique elliptic function. However, we shall not make use of this fact in the following in order to keep the discussion elementary.

7.4 The solar system

The proof of $c > \pi/2$ is broken into several steps: (a) $f'' < 0$ on $(-c, c)$. (b) f is an even function with a zero derivative exactly at 0. [Recall that f even means $f(\phi) = f(-\phi) \forall \phi \in (-c, c)$.] (c) f attains its absolute maximum at 0 and $f' > 0$ on $(-c, 0)$, $f' < 0$ on $(0, c)$.

In the physical context, (a) to (c) are intuitively obvious. Indeed, we have already made the Newtonian interpretation of $\tilde{\gamma}: (-c, c) \to (a, \infty)$ as the parametrization by azimuthal angle of the radial distance of the photon from the solar center. The assertions (a) to (c) about $f = \mu/\tilde{\gamma}$ may therefore be immediately read off from Figure 7.4.6a. For instance, (c) is just the statement that the photon attains a unique absolutely minimal distance from the solar center at $q = \tilde{\gamma}0$, and that on either side of q, the photon either monotonely approaches the sun or monotonely recedes from it. However, this kind of argument is not a substitute for the proofs given below since it is based on Newtonian physics.

Proof of (a). Let I be the open subset of $(-c, c)$ on which $b - f^2 + 2f^3 \neq 0$. In I, the right-hand side of (*) is then differentiable so that $f'' = f(3f - 1) < 0$ by assumption (i). If $I \neq (-c, c)$, let $\xi \in \partial I$, the boundary of I. Then $b - f(\xi)^2 + 2f(\xi)^3 = 0$, $\Rightarrow f(\xi) \neq 0$ by assumption (iii). By assumption (ii), $f''(\xi) = \lim_{x \to \xi} f''(x) = f(\xi)[3f(\xi) - 1] \neq 0$. Thus $f'' < 0$ on the closure I^- of I. Suppose $I^- \neq (-c, c)$; let J be the complement of I in $(-c, c)$. J is open and, by the definition of I, $b - f^2 + 2f^3$ vanishes identically on J. Hence $f' = 0$ on J because of (*) and hence also $f'' = 0$ on J. Continuity of f'' then implies that $f'' = 0$ on ∂J. Thus for $\xi \in \partial J \cap \partial I$ ($= \partial J = \partial I$), $f''(\xi)$ is both zero and nonzero, a contradiction. Hence $I^- = (-c, c)$ and $f'' < 0$ everywhere. □

Proof of (b). First suppose $f'(0) \neq 0$. Then $b - f(0)^2 + 2f(0)^3 \neq 0$ so that (*) has a Lipschitzian right-hand side near $[0, f(0)]$. Define a function g by $g(\phi) = f(-\phi)$; then $g(0) = f(0)$, and g also satisfies (*). The uniqueness theorem on ordinary differential equations then implies that $f = g$, or equivalently, that f is even. But every differentiable even function must have zero derivative at 0, and this contradicts $f'(0) \neq 0$. Hence $f'(0) = 0$. By (a), 0 is the only zero of f'.

Now $f'' < 0$ and $f'(0) = 0$ together imply that f has a strict maximum at 0. Hence by the continuity of f at 0, $\forall$ sequence of negative numbers $\{a_n\}$ converging monotonely to 0, there exists a sequence of positive numbers $\{b_n\}$ such that $b_n \downarrow 0$ and $f(b_n) = f(a_n)$ for all large n. Discarding a finite number of terms if necessary, we may assume the preceding to be true $\forall n$. Define now $\forall n$ two functions

$$k_n, h_n: [0, c - \max\{-a_n, b_n\}) \to (0, \mu/a)$$

such that $k_n(\phi) = f(a_n - \phi)$ and $h_n(\phi) = f(b_n + \phi)$. Then $k_n(0) = h_n(0)$ and both k_n and h_n satisfy (*). Since f' is nonzero at a_n and b_n, the right-hand side of (*) is Lipschitzian at $(0, k_n(0)) = (0, h_n(0))$. Thus the same uniqueness theorem of differential equations implies that $h_n = k_n$ in their common domain of definition. Letting $n \to \infty$, we get $f(\phi) = f(-\phi) \forall \phi \in (-c, c)$. □

Proof of (c). Immediate from (a) and (b). □

7 Further applications

To conclude the proof of $c > \pi/2$, we observe that by (b), we may restrict attention to $(-c, 0)$. On this interval, (c) implies that f is strictly increasing and has a differentiable inverse, which we shall also denote for convenience by ϕ. Thus $\phi: (0, f(0)) \to (-c, 0)$ and $\phi \circ f =$ identity on $(-c, 0)$. From calculus, $f' = 1/\phi' \circ f$. Thus (*) implies:

$$\phi' \circ f = (\mathbf{b} - f^2 + 2f^3)^{-1/2}.$$

In addition, since $0 = [f'(0)]^2 = b - f(0) + 2f(0)^3$, we have:

$$\phi' \circ f = [f(0)^2 - f^2 - 2f(0)^3 + 2f^3]^{-1/2}.$$

Multiplying both sides by f' and integrating from $-c$ to 0, we get the basic formula:

$$c = \int_0^{f(0)} [f(0)^2 - t^2 - 2f(0)^3 + 2t^3]^{-1/2} dt.$$

Although this is an elliptic integral, it is a simple matter to get the qualitative estimate $c > \pi/2$. In fact, $\forall t \in (0, f(0))$, $[-2f(0)^3 + 2t^3] < 0$. Thus upon dropping this term from the integrand, we obtain:

$$c > \int_0^{f(0)} [f(0)^2 - t^2]^{-1/2} dt = \frac{\pi}{2}. \qquad \square$$

For observational purposes, it is important to be able to make a quantitative estimate of the lower bound of c. We proceed as follows. For $t \in (0, f(0))$, let $X = f(0)^2 - t^2$ and $Y = f(0)^3 - t^3$; note that $X > 0$, $Y > 0$. Since in the case of the solar system, $f \simeq 2 \times 10^{-6}$, we have $(2Y/X) < 1$. Thus we may take only the first two terms of the binomial expansion of $[1 - (2Y/X)]^{-1/2}$ to get

$$\left[1 - \left(\frac{2Y}{X}\right)\right]^{-1/2} > 1 + \left(\frac{Y}{X}\right) = 1 + t + \frac{f(0)^2}{t + f(0)},$$

where the inequality is because each term in this binomial series is positive. Since the integrand in the above formula for c equals $X^{-1/2}[1 - (2Y/X)]^{-1/2}$, we obtain finally:

$$c > \int_0^{f(0)} [f(0)^2 - t^2]^{-1/2}(1 + t + f(0)^2[t^2 + f(0)]^{-1}) dt$$

$$= \frac{\pi}{2} + 2f(0).$$

This admits the physical interpretation that when a light signal passes by the sun, it gets bent by a total angle of $2c - \pi > 4f(0) = (4\mu/r_{\min})$. This effect has been measured for light signals from stars during eclipses (most recently in 1973) and for light signals from quasars by radio astronomy techniques. The results are consistent with the theory. The quasar observations are considerably more accurate; their estimated observational accuracy is about 1%.

EXERCISE 7.4.7

(*Hard*) One of the most sensitive tests of general relativity is the measurement of what is called planetary perihelion precession, particularly for Mercury. Let (γ, m) be a freely falling particle with rest-mass $m \neq 0$, total angular momentum $J \neq 0$, and total energy $E < m$. Assume the domain $\mathscr{E}$ of γ is $\mathscr{E} = \mathbb{R}$ and that (γ, m) does not circle (Section 7.3.8). Let $\tilde{\gamma}$ be the orbit function. Show that $\tilde{\gamma}$ is periodic. Write the period P as $P = 2\pi + \Delta\Phi$; $\Delta\Phi$ is called the *perihelion precession*. Compute $\Delta\Phi$ for parameters appropriate to Mercury using appropriate approximations. If N is the number of times Mercury goes around the sun in one earth century, show $N\Delta\Phi \cong 43$ seconds of arc.

For a brief general review of current solar system data see Thorne [1]. A more detailed discussion is in Weinberg [1].

7.5 Black holes

We now discuss the simplest spacetimes used in analyzing complete gravitational collapse. The reader may wish to review the background material in Section 7.1.2 on stellar end-states before embarking on the technical details. We shall define and analyze a maximal, vacuum spacetime (K, g) that contains two normal Schwarzschild spacetimes and two Schwarzschild black holes. The main interest in K lies in these open submanifolds. We give a very brief review of current ideas on black holes in general at the end of the section.

As in Sections 7.3 and 7.4, suppose $\mu \in (0, \infty)$. Let $e = 2.7\cdots$ be the base of natural logarithms and $\mathscr{Q} \subset \mathbb{R}^2$ be the region determined by $(u^1 q)(u^2 q) < 2\mu/e \ \forall q \in \mathscr{Q}$. Define $K = \mathscr{Q} \times \mathscr{S}^2$; let $Q: K \to \mathscr{Q}$ and $P: K \to \mathscr{S}^2$ be the projections. Define functions $u = u^2 \circ Q$ and $v = u^1 \circ Q$ on K. K is connected and orientable; assign K the orientation determined by $du \wedge dv \wedge P^*\zeta$.

We claim that $(r - 2\mu) \exp[(r - 2\mu)/2\mu] = -uv$ defines, implicitly, a C^∞ function $r: K \to (0, \infty)$. In fact, note that for $a \in (0, \infty)$ the C^∞ function $\mathbb{R} \to \mathbb{R}$ given by $u \to u \exp(u/a)$ has positive derivative $\forall u > -a$; the result follows. With r thus defined

$$g = -8\mu^2 \left(\frac{1}{r}\right) \exp\left[\frac{(2\mu - r)}{2\mu}\right](du \otimes dv + dv \otimes du) + r^2 P^*h$$

is a Lorentzian metric on K. Let U be the vector field on K physically equivalent to $-du$. Then $\forall x \in K$, $Ux \neq 0$ and $g(Ux, Ux) = 0$. Thus U determines a time orientation for (K, g). Thus time-oriented by U, (K, g) is a spacetime called the *Kruskal spacetime of active mass* $8\pi\mu$. Some features of a Kruskal spacetime are indicated in Figure 7.5.1. Note that while r is C^∞, $dr = 0$ at $u = 0 = v$, so r is not a coordinate function on K.

Now there is an isometry $K \to K$ determined by $u \to -u$, $v \to -v$. The existence of this isometry will enable us to focus attention on the region $\mathscr{A} \subset K$ determined by $u \geq 0$ without essential loss of generality. Moreover,

7 Further applications

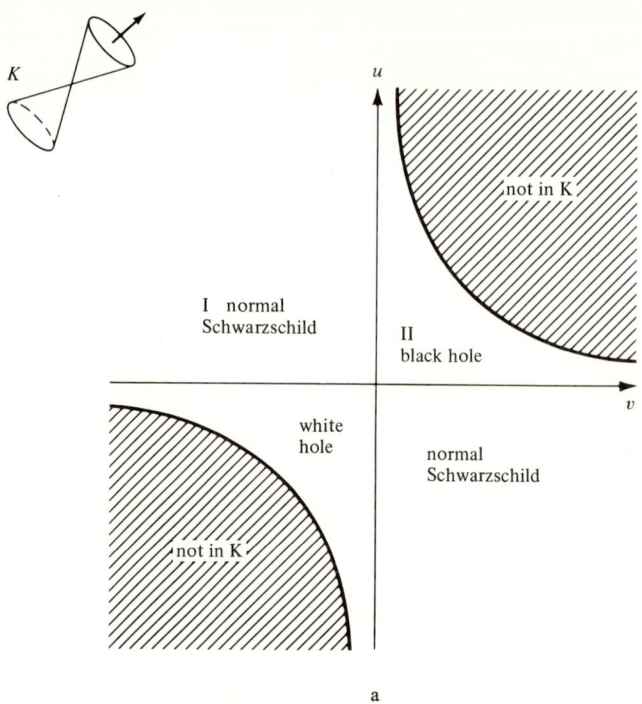

a

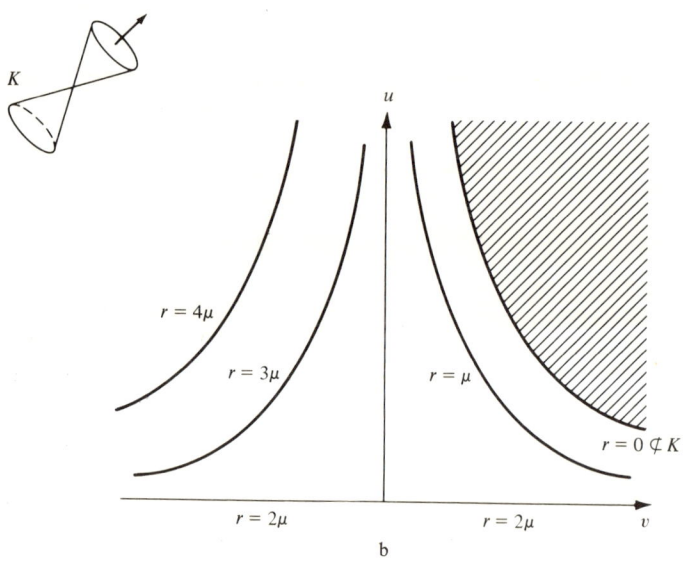

b

7.5 Black holes

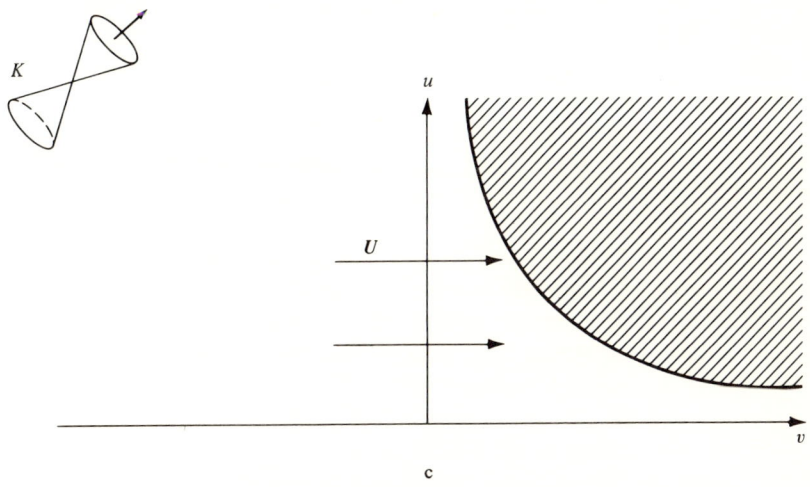

Figure 7.5.1. Some properties of a Kruskal spacetime. K is diffeomorphic to $\mathscr{S}^2 \times$ {the two-dimensional region shown in a }. In b and c only the "top half" of K is sketched.

we have $\mathscr{A} = \text{closure}(M_\text{I} \cup M_\text{II})$, where M_I and M_II are the open submanifolds determined respectively by: $(u > 0, v > 0)$ and $(u > 0, v < 0)$. We now show in what sense K glues together normal Schwarzschild spacetimes and Schwarzschild black holes. The reader is asked to review the diagrams in Example 1.4.2 and Section 7.3 at this point.

Proposition 7.5.2. $(M_\text{I}, g|_{M_\text{I}})$ *is isometric to the normal Schwarzschild spacetime of active mass* $8\pi\mu$.

PROOF. Let $t = 2\mu[\ln u - \ln(-v)]|_{M_\text{I}}$. Then t is C^∞ and maps M_I onto $(-\infty, \infty)$. Let $\psi: M_\text{I} \to (-\infty, \infty) \times (2\mu, \infty) \times \mathscr{S}^2$ be given by $\psi n = (tn, rn, Pn)$ for every $n \in M_\text{I}$. ψ is onto and there is a C^∞ inverse determined by $u = (r - 2\mu) \exp[(r - 2\mu + t)/4\mu]$, $v = (r - 2\mu) \exp[(r - 2\mu - t)/4\mu]$. Thus ψ is a diffeomorphism. Moreover, $dt = 2\mu(dv/v - du/u)$ and $dr = 2\mu(1 - 2\mu/r)(dv/v + du/u)$ on M_I so that

$$-(1 - 2\mu/r)dt \otimes dt + (1 - 2\mu/r)^{-1} dr \otimes dr$$
$$= -8\mu^2 r^{-1} \exp[(2\mu - r)/2\mu](du \otimes dv + dv \otimes du)$$

on M_I. Hence ψ is an isometry. □

Proposition 7.5.3. $(M_\text{II}, g|_{M_\text{II}})$ *is isometric to the Schwarzschild black hole of active mass* $8\pi\mu$.

PROOF. Define $t = 2\mu(\ln u - \ln v)$ on M_II. The calculation now duplicates that of Proposition 7.5.2 except for the change in domains. □

Define $f: K \to \mathbb{R}$ by $f = R_{ijkl}R^{ijkl}$.

7 Further applications

Corollary 7.5.4.

(a) $f = 144\mu^2/r^6$.
(b) *K is vacuum, not flat, and not stationary.*

PROOF. By Corollary 7.3.4 and Proposition 7.5.2, f and $144\mu^2/r^6$ coincide on the open submanifold M_I of K. Since both are real-analytic functions on K, they coincide. That K is vacuum and not flat follows by a similar argument.

Now suppose $X: K \to TK$ is Killing. Using Proposition 7.5.2, the proof of Proposition 7.3.5 already implies that $dr(X) = 0$ in M_I. By the same analyticity argument as above, $dr(X) = 0$ in K. But by the proof of Proposition 7.5.2, dr is timelike on M_{II}. Thus X must be spacelike on M_{II}. Thus K is not stationary (and *a fortiori* not static). □

Roughly, the gravity in the black hole region M_{II} is so large that even a photon going directly outward cannot escape to infinity (cf. Section 7.1.2). Specifically, suppose $\gamma: \mathscr{E} \to K$ is a particle or observer such that for some $s_0 \in \mathscr{E}$, $\gamma s_0 \in M_{II}$. Define $\hat{\mathscr{E}} = \{s \in \mathscr{E} \mid s \geq s_0\}$.

Proposition 7.5.5. $\gamma\hat{\mathscr{E}} \subset M_{II}$.

PROOF. Let V be the vector field physically equivalent to $-dv$. Then $g(V, V) = 0$ and $g(U, V) < 0$, so V is future pointing, lightlike, and nowhere parallel to U. We compute $(d/ds)(r \circ \gamma)$: On M_{II}, $dr = 2\mu(1 - 2\mu/r) \times (du/u + dv/v)$, where $u > 0$, $v > 0$ and $(1 - 2\mu/r) < 0$. Thus $(d/ds)(r \circ \gamma) = dr(\gamma_*) = -2\mu(1 - 2\mu/r \circ \gamma)[(1/u \circ \gamma)g(U, \gamma_*) + (1/v \circ \gamma)g(V, \gamma_*)] < 0$, and therefore $r \circ \gamma$ is strictly decreasing. Since the boundary of $M_{II} \subset K$ is determined by $r = 2\mu$ and $M_{II} \subset \{r < 2\mu\}$, the result follows. □

On the other hand, it is quite easy to enter M_{II}. For example, each maximal integral curve of U in $\{u > 0\}$ is a freely falling photon which starts from infinity (in the sense of Section 7.3), goes through the normal Schwarzschild region M_I, and then enters M_{II}. Observers can also fall into M_{II}. Figure 7.5.6 shows a typical situation.

γ and $\hat{\gamma}$ are observers hovering above the north pole in the normal Schwarzschild spacetime region M_I at $r = 5\mu$. At x, γ turns off his motor. The freely falling photons 1 to 5 carry the following conversation.

1. "Why are you wasting fuel?"
2. "Put your motor back on, you idiot!"
3. "I put my motor back on, but I seem to be in a funny gravitational field."
4. "Stop!"
5. "Help!"

Photon 4 will never reach γ; no one outside M_{II} will ever receive photon 5. In this sense K is an excellent model for the complete history of the twentieth century.

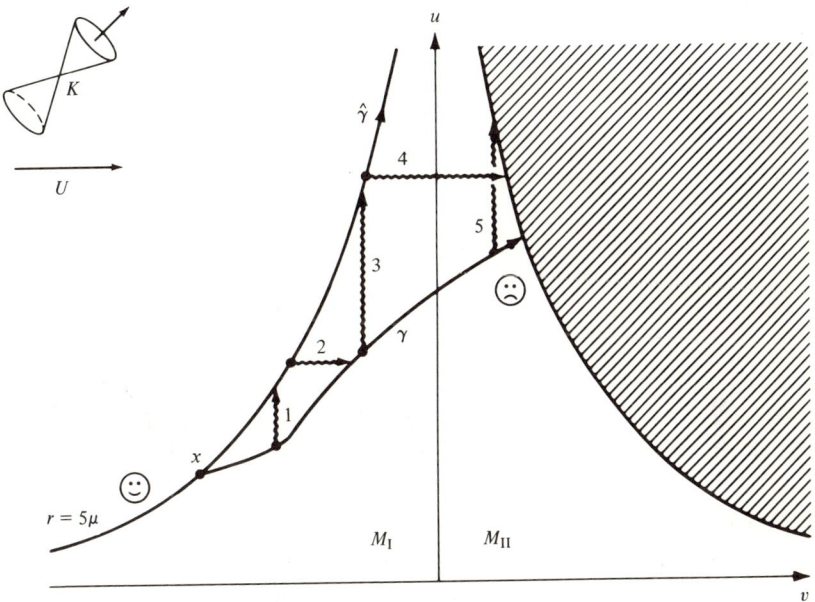

Figure 7.5.6

In more realistic models of uncharged, spherically symmetric black holes, the entire region below the v axis in Figure 7.5.1a, and part of the region above the v axis, is replaced by a nonvacuum spacetime. Attention is focused on the behavior of particles in the region M_I. For example, such particles may undergo violent collisions near M_{II} and create photons that (barely) escape to infinity in the sense of Section 7.3. As we write there is a possibility that Cygnus X1 may be an observed black hole. Roughly, the active mass μ is judged from the gravitational field in the region M_I, which influences a companion star. Photons are observed from the system, which may correspond to particles flowing from the companion star to the black hole, and colliding as above.

R. Kerr, E. Newman, W. Israel, B. Carter, and many other physicists have developed a theory for black holes that have a net electric charge and/or angular momentum but are, like the black holes modeled by K above, "stationary near infinity." R. Penrose, S. Hawking, and others have developed an extremely clever global theory for more general black hole spacetimes without the restriction that there be a timelike Killing vector field on some open submanifold. A comprehensive survey of current theory and observations of black holes is given in DeWitt and DeWitt [1].

7 Further applications

EXERCISE 7.5.7

Let γ be a freely falling observer in a Kruskal spacetime. Suppose γ starts at $r = r_0 > 2\mu$ in M_I and goes directly in with some energy E (cf. Section 7.3). How long will γ live before he gets destroyed "at" $r = 0$ by infinite "tidal forces"?

EXERCISE 7.5.8

Show that a Kruskal spacetime is maximal.

EXERCISE 7.5.9

Let K be a Kruskal spacetime and $S = r^{-1}a \subset K$ be the level hypersurface of r for $a \in (0, \infty)$.

(a) Show that S can be regarded as an immersed 3-submanifold which is imbedded iff $a \neq 2\mu$.
(b) Show that S can contain a photon worldline iff $a = 3\mu$ or $a = 2\mu$. For $a = 2\mu$, we can regard the photon as one which runs outward as fast as it can in order to remain still.
(c) In (a) let $a = 2\mu$ and suppose $\psi: \mathscr{S}^2 \to S$ is any imbedding. Show ψ is spacelike and that the intrinsic 2-area of $\psi\mathscr{S}^2 \subset K$ is $4\pi\mu^2$ no matter what shape $\psi\mathscr{S}^2$ has! [*Hint:* compare Exercise 1.4.9 for part (c).]

7.6 Gravitational plane waves

We now analyze spacetimes that model gravitational radiation in regions far from the source of the radiation. The spacetimes are very simple formally, but the key role played by lightlike quantities makes the interpretations tricky.

Throughout this section: $(\mathbb{R}^4, h)$ is Minkowski space (Example 1.4.1) $f: \mathbb{R} \to \mathbb{R}$ and $g: \mathbb{R} \to \mathbb{R}$ are C^∞ functions such that $f^2 + g^2$ is not identically zero; $\phi: \mathbb{R}^4 \to \mathbb{R}$ is $\phi = u^4 - u^3$; and $F: \mathbb{R}^4 \to \mathbb{R}$ is

$$F = \tfrac{1}{2}(f \circ \phi) \cdot [(u^1)^2 - (u^2)^2] + (g \circ \phi) \cdot u^1 u^2.$$

For comparison, we first define an electromagnetic field χ on Minkowski space by $\chi = d\alpha$, where $\alpha = -[(f \circ \phi)u^1 + (g \circ \phi)u^2]d\phi$. The reader may verify the following two points: (a) χ is a linearly polarized, electromagnetic plane wave (Example 3.7.3) iff f and g are linearly dependent over $\mathbb{R}$ (i.e., iff there exist $a, b \in \mathbb{R}$ such that $af + bg = 0$). (b) $((\mathbb{R}^4, h), \chi, 0)$ obeys Maxwell's equations. In what follows, α and χ should be compared to the metric g and the curvature tensor $\hat{R}$, respectively.

Define $g = h + 2F d\phi \otimes d\phi$ and $Y = \partial_4 + \partial_3$ on $\mathbb{R}^4$. We claim g is a Lorentzian metric and Y is lightlike on $(\mathbb{R}^4, g)$. In fact, define on $\mathbb{R}^4$ the vector fields $X_1 = \partial_1$, $X_2 = \partial_2$, $X_3 = Y$, $X_4 = (1/2)(\partial_4 - \partial_3) + FY$. Then

$$[g(X_i, X_j)] = \begin{bmatrix} 1 & 0 & 0 & 0 \\ 0 & 1 & 0 & 0 \\ 0 & 0 & 0 & -1 \\ 0 & 0 & -1 & 0 \end{bmatrix}$$

242

by direct computation; this immediately yields both results. Orient $\mathbb{R}^4$ as usual and time orient $(\mathbb{R}^4, g)$ by Y. Then $(\mathbb{R}^4, g)$ is a spacetime, called a *gravitational plane wave* spacetime. Such a spacetime is *linearly polarized* iff f and g are linearly dependent over $\mathbb{R}$. It is *monochromatic* iff there exist real numbers a, b, c, d, ω such that $f(t) = a \cos \omega t + b \sin \omega t$ and $g(t) = c \cos \omega t + d \sin \omega t$, $t \in \mathbb{R}$; in this case ω is its *frequency*.

In Newtonian physics, if a sound wave is detected at a location many wavelengths away from a small source such as a whistle, it is usual to approximate the "wavefronts" as 2-planes and the wave as a plane sound wave (cf. Examples 1.1.6 and 3.7.3). According to general relativity, a collapsing star will send out gravitational waves with wavelengths typically not much larger than the active mass of the star. Thus if we observe gravitational radiation from a star of active mass 10^{-4} seconds located at a distance of perhaps 10^{10} seconds (about 300 light years), $(\mathbb{R}^4, g)$ should be a reasonable model for the wave near us.

To get at intrinsic properties of $(\mathbb{R}^4, g)$ we start, as usual, by computing curvatures.

Proposition 7.6.1. $(\mathbb{R}^4, g)$ *is vacuum but not flat.*

PROOF. Let $\{\omega^i\}$ be the basis dual to $\{X_i\}$ above. By algebra, $\omega^1 = du^1$, $\omega^2 = du^2$, $\omega^3 = \frac{1}{2}(du^4 + du^3) - Fd\phi$, $\omega^4 = d\phi$. To compute the connection forms note that $(\omega^1, \omega^2, (\omega^3 - \omega^4)/2, (\omega^3 + \omega^4)/2)$ is an orthonormal basis; Section 1.0.3 now shows that the only nonvanishing connection forms for the $\{\omega^i\}$ basis are $\omega_A{}^3 = -(\partial_A F)\omega^4 = \omega_4{}^A$ $A \in (1, 2)$. Thus, in the notation of Section 3.6.1, $\omega_j{}^i \wedge \omega_k{}^j = 0$ and the curvature tensor is $\mathbf{R} = -2X_i \otimes \omega^j \otimes d\omega_j{}^i = -2 \sum_{A,B=1}^2 [\partial_A(\partial_B F)](X_A \otimes \omega^4 + X_3 \otimes \omega^A) \otimes (\omega^B \wedge \omega^4)$. $R(\omega^1; X_4, X_1, X_4) = -\partial_1^2 F = -f \circ \phi$ and $R(\omega^1; X_4, X_2, X_4) = -g \circ \phi$, so $\mathbf{R}$ is not identically zero and $(\mathbb{R}^4, g)$ is not flat. On the other hand, $\mathbf{Ric} = -[\partial_1(\partial_1 F) + \partial_2(\partial_2 F)](\omega^4 \otimes \omega^4) = 0$, so the Einstein tensor vanishes and $(\mathbb{R}^4, g)$ is vacuum. □

Let $\hat{\mathbf{R}}$ be the $(0, 4)$ tensor field physically equivalent to $\mathbf{R}$.

Corollary 7.6.2

$$-\hat{\mathbf{R}}/4 = (f \circ \phi)[(\omega^1 \wedge \omega^4) \otimes (\omega^1 \wedge \omega^4) - (\omega^2 \wedge \omega^4) \otimes (\omega^2 \wedge \omega^4)]$$
$$+ (g \circ \phi)[(\omega^1 \wedge \omega^4) \otimes (\omega^2 \wedge \omega^4) + (\omega^2 \wedge \omega^4) \otimes (\omega^1 \wedge \omega^4)].$$

PROOF. Algebra. □

Note that the electromagnetic field χ above is $\chi = (f \circ \phi)\omega^4 \wedge \omega^1 + (g \circ \phi)\omega^4 \wedge \omega^2$. This formal similarity and the interpretations assigned in Example 3.7.3 suggest calling f and g the *amplitudes* of $(\mathbb{R}^4, g)$. We will justify this term in more detail presently and will use it henceforth.

7 Further applications

The formal similarity between $(\mathbb{R}^4, g)$ and $((\mathbb{R}^4, h), \chi)$ has a deeper reason. Let $\mathscr{G}M$ be the isometry group of $(\mathbb{R}^4, g)$, and let $\mathscr{G}$ be the natural automorphism group $\mathscr{G} = \{\psi: \mathbb{R}^4 \to \mathbb{R}^4 \mid \psi^*h = h \text{ and } \psi^*\chi = \chi\}$. Unless f and g above are chosen in a very special way, the following results hold: $\mathscr{G}M$ is isomorphic to $\mathscr{G}$; $\mathscr{G}M$ is a certain five-parameter Lie group; and each orbit of $\mathscr{G}M$ in $(\mathbb{R}^4, g)$, like each orbit of $\mathscr{G}$ in $(\mathbb{R}^4, h)$, is 3-dimensional and lightlike (cf. Optional exercise 8.4.8). Moreover, each maximal vacuum spacetime whose isometry group is $\mathscr{G}M$ is isometric to a gravitational plane wave.

Thus, conceptually, a gravitational plane wave is a vacuum spacetime with "plane wave symmetry" where the familiar case of electromagnetic plane waves on Minkowski spacetime can be used to define the concept of plane wave symmetry.

We can now characterize the "travel direction" and "wavefronts" of $(\mathbb{R}^4, g)$. Both are lightlike.

Proposition 7.6.3. *A vector field X is parallel iff $X = aY$ for some $a \in \mathbb{R}$.*

PROOF. Suppose $X = aY$. Then $\forall (z, Z) \in TM$,

$$D_Z X = a\{Z[\omega^i(Y)] + \omega_j{}^i(Z)\omega^j(Y)\}X_i = a\omega_3{}^i(Z)X_i = 0,$$

where we have used Section 3.6.1e and the proof of Proposition 7.6.1. Conversely, suppose $DX = 0$. Then $\hat{R}(\cdot, X, \cdot, \cdot) = 0$ by the definition of a curvature tensor (Section 1.0.2). There exists an open connected region $\mathscr{U} \subset M$ such that $f^2 + g^2$ is nowhere zero on $\mathscr{U}$. Algebra and Corollary 7.6.2 then show that on $\mathscr{U}$, $X = jY$ where $j: \mathscr{U} \to \mathbb{R}$ is a C^∞ function. Now on $\mathscr{U}$, $0 = DX = jDY + Y \otimes dj = Y \otimes dj$. Thus $dj = 0$ and j is a constant, say $j = a$. We now have that $X - aY$ is parallel and $(X - aY)|_\mathscr{U} = 0$; consequently by Section 3.6.3d and Exercise 7.3.14, $X = aY$. □

Since Y is parallel, its integral curves are (future-pointing, lightlike) geodesics and Y is Killing (Section 3.6.3d). We call Y the *spacetime propagation direction*. Since there is in general no natural distinction between Y and aY $\forall a \in (0, \infty)$, this term would more properly apply to the set $\{aY \mid a \in (0, \infty)\}$, but we have followed standard terminology. Now note that $\forall (z, Z) \in TM$, $g(Z, Y) = 0$ iff $Z\phi = 0$. Thus the level hypersurfaces of ϕ are intrinsically determined as those maximal imbedded submanifolds which are everywhere orthogonal to some nonzero, parallel vector field. Precisely one such level hypersurface, say $\mathscr{W}$, contains any given $z \in \mathbb{R}^4$. We call $\mathscr{W}$ the *spacetime wavefront* through z. $Y|_\mathscr{W}$ is orthogonal to $\mathscr{W}$ and is lightlike, so $\mathscr{W}$ is lightlike (cf. Section 1.2). On the other hand, $Y\phi = 0$, so $Y|_\mathscr{W}$ is also tangent to $\mathscr{W}$ (see figure below).

The ambiguity in the positive constant a of $\{aY\}$ can be eliminated only by an *ad hoc* choice for each discussion. The reader may have noted that the definitions of amplitude and frequency given earlier in this section as well as all other associated quantities are in fact also

7.6 Gravitational plane waves

indeterminate up to (a function of) this constant a. This ambiguity corresponds to the fact that for a monochromatic gravitational plane wave, one cannot speak of "the" frequency without introducing some observer (cf. Exercise 7.6.5). Similarly, one cannot discuss "the" energy of a photon unless an instantaneous observer is specified (Sections 3.1 and 5.1).

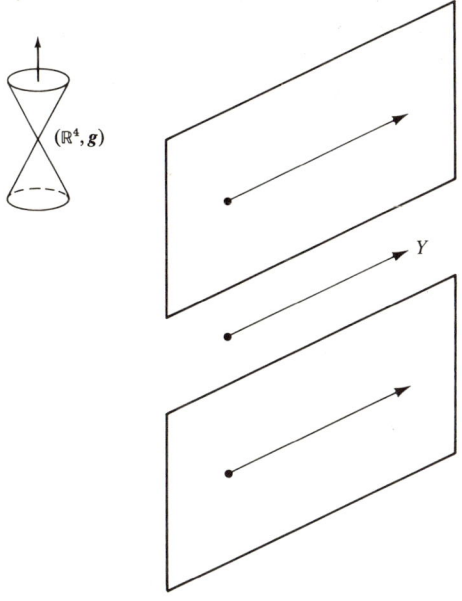

The term "spacetime propagation direction" can be motivated by Newtonian and special relativistic analogues (cf. Examples 1.1.6 and 3.7.3) or directly by the following argument. Suppose $z, \hat{z} \in \mathbb{R}^4$ lie on the same integral curve of Y, and $\mathcal{U}$ is an open neighborhood of z. Since Y is Killing there exists an open neighborhood $\hat{\mathcal{U}}$ of $\hat{z}$ and an isometry $\psi: \mathcal{U} \to \hat{\mathcal{U}}$ (cf. Section 3.6.3). Thus any "information" the amplitudes f and g "register" on $\hat{\mathcal{U}}$ is registered identically on $\mathcal{U}$. In this sense "information propagates along Y."

The Newtonian analogue (Example 1.1.6) of a spacetime wavefront $\mathcal{W}$ is a Euclidean 2-plane travelling at the speed of light in the Euclidean direction perpendicular to itself. $\mathcal{W}$ contains, and is orthogonal to, integral curves of Y. $\mathcal{W}$ might be called a "hypersurface of constant phase" since $\phi|_\mathcal{W} = $ constant, or a "hypersurface of constant amplitudes" since $f \circ \phi$ and $g \circ \phi$ are constant on $\mathcal{W}$. With $\mathcal{G}M$ as in the previous fine-print comment, each $\mathcal{W}$ is an orbit of $\mathcal{G}M$. Appropriate observers thus observe "homogeneity" for $\mathcal{W}$. On balance, such (rather difficult) intrinsic interpretations are better than the reference frame dependent interpretations of Example 3.7.3.

In order to try to detect a gravitational wave, one sometimes uses a metal bar and tries to observe how the wave excites vibrations of the metal bar.

7 Further applications

The histories of the molecules of this metal bar can be regarded as a family of observers filling up a small open subset N of the spacetime; alternately, one may regard these observers as the integral curves of a reference frame defined on N. For this reason, a detector is modeled by a reference frame defined on a small open submanifold of spacetime. We first discuss the geodesic case for simplicity. Thus consider a *local geodesic reference frame* Z—that is, there exists an open submanifold N of M such that Z is a geodesic reference frame on $(N, g|_N)$. We want to compute the 3-accelerations of the neighbors of an observer in Z (Section 2.3). Roughly, these 3-accelerations correspond to the vibrations in the detector due to the excitation by the gravitational plane wave (M, g).

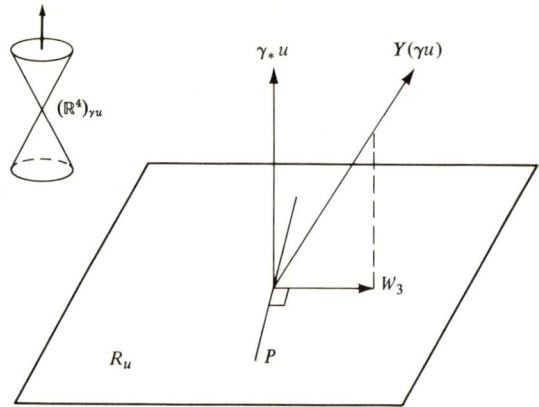

Let γ be an observer in Z, let R_u be his local rest space at proper time u, and let $\mathcal{W}$ be the spacetime wavefront that contains γu. Let $P = \mathcal{W}_{\gamma u} \cap R_u$; P is a (2-dimensional) plane in R_u. Let $W_3 \in R_u$ be the unique unit vector obtained by normalizing the projection of $Y(\gamma u)$ into R_u (see figure). A simple algebraic computation shows that W_3 and P are orthogonal. Roughly, γ sees the wave moving in the spatial direction W_3 with 2-dimensional wavefront P (cf. Example 1.1.6 and 3.7.3). We shall now compute relative accelerations, using a basis suggested by W_3 and P. First define $\mathsf{E} = -g(\gamma_* u, Y)$; $\mathsf{E} > 0$.

> Although general relativity cannot at present be systematically combined with quantum theory, physicists often think about "gravitons." Roughly, a graviton is related to a gravitational plane wave as a photon is related to an electromagnetic plane wave (cf. Exercise 3.7.7). Now suppose $(\mathbb{R}^4, g)$ is monochromatic as defined at the beginning of this section. Then one can think of $(\mathbb{R}^4, g)$ roughly as a graviton particle flow with energy-momentum vector field $P = aY$, where $a \in (0, \infty)$ depends on ω. For $a = 1$, E above is roughly the energy of a graviton observed by $(\gamma u, \gamma_* u)$, corresponding to the observed energy of a particle with energy-momentum $Y(\gamma u)$.

7.6 Gravitational plane waves

Define a basis $\{W_\alpha\}$ for R_u by $W_A = \partial_A(\gamma u) - \mathsf{E}^{-1}g(\gamma_* u, \partial_A)Y(\gamma u)$, $A \in (1, 2)$, and $W_3 = -\mathsf{E}^{-1}Y(\gamma u) + \gamma_* u$. Algebra shows that $\{W_\alpha\}$ is orthonormal for $(R_u, g|_{R_u})$ and that $P = \text{span}\{W_1, W_2\}$. Let ψ_u be the linear transformation $R_u \to R_u$, which assigns to each neighbor W of γ in Z the relative 3-acceleration $\psi_u W \in R_u$ (Proposition 2.3.3). More algebra using Corollary 7.6.2 shows that the matrix of ψ_u in the $\{W_\alpha\}$ basis is

$$[\psi_u] = \begin{pmatrix} A & B & 0 \\ B & -A & 0 \\ 0 & 0 & 0 \end{pmatrix} \quad \begin{array}{l} A = \mathsf{E}^2(f\phi\gamma)u \in \mathbb{R} \\ B = \mathsf{E}^2(g\phi\gamma)u \in \mathbb{R} \end{array}$$

Thus the observer measures a "transverse" wave: all relative accelerations are orthogonal to W_3, lying in the plane P. He does not directly measure the amplitudes f and g, but measures $\mathsf{E}^2 f$ and $\mathsf{E}^2 g$ instead. However, note that since Y is Killing and γ is geodesic, E is constant along γ's world line (Section 3.6.3e). Moreover, $\phi \circ \gamma$ is a linear function of proper time u, since $(d/du)(\phi \circ \gamma) = -g(\gamma_* u, Y) = \mathsf{E}$. Thus the observer can measure the graphs of f and g, with a constant stretching of ordinate and abscissa that depends on E. f and g are distinguished from each other by the directions of the observed relative accelerations. We plot the relative acceleration patterns in the principal directions of ψ_u for two special cases:

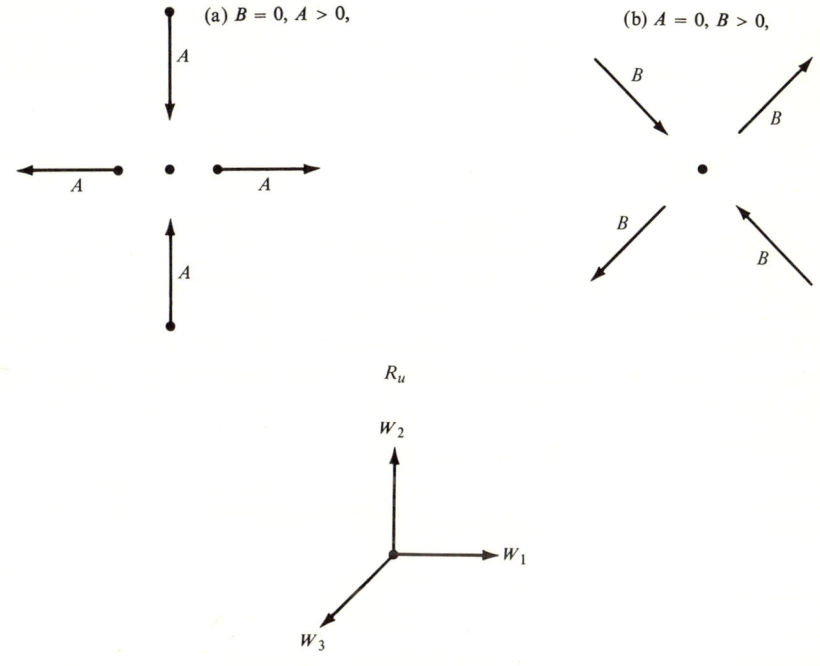

Figure 7.6.4

7 Further applications

The fact that ψ_u has trace zero is implied by the Einstein field equation (cf. Proposition 4.2.2). In Figure 7.6.4, trace $\psi_u = 0$ shows up via the fact that the relative accelerations of neighbors whose directions differ by 90° have opposite signs. The figure suggests, correctly, that the wave will excite "shear modes" of a detector, rather than "expansion modes." For example, a spherically symmetric detector would be set in a vibration that distorts its shape in the course of the vibration ("shear") rather than a spherically symmetric pulsation in which only the radius changes ("expansion"). In principle, this prediction is a sharp test of general relativity.

> The fact that ψ_u is a tensor, and that the patterns in Figure 7.6.4 are more complicated than the corresponding diagrams for an electric field of an electromagnetic plane wave (Example 3.7.3), corresponds to the quantum theoretical statement that a "graviton has spin 2," as opposed to "spin 1" for a photon. There are only two amplitudes f and g, and two "polarization modes" (Figure 7.6.4a, b). The corresponding quantum result is that every rest-mass zero field has two possible spin states, regardless of the total spin (Messiah [1]).

Some detectors must be modeled by reference frames that are not geodesic, since each observer (molecule) is subject to nongravitational forces from neighbors. However, once the response (Figure 7.6.4) of a geodesic reference frame is known, the response of an actual detector can be inferred from standard Newtonian physics and solid state physics (Weber [1]).

A vigorous controversy on whether or not gravitational waves have been detected is taking place as we write. It seems likely that once a highly sensitive, supercooled detector being designed at Stanford becomes operative, some waves should be observed, but even this is by no means certain.

> To design a detector one needs a guess as to the typical frequencies and amplitudes one might expect. Suppose the source is a star with active mass $\bar{m} \simeq 10^{-4}$ seconds at a distance L. A typical frequency might be $\omega \simeq 1/m$ and a typical amplitude might be $f \simeq 1/(100mL)$ under the most favorable circumstances.

Throughout the following exercises, (M, g) is a gravitational plane wave spacetime with amplitudes f and g. The first two exercises are designed to demonstrate in a quantitative way the fact that amplitude and frequency are observer-dependent.

Exercise 7.6.5

Suppose $A \in \mathbb{R}$, $\omega \in (0, \infty)$, $f(t) = A \cos \omega t$, $g = 0$, and $\gamma: \mathscr{E} \to M$ is a freely falling observer. Show that $(f\phi\gamma)u = A\mathsf{E}^2 \cos(\mathsf{E}\omega u + \phi_0) \forall u \in \mathscr{E}$, where $\phi_0 \in \mathbb{R}$.
Remark: Thus if relative to γ, $\mathsf{E} = 1$, then the frequency of this monochromatic gravitational plane wave is ω by definition; if, however, $\mathsf{E} = 7$ relative to γ, then the frequency is 7ω.

7.6 Gravitational plane waves

EXERCISE 7.6.6

Say $(z, \hat{Z})$ = us-now observes the wave in Exercise 7.6.5 and obtains a measurement of $\hat{E}\omega = 10^4$ (seconds)$^{-1}$. An instantaneous observer (z, Z) moving directly toward the source of the wave measures instead $E\omega = 10^{10}$ (seconds)$^{-1}$. What Newtonian speed do we observe for (z, Z) (cf. Section 2.1.6)?

EXERCISE 7.6.7

In Exercise 7.6.5 show $X \equiv \partial_1$ has the following properties. (a) $|X| = 1$. (b) $\forall z \in M$, $\mathscr{P}_z \equiv \text{span}\{Xz, Yz\}$ is lightlike. (c) $D(X \wedge Y) = 0$. (d) $\forall$ geodesic observer $\gamma: \mathscr{E} \to M$, and $\forall u \in \mathscr{E}$, a principal direction of the relative acceleration transformation ψ_u is the line $(\mathscr{P}_{\gamma u}) \cap R_u \subset R_u \subset M_{\gamma u}$. *Remark:* Here (M, g) is *linearly polarized*; (d) identifies $\mathscr{P}_{\gamma u} \subset M_{\gamma u}$ as the *spacetime polarization plane*; and (c) shows this plane is "constant" (parallel).

EXERCISE 7.6.8

Let $\gamma: \mathscr{E} \to \mathbb{R}^4$ be a geodesic observer, and let W be a unit vector field over γ such that Wu is a principal direction for ψ_u $\forall u \in \mathscr{E}$. Show W is parallel $\forall$ such observer iff (M, g) is linearly polarized.

EXERCISE 7.6.9

Suppose f has compact support and $g = 0$. Show (M, g) is complete (and hence maximal).

8 Optional exercises: relativity

In the previous chapters we have tried to stick to topics that are essential, require no mathematical background beyond the second-year graduate level, and require no physics background beyond the freshman level. This and the next chapter are primarily intended to round out our presentation somewhat by outlining some topics that fail to meet one or more of these criteria. (M, g) is a spacetime throughout this chapter.

8.1 Lorentzian algebra

This section covers some details on Lorentzian vector spaces.

8.1.0 Notation

Throughout this section: Z is a reference frame on M; $z \in M$; $Z = Zz$; $\zeta \in M_z^*$ is physically equivalent to $-Z$; $g = gz \in T_2^0(M_z)$; and $\mathscr{T}_z^+ = \{X \in M_z \mid X \text{ is timelike and } \zeta(X) > 0\}$ (cf. Exercise 1.1.9). Thus (M_z, g) is a Lorentzian vector space (Sections 0.0.3 and 1.1) and Z is a unit vector in $\mathscr{T}_z^+$. We shall be concerned primarily with the algebraic properties of $(M_z, g, \mathscr{T}_z^+, Z)$.

8.1.1 References

We give some cross-references and references that indicate how the algebra below is applied. The concepts of a wider metric (Section 8.1.2) and an Alexandrov basis (Section 8.1.3) are used in the study of topology and causality (cf. Sections 8.2, 8.3, and the further references given there). The algebra of stress-energy tensors (Sections 8.1.4 to 8.1.8), reference frames (Sections 8.1.9 and 8.1.10), and lightlike geodesic vector fields (Section 8.1.12) is important in the study of collapse (Section 4.3; Hawking–Ellis [1]) of cosmology (Chapter 6; Ellis [1]) and of theoretical models for black holes (cf. DeWitt and DeWitt [1]). The algebra of curvature tensors (Section 8.1.13) is important in various contexts, such as analyzing groups of isometries (cf. MacCallum [1]) and discussing "asymptotically flat" spacetimes (cf. Penrose–Rindler [1]).

We remark that there is a powerful machine, the two-component spinor formalism, for doing Lorentzian algebra neatly (Penrose and Rindler [1]).

8.1.2 Inner products

A Lorentzian inner product $\bar{g}$ on M_z is defined as *wider* than g iff $\bar{g}(Y, Y) < 0\,\forall$ causal $Y \in M_z$. Let $h = g + a\zeta \otimes \zeta$ with $a \in \mathbb{R}$ and ζ as in Section 8.1.0. Show that on M_z: (a) h is a positive definite inner product iff $a > 1$; (b) h is a Lorentzian inner product narrower than g iff $0 < a < 1$; (c) h is a Lorentzian inner product wider than g iff $0 > a$.

8.1.3 Topology

(a) $\forall X, Y \in M_z$ define $\langle X, Y \rangle = \{V \in M_z \,|\, (Y - V) \in \mathcal{T}_z^+ \text{ and } (V - X) \in \mathcal{T}_z^+\}$. Show $\{\langle X, Y \rangle\}$ is a basis for the standard topology of M_z; it is called the *Alexandrov basis*. (b) $\forall \varepsilon > 0$ and $\forall Y \in M_z$ define $B_\varepsilon Y = \{X \in M_z \,|\, |Y - X| < \varepsilon\}$. Show $\{B_\varepsilon Y\}$ is not a basis for the standard topology of M_z.

8.1.4 Self-adjoint linear transformations

Suppose $T \in T_0^2(M_z)$ is symmetric; thus the physically equivalent linear transformation $\tilde{T}: M_z \to M_z$ is self-adjoint with respect to g (cf. Exercise 1.0.6 and Section 8.1.0). Recall that a vector subspace $W \subset M_z$ is an *invariant subspace of* $\tilde{T}$ iff $\tilde{T}W \subset W$ and an *eigenvalue* of $\tilde{T}$ is any solution (real or not) of the characteristic equation. In this and the following exercise we work with $\mathbb{R}$, rather than the complex numbers $\mathbb{C}$, unless explicitly indicated otherwise. Show the following results. (Hint: the results extend to a Lorentzian vector space of any finite dimension except that (h) can fail if the dimension is less than 4.)

(a) W is lightlike iff $W \cap W^\perp$ is lightlike iff dimension $(W \cap W^\perp) \geq 1$ iff dimension $(W \cap W^\perp) = 1$.
(b) If dimension $W \geq 2$ then W is timelike iff W contains two linearly independent lightlike vectors.
(c) W is an invariant subspace of $\tilde{T}$ iff $W^\perp$ is.
(d) If W is a spacelike invariant subspace of $\tilde{T}$ then $\tilde{T}|_W$ is diagonalizable.
(e) If $\tilde{T}Y \in \mathcal{T}_z^+$ $\forall$ causal $Y \in M_z$ then: $T(\omega, \omega) < 0$ $\forall$ causal $\omega \in M_z^*$; moreover $\tilde{T}$ is diagonalizable—that is, each eigenvalue of $\tilde{T}$ is real and there exists an orthonormal basis of eigenvectors.
(f) If W is a lightlike invariant subspace, $\tilde{T}|_W$ has a lightlike eigenvector.
(g) If $a + \sqrt{-1}b$ is an eigenvalue of $\tilde{T}$ and $X + \sqrt{-1}Y$ is a corresponding eigenvector, span $\{X, Y\}$ is an invariant subspace of $\tilde{T}$.
(h) $\tilde{T}$ has a spacelike eigenvector. [Hint: use (d)–(g).]

8 Optional exercises: relativity

8.1.5 Normal forms

In Section 8.1.4 show exactly one of the following holds. (Hint: one proof uses Jordan normal forms for the endomorphisms of a vector space over $\mathbb{C}$.)

(a) There is an orthonormal basis (X_1, X_2, X_3, X_4) for M_z such that $\hat{T} = \sum_{\alpha=1}^{3} p_\alpha X_\alpha \otimes X_\alpha + \rho X_4 \otimes X_4$, where $p_1, p_2, p_3, \rho \in \mathbb{R}$.

(b) There is a basis as in (a) such that
$$\hat{T} = \sum_{i=1}^{4} \lambda_i X_i \otimes X_i + q(X_3 \otimes X_4 + X_4 \otimes X_3),$$
where $\lambda_i, q \in \mathbb{R}$ and $(\lambda_4 - \lambda_3)^2 < 4q^2$.

(c) There is an ordered basis (X_1, X_2, Y_1, Y_2) such that: $\forall A \in \{1, 2\} X_A$ is unit spacelike and Y_A is lightlike; X_1, X_2, and span $\{Y_1, Y_2\}$ are pairwise orthogonal; $g(Y_1, Y_2) = \pm 1$; and the matrix representing $\tilde{T}$ in this basis is

$$\tilde{T} \sim \begin{bmatrix} \lambda_1 & 0 & 0 & 0 \\ 0 & \lambda_2 & 0 & 0 \\ 0 & 0 & \lambda_3 & 1 \\ 0 & 0 & 0 & \lambda_3 \end{bmatrix} \qquad \lambda_1, \lambda_2, \lambda_3 \in \mathbb{R}.$$

(d) There is a basis as in (c) such that

$$\tilde{T} \sim \begin{bmatrix} \lambda_1 & 0 & 0 & 0 \\ 0 & \lambda_2 & 0 & 1 \\ 0 & 1 & \lambda_2 & 0 \\ 0 & 0 & 0 & \lambda_2 \end{bmatrix} \qquad \lambda_1, \lambda_2 \in \mathbb{R}.$$

8.1.6 Stress-energy tensors

In Section 8.1.5 show $\hat{T}$ has the basic algebraic property of a stress-energy tensor—that is, $\hat{T}(\omega, \omega) \geq 0 \; \forall$ causal $\omega \in M_z^*$, iff either: (a) case 8.1.5a holds with $\rho \geq 0$ and $\rho \geq -p_\alpha \; \forall \alpha \in (1, 2, 3)$; or (b) case 8.1.5c holds with $\lambda_3 \leq 0$, $\lambda_3 - \lambda_1 \leq 0$ and $\lambda_3 - \lambda_2 \leq 0$.

8.1.7 Conditions on stress-energy tensors

In Sections 8.1.4 to 8.1.6, suppose $\hat{T} = \tilde{T}z$, where $\hat{T}$ is a stress-energy tensor on M. $\hat{T}$ is said to obey the *strong energy condition* iff $\hat{T}(\omega, \omega) \geq -\frac{1}{2}(\text{trace } \tilde{T})|\omega|^2 \; \forall$ causal $\omega \in M_z^*$, and is said to obey the *dominant energy condition* iff $\tilde{T}Y$ is causal $\forall$ causal $Y \in M_z$.

(a) Show that if $\hat{T}$ obeys the timelike convergence condition (Exercise 4.3.7), then $\hat{T}$ obeys the strong energy condition.
(b) Assume case 8.1.6a holds. Show $\hat{T}$: obeys the strong energy condition iff $\rho \geq -\sum_{\alpha=1}^{3} p_\alpha$; obeys the timelike convergence condition iff it obeys the strong energy condition and is nonzero; obeys the dominant energy condition iff $\rho > 0$ and $\rho \geq |p_\alpha| \forall \alpha \in \{1, 2, 3\}$; and is normal iff $\rho > |p_\alpha| \forall \alpha \in \{1, 2, 3\}$.
(c) Give corresponding results for case 8.1.6c.
(d) Show the stress-energy tensor for a perfect fluid obeys all four conditions (strong and dominant energy, timelike convergence, and normalcy).

8.1.8 Electromagnetic stress energy tensors

Let: F be an electromagnetic field on M; $\hat{E}$ be its stress-energy tensor; $F = Fz \in T_2{}^0(M_z)$; $\hat{T} = \hat{E}z \in T_0{}^2(M_z)$; and E (respectively, B) be the electric (respectively, magnetic) field measured by (z, Z). Compare Section 3.4. Suppose $F \neq 0$; show:

(a) $g(E, B) = 0$ iff $F \wedge F = 0$;
(b) $g(E, E) = g(B, B)$ iff $F^{ij}F_{ij} = 0$ (cf. Section 3.6.1);
(c) $\tilde{T}$ is diagonalizable iff $\hat{T}$ obeys the dominant energy condition iff either $g(E, B) \neq 0$ or $g(E, E) \neq g(B, B)$;
(d) $\hat{T}$ is not normal but obeys the timelike convergence condition and the strong energy condition.

8.1.9 Decompositions of (2, 0) tensors

We use the notation of Section 8.1.0. Suppose $S \in T_0{}^2(M_z)$; $\hat{h} = \hat{g} + Z \otimes Z$ is physically equivalent to the projection tensor (Section 2.1.5); and R is the rest space $Z^\perp$.

(a) Show there exist: unique $a, \theta \in \mathbb{R}$; unique $X, Y \in R$; a unique symmetric tensor $\sigma \in R \otimes R$, trace-free with respect to $g|_R = h|_R$; and a unique skew-symmetric tensor $\omega \in R \otimes R$ such that the following decomposition of S holds.
$$S = aZ \otimes Z + Y \otimes Z + Z \otimes X + \sigma + \omega + \tfrac{1}{3}\theta\hat{h}.$$

(b) Write out the components in (a) using an orthonormal basis (X_1, X_2, X_3, Z) and its dual.

(c) Show the quantities in (a) can equivalently be defined by the following conditions. $\forall \alpha, \beta \in M_z{}^*$ such that $\alpha(Z) = 0 = \beta(Z)$:
$$a = S(\zeta, \zeta), \qquad \theta = \text{trace } S + a;$$
$$\alpha(X) = S(\zeta, \alpha), \qquad \alpha(Y) = S(\alpha, \zeta);$$
$$\omega(\alpha, \beta) = \tfrac{1}{2}[S(\alpha, \beta) - S(\beta, \alpha)];$$
$$\sigma(\alpha, \beta) = \tfrac{1}{2}[S(\alpha, \beta) + S(\beta, \alpha)] - \tfrac{1}{3}\theta\hat{h}(\alpha, \beta).$$

8.1.10 Acceleration, expansion, rotation and shear of a reference frame

Let S be the (2, 0)-tensor physically equivalent to $(DZ)z$ (cf. Section 8.1.0).

(a) Show S algebraically determines the relative velocity linear transformation $(A_Z)z\colon Z^\perp \to Z^\perp$ (cf. Proposition 2.3.4).
(b) Show that in Section 8.1.9 $S = X \otimes Z + \sigma + \omega + \tfrac{1}{3}\theta\hat{h}$ with $-X = (D_Z Z)z$ (the acceleration) and $\theta = (\text{div } Z)z$. In this case θ is called the *expansion*, σ the *shear*, and ω the *rotation* of the reference frame Z at z. Now show:
(c) Z is irrotational at z iff $\omega = 0$;
(d) Z is rigid at z iff $\sigma = 0 = \theta$;
(e) If Z is proportional to a Killing vector field, Z is rigid.
(f) Show by an example that in general there is no shear free reference frame on any open neighborhood of a given point in a given spacetime. (Hint: in local coordinates, the requirement $\sigma = 0$ becomes five independent conditions on the components of Z.)

(g) Imagine an "infinitesimal" solid sphere in $Z^\perp$. Sketch the intrinsic shape of the slightly distorted sphere after an "infinitesimal" proper time, assuming it is dragged along by the flow of Z (cf. Sections 2.0.3 and 2.3); identify θ, ω, and σ on your figure. (You may assume $X = 0$ for simplicity.)

(h) Show $Z^{i|j}Z_{j|i} = \sigma^{ij}\sigma_{ij} - \omega^{ij}\omega_{ij} + \frac{1}{3}\theta^2$ (cf. Section 3.6.1).

(i) In (h) show $\sigma^{ij}\sigma_{ij} = 0$ implies $\sigma_{ij} = 0$ and $\omega^{ij}\omega_{ij} = 0$ implies $\omega_{ij} = 0$.

(j) Use component notation (Section 3.6.1) and the above to simplify the proof of Lemma 4.3.3. (Hint: in component notation the curvature tensor R obeys $Z^i_{|j|k} - Z^i_{|k|j} = R^i_{mjk}Z^m$, as follows from its definition in Section 1.0.2.)

(k) Show the linear transformation $R \to R$ physically equivalent to σ (respectively, ω) is self-adjoint (respectively, skew-adjoint).

8.1.11 Averages over directions

Let $\mathscr{S}^2$ be the unit sphere in the rest space R (cf. Section 8.1.0).

(a) Show $\mathrm{Ric}(Z, Z) = (3/4\pi)\int_{X \in \mathscr{S}^2} K_X \zeta$, where ζ is the standard volume element and K_X is the sectional curvature (Bishop–Goldberg 5.14) of span $\{X, Z\}$.

(b) For $X \in R$, let α_X be the relative 3-acceleration of that neighbor W of an observer γ in Z through z such that $W\gamma^{-1}z = X$ (cf. Section 2.3). Let $\alpha: M \to \mathbb{R}$ be the mean relative-acceleration of Z (cf. Section 4.2). Show $\alpha z = (1/4\pi)\int_{X \in \mathscr{S}^2} g(\alpha_X, X)\zeta$.

(c) Using the relative velocity linear transformation of Section 2.3, express $(\mathrm{div}\, Z)z$ as an average similar to those in (a) and (b).

8.1.12 Lightlike geodesics

(Hard) Let Y be a future-pointing lightlike geodesic vector field on M. Give a decomposition of $(DY)z$, similar to that of Section 8.1.10, as follows: Introduce a projection operator that projects orthogonally into the 2-space $[\mathrm{span}\{Yz, Z\}]^\perp$ and there decompose into a shear, rotation, and expansion. Analyze in detail to what extent the various bits of the decomposition depend on the arbitrarily chosen instantaneous observer (z, Z). Give geometric and physical interpretations for each intrinsic bit. (Cf. Ellis [1] and Hawking–Ellis [1]; there is a very large literature on such vector fields Y and such decompositions.)

8.1.13 Curvature tensors

Suppose $\mathrm{Ric} = 0$. (M, g) is defined to have the *Petrov type N* at z iff the curvature tensor Rz is nonzero and there exists a nonzero $L \in M_z$ such that $R(\tau, L, X, Y) = 0 \,\forall \tau \in M_z^* \,\forall X, Y \in M_z$. Show that then L is lightlike. (For a systematic discussion of Petrov types, cf. Penrose–Rindler [1].)

8.2 Differential topology and geometry

When standard methods of differential topology and geometry are applied to Lorentzian manifolds, special features occur. This section and the next outline some.

To work all the exercises in Sections 8.2 to 8.3, particularly to find the various examples and counterexamples, the reader will probably need to consult Penrose [1] and Hawking–Ellis [1]. Many of the recent results are due to Penrose, Hawking, or Geroch.

8.2.1 PARACOMPACTNESS

In this exercise, and only in this exercise, we take "manifold" to mean manifold in the sense of Kobayashi–Nomizu [1]—that is, second countability is dropped from the list of defining axioms. (Contrast the Bishop–Goldberg [1] definition of manifold and Section 0.0.4). Let N be a connected, Hausdorff, real, C^∞ manifold of finite dimension ≥ 2.

(a) Show N is second countable iff N is Lindeloff iff N is paracompact iff N is metrizable iff there is a Riemannian metric on N.
(b) Suppose N is a Lie group. Show there is a Riemannian metric on N.
(c) Suppose there is a Lorentzian metric g on N. Let ON be the bundle of Lorentzian orthonormal frames above (N, g). Show there is a Riemannian metric on ON. (Hint: See Sections 8.1.2a and 8.2.1b).
(d) Show that if there is a Lorentzian metric on N then N is paracompact.

8.2.2 LORENTZIAN MANIFOLDS

Let N be a manifold (Section 0.0.4). Thus N is real, Hausdorff, finite-dimensional, paracompact and C^∞.

(a) Show there exists a Lorentzian metric on N iff there exists a nowhere zero vector field on N iff there exists a Lorentzian metric g on N such that (N, g) is time orientable.
(b) Suppose N is connected, orientable, and 4-dimensional. Show there exists a spacetime (N, g) iff there is a nowhere zero vector field on N.
(c) Suppose N is connected, orientable, not compact and 4-dimensional. Show there exists a spacetime (N, g).

8.2.3 TIME ORIENTABILITY

Let (N, g) be a connected Lorentzian manifold. Show:

(a) (N, g) is time orientable iff there is a causal vector field on (N, g) iff there is a timelike vector field on (N, g) iff, $\forall$ curve $\gamma: [0, 1] \to M$ with $\gamma 0 = \gamma 1$, parallel transport with respect to the Levi–Civita connection of any timelike vector in $M_{\gamma 0}$ along γ gives a vector in the same connected component of $\mathcal{T}_{\gamma 0} \subset M_{\gamma 0}$ (cf. Section 1.1).
(b) If N is simply connected, (N, g) is time orientable.

8.2.4 SOME COUNTEREXAMPLES

Let (N, g) be a connected Lorentzian manifold. Show by examples that each of the following is possible:

(a) N is orientable but (N, g) is not time orientable.
(b) (N, g) is time orientable but N is not orientable.
(c) N is not orientable and (N, g) is not time orientable.

8 Optional exercises: relativity

(d) (N, g) is time orientable but there is a timelike curve into N which is not $1-1$.
(e) Each timelike curve into N is $1-1$ but (N, g) is not time orientable.

8.2.5 SPACETIMES

Let (M, g) be a spacetime. Show:

(a) There is a reference frame Z on M.
(b) If M is compact, M is not simply connected. (Hint: use Poincaré duality. Remarks: thus one can work with a noncompact covering space; indeed, compact spacetimes are almost never considered in relativity.)
(c) If ω is a 1-form physically equivalent to a reference frame on M; $g + 2\omega \otimes \omega$ is a Riemannian metric on M.
(d) There is a volume element χ on M such that $\int_M \chi = 1$. (Remarks: χ is not unique; for suitable $A \in (0, \infty)$ and $M =$ Minkowski space,

$$\chi = A \exp\left[-\sum_{i=1}^{4}(u^i)^2\right] du^1 \wedge \cdots \wedge du^4$$

is an example.)

8.2.6 MORE COUNTEREXAMPLES

Let (M, g) be a spacetime. Show by examples that each of the following is possible.

(a) (M, g) is compact but not complete.
(b) There exists an inextendible lightlike geodesic $\gamma: (-\infty, 0) \to M$ whose image is the (compact) set $\gamma[-2, -1]$.
(c) (M, g) is complete but there are distinct points $x, y \in M$ such that no geodesic goes from x to y.
(d) Each inextendible causal geodesic has domain $\mathbb{R}$ but (M, g) is not complete.
(e) There is no lightlike vector field on (M, g).

8.2.7 SECOND FUNDAMENTAL FORMS

Let B be a 3-manifold and let $\beta: B \to M$ be a spacelike imbedding. Thus with $b = \beta^* g$, (B, b) is a Riemannian manifold. (a) Show B is orientable. (b) Show there exists a unique symmetric $(0, 2)$-tensor field k on B with the following property. Let $\mu: \mathscr{E} \to B$ be a geodesic of (B, b), $\nu = \beta \circ \mu$. Then $D_{\nu_*}\nu_* = k(\mu_*, \mu_*)Z$, where Z is a future-pointing unit timelike vector field over ν which is orthogonal to βB. k is defined as the *second fundamental form of β*. Note that k measures, roughly speaking, how much geodesics of B are curved in M, which is an indication of how B is inserted into M; k is thus sometimes called the *extrinsic curvature of β*; we are using the sign convention standard in physics tests.

(c) Let Z be a reference frame on M such that the physically equivalent 1-form ζ obeys $\beta^*\zeta = 0$. Show that $k = -\beta^*(D\zeta)$. Thus k also measures, roughly speaking, how much the unit normal to B tilts as we move around on B. (d) Let s be the $(0, 2)$-tensor field on M such that sz is physically equivalent to the shear of Z at z $\forall z \in M$; θ be the expansion of Z. Show $-k = \beta^*s + \frac{1}{3}(\theta \circ \beta)b$. (e) Now assume Z is geodesic. Show $k = -\beta^* L_Z(g + \zeta \otimes \zeta)$, where L_Z is the Lie derivative. (f) Interpret (d) and (e).

8.3 Chronology and causality

We outline some properties of the causality and chronology relations. Penrose [1] gives a systematic presentation. Hawking–Ellis [1] and DeWitt–DeWitt [1] give the main applications to black hole theory, collapse theorems, cosmology, and other areas. This section merely states a few of the main definitions and standard lemmas.

8.3.1 BASIC PROPERTIES

(a) Show $\ll \circ \ll = \ll$, where "$\circ$" denotes composition of relations. (Thus, in particular, $\ll$ is transitive.)
(b) By definition (M, g) *obeys the chronology condition* iff no point in M chronologically precedes itself. (Roughly: ancestorcide is impossible.) Show a Kruskal spacetime obeys the chronology condition.
(c) Find a spacetime such that $x \ll x\ \forall x \in M$.
(d) Show $\leq\ =\ \ll \circ \leq\ =\ \leq \circ \ll\ =\ \leq \circ \leq$.

8.3.2 CHRONOLOGY, CAUSALITY AND TOPOLOGY

(a) Denote the chronological (respectively, causal) past of $z \in M$ by $I^-\{z\}$ (respectively, by $J^-\{z\}$). Show $I^-\{z\}$ is perfectly open—that is,
$$\text{Interior [Closure } I^-\{z\}] = I^-\{z\}.$$
Show Closure $J^-\{z\}$ = Closure $I^-\{z\}$ = $\{x \in M \mid I^-\{x\} \subset I^-\{z\}\}$. Give an example where $J^-\{z\}$ is neither open nor closed.
(b) $\forall x, y \in M$, define $\langle x, y \rangle = \{z \in M \mid x \ll z \ll y\}$. Show $\{\langle x, y \rangle\}$ is a base for a topology on M. (This topology is called the *Alexandrov topology*; cf. Section 8.1.3a.)
(c) Suppose M is compact in the Alexandrov topology; show (M, g) does not obey the chronology condition.
(d) Suppose M is compact in its ordinary topology; show (M, g) does not obey the chronology condition.
(e) (Penrose) Show the Alexandrov topology is the ordinary topology iff the Alexandrov topology is Hausdorff. Spacetimes that obey this condition are defined as *strongly causal*.
(f) Show a strongly causal spacetime obeys the chronology condition; give an example to show the converse can fail.

8.3.3 PERTURBED METRICS

A Lorentzian metric w on M is defined as *wider* than g iff wx is wider than $gx\ \forall x \in M$ (cf. Section 8.1.2).

(a) Show there exist spacetimes (M, w) and (M, n) such that w is wider than g and g is wider than n.
(b) Suppose (M, g) obeys the chronology condition; show in (a) that (M, n) obeys the chronology condition.

8 Optional exercises: relativity

(c) Show by an example that (M, g) can obey the chronology condition even if (M, w) violates the chronology condition $\forall w$ wider than g. (Roughly, this shows that the chronology condition does not have the kind of stability which one usually demands for physically acceptable restrictions; the next exercise discusses a more plausible restriction.)

8.3.4 Stable causality

(Hawking) (M, g) is defined as *stably causal* iff there is a spacetime (M, w) such that w is wider than g and (M, w) obeys the chronology condition.

(a) Show a stably causal spacetime is strongly causal but the converse can fail.
(b) Choose any of the spacetimes defined in Section 1.4 or Chapter 7 and show it is stably causal.

8.3.5 C^0 Global time functions

A C^0 function $T: M \to \mathbb{R}$ is a *global time function* iff T is monotonic increasing along each future-pointing curve. Show (M, g) is stably causal iff there is a C^0 global time function on M.

8.3.6 Causal continuity

Let χ be a volume element on M such that $\int_M \chi = 1$ (cf. Section 8.2.5d). Define two functions, not necessarily C^0, $T_\chi^\pm : M \to (0, 1]$ by $T_\chi^\pm x = \pm \int_{I^\pm\{x\}} \chi \, \forall x \in M$.

(a) By removing a closed subset of Minkowski space, give an example where $T_\chi^\pm$ is not C^0.
(b) (M, g) is defined as *causally continuous* iff there exists a χ as above such that T_χ^+ and T_χ^- are C^0 global time functions. Show (M, g) is causally continuous iff (M, g) is strongly causal and, $\forall x, y \in M$, $x \in$ Closure $I^-\{y\}$ iff $y \in$ Closure $I^+\{x\}$. Show that then $T_\chi^\pm$ is a C^0 global time function $\forall \chi$ as above.
(c) Show that a causally continuous spacetime is stably causal but the converse need not hold.

8.3.7 Global hyperbolicity

(M, g) is defined as *globally hyperbolic* iff there exists a subset $\mathscr{B} \subset M$ such that each endless timelike curve intersects $\mathscr{B}$ exactly once. $\mathscr{B}$ is then called a *Cauchy surface*. Let (M, g) be a globally hyperbolic spacetime, $\mathscr{B} \subset M$ be a Cauchy surface. Show:

(a) $\mathscr{B}$ is a C^0 imbedded submanifold.
(b) Each inextendible lightlike geodesic intersects $\mathscr{B}$ in a compact set.
(c) (M, g) is causally continuous, but a causally continuous spacetime need not be globally hyperbolic.
(d) M is homeomorphic to $\mathscr{B} \times \mathbb{R}$ and each Cauchy surface is homeomorphic to $\mathscr{B}$.

8.3.8 EXAMPLES

(a) Show Minkowski space, Einstein–de Sitter spacetime, and any Kruskal spacetime are globally hyperbolic.
(b) Find a complete spacetime that is not globally hyperbolic.
(c) Show that in Minkowski space there is a 3-dimensional, spacelike imbedded submanifold that is complete in its induced metric but is not a Cauchy surface.

8.3.9 PROPER TIME

$\forall$ timelike future pointing $\gamma: [a, b] \to M$ let $L_\gamma = \int_a^b |\gamma_* u|\, du$ be the arclength; thus $L_\gamma > 0$. Define the *proper time function* $t: \ll \to (0, \infty]$ by: $t(p, q) =$ lub $\{L_\gamma | \gamma$ is a future-pointing timelike curve from p to $q\}$ $\forall p, q \in M$ such that $p \ll q$.

(a) Give an example to show t need not be continuous.
(b) Show t obeys the global version of the wrong-way triangle inequality (Exercise 1.2.4b)—that is, $t(x, z) \geq t(x, y) + t(y, z)\, \forall x, y, z \in M$ such that $x \ll y \ll z$.

8.3.10 EXISTENCE OF GEODESICS

Let (M, g) be a globally hyperbolic spacetime, and let t be the proper time function.

(a) Show $t(p, q)$ is finite $\forall (p, q) \in \ll$.
(b) Show that $t_p: I^+\{p\} \to (0, \infty)$, defined by $t_p q = t(p, q)$, is continuous.
(c) Show that $\forall (p, q) \in \ll$ there is a future-pointing timelike geodesic γ from p to q such that $L_\gamma = t(p, q)$.

8.3.11 COLLAPSE THEOREMS

Let $(M, \mathcal{M}, F)$ be a relativistic model such that the Einstein field equation (Definition 4.1.1) holds and the matter stress-energy tensor obeys the dominant energy condition at each point in M. Suppose there is an imbedded, spacelike Cauchy surface such that the trace of the second fundamental form of the imbedding is everywhere less than some fixed negative number. Show the spacetime is not complete. (Roughly, this says that if the galaxies are now diverging there must have been a big bang.)

8.4 Isometries and characterizations of gravitational fields

How can one, starting from some intuitive physical picture, find an appropriate spacetime model? How can one characterize a gravitational field intrinsically? In general, neither of these two related questions has a straightforward answer. This section indicates the main ideas used to obtain and characterize the examples used in the text. Throughout the section: $\mathcal{G}M$ is the isometry group of (M, g) and $\mathcal{H} \subset \mathcal{G}M$ is a subgroup. For each $x \in M$

8 Optional exercises: relativity

the *orbit* of $\mathscr{H}$ through x is $\{y \in M \mid \text{there exists } \phi \in \mathscr{H} \text{ such that } \phi x = y\}$. Each orbit of $\mathscr{H}$ is a submanifold (Kobayashi–Nomizu [1]).

8.4.1 Minkowski space

Show (M, g) represents the trivial gravitational field iff (M, g) is simply connected, complete, and flat (cf. the Ambrose–Hicks theorem—e.g., in Wolf [1]; or use the results of Bishop–Goldberg sections 5.6–5.12).

8.4.2 The Poincaré group

(a) Let (M, g) be Minkowski space. Show $\mathscr{G}M$ is a ten-dimensional Lie group with four connected components; this $\mathscr{G}M$ is called the *Poincaré group*; it is often considered the most basic object in theoretical physics.
(b) In (a), suppose $x \in M$ and let $\mathscr{H} = \{\phi \in \mathscr{G}M \mid \phi x = x\}$. Show $\mathscr{H}$ is a six-dimensional Lie group; it is called the ("homogeneous") *Lorentz* group.
(c) Show the spacetime (M, g) is isometric to Minkowski space iff $\mathscr{G}M$ is isomorphic to the Poincaré group.

8.4.3 Remarks

We shall now mention some characterizations similar to Section 8.4.2c. The use of isometry groups corresponds to the physicist's concept of symmetry. A really detailed, realistic spacetime model normally has no symmetry: $\mathscr{G}M$ is the trivial group. But models simple enough to analyze explicitly usually have a nontrivial isometry group. The characterizations usually also involve algebraic conditions on the Ricci or Einstein tensor. These correspond to assuming particular matter models and assuming the Einstein field equation (cf. Chapters 3 and 4). The main trick required for the exercises below is the introduction of local coordinates adapted to group orbits, similar to the coordinates adapted to an integrable distribution (cf. Bishop–Goldberg 3.11 and 3.12).

8.4.4 The simplest example

Let (N, h) be a 2-dimensional connected Lorentzian manifold. Suppose the isometry group contains a subgroup $\mathscr{H}$ isomorphic to the additive group $\mathbb{R}$, and suppose each orbit of $\mathscr{H}$ is one-dimensional and spacelike.

(a) Show there exists a smooth function $\mathscr{R}: \mathbb{R} \to (0, \infty)$ such that (N, h) is locally isometric to $(\mathbb{R}^2, (\mathscr{R}^2 \circ u^2) du^1 \otimes du^1 - du^2 \otimes du^2)$. (Hint: use coordinates such that, locally, each orbit is a level hypersurface of u^2 and consider the orthogonal trajectories to the orbits.)
(b) Show that N is diffeomorphic either to $\mathbb{R}^2$ or to $\mathbb{R} \times \mathscr{S}^1$ or to $\mathscr{S}^1 \times \mathscr{S}^1$.

8.4.5 Simple cosmological spacetimes

Suppose that $\mathscr{G}M$ is isomorphic to the (6-dimensional Lie) isometry group of Euclidean 3-space and that each orbit of $\mathscr{G}M$ is 3-dimensional spacelike. Show (M, g) is locally isometric to a simple cosmological spacetime.

8.4 Isometries and characterizations of gravitational fields

8.4.6 SPHERICAL SYMMETRY

Define (M, g) as *spherically symmetric* iff $\mathscr{G}M$ contains a subgroup $\mathscr{H}$ with the following two properties: $\mathscr{H}$ is isomorphic to the group $\mathscr{SO}^3$ of proper rotations of Euclidean 3-space, and each orbit of $\mathscr{H}$ is spacelike and at most 2-dimensional.

(a) Show that then an orbit diffeomorphic to the 2-sphere has the intrinsic geometry of a 2-sphere in Euclidean space.
(b) Show that then no orbit of $\mathscr{H}$ is one-dimensional.
(c) Show that each spacetime defined in Section 1.4 is spherically symmetric.
(d) Suppose (M, g) is Einstein–de Sitter spacetime and $\mathscr{H}$ is one of the subgroups with the above properties. Show there are 0-dimensional orbits, and that all these orbits lie on one integral curve of ∂_4. (Roughly: each such orbit is a "center of rotation.")
(e) Suppose (M, g) is a normal Schwarzschild spacetime or a Schwarzschild black hole, and let $\mathscr{H}$ be as above. Show there is no 0-dimensional orbit (no "center"; in spherically symmetric models of stable stars one replaces part of a normal Schwarzschild spacetime with a different spacetime, using some appropriate matter model; then there is a timelike geodesic—the history of the center of the star—each point on which is a zero-dimensional orbit).
(f) Show by an example that there exists a spherically symmetric spacetime such that each orbit of $\mathscr{H}$ is diffeomorphic to a projective plane. (This case is sometimes excluded by definition.)
(g) Show by an example there is a spacetime such that $\mathscr{G}M$ is isomorphic to $\mathscr{SO}^3$ but each orbit is 3-dimensional. (Hint: consider $\mathscr{SO}^3 \times \mathbb{R} = M$.)

8.4.7 SCHWARZSCHILD AND KRUSKAL SPACETIMES

Let (M, g) be a Ricci-flat, spherically symmetric spacetime that is not flat.

(a) Show (M, g) is locally isometric to a Kruskal spacetime.
(b) Give a global version of (a).

8.4.8 PLANE WAVES

Suppose: (M, g) is Ricci flat but not flat; there are subgroups $\mathscr{H}$, $\mathscr{J}$, such that $\mathscr{H} \supset \mathscr{J}$; $\mathscr{J}$ is isomorphic to the additive group $\mathbb{R}$ and has one-dimensional lightlike orbits; and $\mathscr{H}$ is Abelian, 3-dimensional, and has 3-dimensional lightlike orbits. (Despite their clumsiness, the conditions on $\mathscr{H}$ and $\mathscr{J}$ merely make precise the idea that we have a wave with plane wavefronts travelling at the speed of light.)

(a) Show (M, g) is locally isometric to a gravitational plane wave.
(b) Suppose (M, g) is complete and diffeomorphic to $\mathbb{R}^4$. Show (M, g) is a plane wave and $\mathscr{G}M$ is a Lie group with dimension ≥ 5.

8.4.9 STATIONARY AND STATIC SPACETIMES

Suppose Z is an absolute stationary reference frame on M. Show:

(a) $\forall x \in M$, there is a coordinate neighborhood $\mathscr{U}$ of x such that on $\mathscr{U}$ Z is proportional to ∂_4 and $\partial_4 g_{ij} = 0$.
(b) (M, g) is static iff $\forall x$ we can find such a coordinate neighborhood with the additional property that $g_{4\alpha} = 0 \; \forall \alpha \in (1, 2, 3)$.

8 Optional exercises: relativity

8.5 The Einstein field equation

The exercises of this section state some results on the Einstein field equation. We emphasize the vacuum condition **Ric** = 0, interpreted in Section 4.2.

8.5.1 THE CONSTRAINT EQUATIONS

Let B be a 3-manifold, let $\beta: B \to M$ be a spacelike imbedding, and let k be the second fundamental form of β; write $b = \beta^* g$. Thus (B, b) is a Riemannian 3-manifold. Denote its scalar curvature by b: $B \to \mathbb{R}$ and let $\tilde{k}$ be the (1, 1)-tensor field on B physically equivalent to k via b. (a) Show that if (M, g) has zero Ricci tensor the triple (B, b, k) obeys the *constraint equations*:

$$b = (\text{trace } \tilde{k})^2 - \text{trace [contraction } \tilde{k} \otimes \tilde{k}];$$

and

$$\text{div } \tilde{k} = d(\text{trace } \tilde{k}).$$

Here the Levi–Civita connection of b is used to form the divergence div; and contraction $\tilde{k} \otimes \tilde{k}$ has components $\sum_{\rho=1}^{3} k_\rho{}^\alpha k_\delta{}^\rho$ in local coordinates. (b) Conversely, suppose the constraint equations hold whenever β is such an imbedding. Show (M, g) is Ricci flat.

> As the next exercise indicates, the constraint equations play for the equation **Ric** = 0 a role analogous to that played for Maxwell's equation by the constraint equations of Theorem 3.11.1.

8.5.2 THE INITIAL VALUE PROBLEM

This exercise (cf. Bruhat [1]) states a "present determines the future" result, similar to those in Chapter 3, for the Einstein vacuum condition. For convenience, we restrict attention to spacetimes that have a compact Cauchy surface. Let (B, b) be a Riemannian 3-manifold with B compact and orientable, k be a symmetric (0, 2)-tensor field on B. Suppose the triple (B, b, k) obeys the constraint equations (Section 8.5.1a).

(a) Show that $\forall$ sufficiently small $a \in (0, \infty)$ there exists a Ricci flat spacetime (M, g) that is the *time development* of (B, b, k) in the following sense. $M = B \times (-a, a)$; the map $\beta: B \to M$ given by $\beta b = (b, 0) \; \forall b \in B$ is a spacelike imbedding; $b = \beta^* g$; and k is the second fundamental form of β. (b) Suppose such a sufficiently small a is given; show the spacetime equivalence class $[(M, g)]$ is uniquely determined.

8.5.3 LINEARIZED EQUATIONS

An enormous amount of work has been done, especially in cosmology and in black hole theory, on comparatively simple linear equations related to the Einstein field equation in roughly the way the equation for the tangent plane to a surface in $\mathbb{R}^3$ at a point is related to the equation for the surface. This exercise outlines some of the ideas involved in the case of the vacuum equations **Ric** = 0.

8.5 The Einstein field equation

(a) Suppose $0 < a < c$ and $\forall u \in (a, c)$ we have a Lorentzian metric g_u on $\mathbb{R}^4$. Assume that $\forall i, j \in (1, \ldots, 4)$ the function $g_{ij}: \mathbb{R}^4 \times (a, c) \to \mathbb{R}$ defined by $g_{ij}(x, u) = g_u(\partial_i x, \partial_j x)$ is C^∞. We may, and shall, regard (M, g_u) as a spacetime $\forall u \in (a, c)$. With $b \in (a, c)$, abbreviate g_b by g. Note that $h = \sum_{i,j=1}^{4} (\partial g_{ij}/\partial u)_{u=b} du^i \otimes du^j$ is a symmetric $(0, 2)$-tensor field on $\mathbb{R}^4$; intuitively speaking, $g + (u - b)h$ is a linear approximation to g_u for $|b - u|$ sufficiently small. Show that if each spacetime $(\mathbb{R}^4, g_u)$ is Ricci flat, h obeys the following equation.
$$g^{ij}(h_{ki|l|j} - h_{ki|l|j} - h_{li|k|j} + h_{ij|k|l}) = 0$$
where we use the notation of Section 3.6.1.

(b) Let (M, g) be a Ricci flat spacetime, and let h be a symmetric $(0, 2)$-tensor field on M. By definition, h *obeys the vacuum Einstein equation linearized about (M, g)* iff h obeys the equation at the end of (a) in each local coordinate system. An example can be constructed as follows. Let (M, g_u) be the Kruskal spacetime of active mass u. By formally differentiating g_u with respect to u, find an h on M which obeys the vacuum Einstein equation linearized about (M, g_1).

In this case we do not really need the linear approximation $g_1 + (u - 1)h$, since we have g_u available. However, in applications it often happens that one can find many solutions h of the linearized vacuum equations without being able to find corresponding collections $\{(M, g_u) \mid u \in (a, c)\}$ ("curves in the space of Lorentzian metrics") explicitly. Then, given h, one should ask whether at least one corresponding curve exists. In certain cases—for example, when M has a compact Cauchy surface—an extremely elegant existence result can be stated and proved using appropriate infinite dimensional manifolds (cf. Fischer–Marsden [1] and Moncrief [1]). In this case the result is, roughly, that there is at least one curve for each solution h iff (M, g) has no Killing vector field other than the zero vector field.

8.5.4 Gauge transformations

In Section 8.5.3a, let $\varepsilon: \mathbb{R}^4 \times (a, c) \to \mathbb{R}^4$ be a C^∞ map such that $\forall u \in (a, c)$ $\varepsilon_u: \mathbb{R}^4 \to \mathbb{R}^4$ defined by $\varepsilon_u(x) = \varepsilon(x, u)$ is a diffeomorphism, with ε_b the identity. Thus with $g_u' = \varepsilon_u^* g_u$, $(\mathbb{R}^4, g_u')$ is a Ricci flat spacetime $\forall u \in (a, c)$. (a) Show there is a unique vector field V on M such that $df(V) = [\partial(f \circ \varepsilon_u)/\partial u]_{u=b}$ $\forall$ function f on M. (b) Defining h' by analogy with Section 8.5.3a, show $h' = h + L_V g$, where L_V is the Lie derivative. (c) In Section 8.5.3b, let V be a vector field on M and define $h' = h + L_V g$. Show h' obeys the vacuum Einstein equation linearized about (M, g) iff h does. Here h' is defined as the *transform of h under the gauge transformation* determined by V. In view of (b) above one regards h' and h as equivalent: they differ merely via diffeomorphisms.

8.5.5 Variational principle

Let $(\mathbb{R}^4, g)$ be a spacetime. Regarding the components $g(\partial_i, \partial_j)$ as fields to be varied, show the Einstein vacuum field equation **Ric** $= 0$ is equivalent to the Euler–Lagrange equations for the action $\int S\Omega$, where S is the scalar curvature and Ω is the metric volume element.

263

8.5.6

Generalize Sections 8.5.1 to 8.5.5 to the case of relativistic models $(M, \mathcal{M}, F)$ which obey the Einstein field equation and Maxwell's equation with $\mathcal{M}$ a complete matter vacuum matter model. (Cf. Misner–Thorne–Wheeler [1] for the generalization of Section 8.5.5.)

8.6 Gases

We outline some of the geometric methods used when discussing gases, in particular photon gases. Throughout the section: $\Pi: TM \to M$ is the projection, $K: TM \to \mathbb{R}$ is the function of Section 5.0.5, $\mathscr{L}^+ \subset TM$ is the future light cone, Ξ is the restriction of Π to $\mathscr{L}^+$, and $\forall x \in M$, Λ_x is the natural volume element of $\mathscr{L}_x^+ = \mathscr{L}^+ \cap M_x$ as defined in Section 5.6.

8.6.1 HORIZONTAL VECTOR FIELDS

Let $I: TM \to TM$ be the identity map. Thus we can regard I as a vector field I over the projection Π (cf. Section 2.0). Similarly we regard the restriction I_0 of I to $\mathscr{L}^+$ as a vector field over Ξ. A vector field H on TM is defined as *horizontal* iff $(\Pi^*D)_H I = 0$ (cf. Bishop–Goldberg [1], Section 5.12). Similarly a vector field $H: \mathscr{L}^+ \to T\mathscr{L}^+$ is *horizontal* iff $(\Xi^*D)_H I_0 = 0$. Let H be a horizontal vector field on TM. (a) Show that $dK(H) = 0$. (b) Suppose $v \in \mathscr{L}^+$. Show $H(v)$ is tangent to $\mathscr{L}^+$. (c) Show there exists a unique horizontal vector field G on TM such that $\Pi_* G = I$. (d) Show: if γ is an integral curve of G, $\Pi \circ \gamma$ is a geodesic; if $\alpha: \mathscr{E} \to M$ is a geodesic, $\alpha_*: \mathscr{E} \to TM$ is an integral curve of G. G is defined as the *geodesic spray*. By (b) we may regard the restriction of G to $\mathscr{L}^+$ as a vector field $L: \mathscr{L}^+ \to T\mathscr{L}^+$. Following physics terminology we will call L the *Liouville operator* on $\mathscr{L}^+$.

8.6.2 LOCAL UNIFORM BOUNDEDNESS

Let $F: \mathscr{L}^+ \to [0, \infty)$ be a function; thus F is C^∞ (Section 0.0.4). Suppose $\forall x \in M$ there exists an open neighborhood $\mathscr{U}_x \subset M$ of x such that, on $\Xi^{-1}\mathscr{U}_x$, $\sup|\bar{\sigma}^N H^L F| < \infty$ whenever: σ is a 1-form on $\mathscr{U}_x$ with $\bar{\sigma}: \mathscr{L}^+ \to \mathbb{R}$ the corresponding function; N is a positive integer; H is a horizontal vector field on $\Xi^{-1}\mathscr{U}_x$ which is Ξ related to some vector field on $\mathscr{U}_x$; L is a nonnegative integer; and H^L denotes the Lth derivative—that is, $H^0 F = F$, $H^1 F = dF(H)$, $H^2 F = \{d[dF(H)]\}(H)$, and so on. (a) Prove F is a photon gas. (b) Show F is conserved on M iff $LF = 0$, where L is the Liouville operator on $\mathscr{L}^+$. (c) Suppose F is conserved on M. Show the particle number density N of F obeys div $N = 0$ and the stress-energy tensor T of F obeys div $T = 0$. [Hint: suppose $x \in M$; the easiest way to show (div $N)(x) = 0$ is to introduce in a neighborhood of x coordinates normal at x and corresponding coordinates in a neighborhood of $M_x \subset TM$.]

8.6.3 CONSERVED PHOTON GASES

This exercise outlines the motivation for the physical interpretation of the conservation condition in Section 5.7. (a) Show there is a unique 3-form Λ on $\mathscr{L}^+$ such that: $i(H)\Lambda = 0$ $\forall$ horizontal vector field H on $\mathscr{L}^+$; and $\Lambda(X, Y, Z) = \Lambda_x(X, Y, Z) \forall X, Y, Z \in \mathscr{L}_x^+$, $\forall x \in M$, where Λ_x is the natural volume element of

$\mathscr{L}_x^+$, X is X regarded as a vector field on $\mathscr{L}_x^+$ (cf. Exercise 0.0.10a), and similarly for Y, Z.

(b) Define the 7-form $\omega = \Lambda \wedge (\Xi^*\Omega)$ on $\mathscr{L}^+$, where Λ is defined in (a), and Ω is the Lorentzian volume element of M. Show ω is a volume element on $\mathscr{L}^+$ with the following property. If ω' is any 7-form on TM whose pullback to $\mathscr{L}^+ \subset TM$ equals ω then $\omega' \wedge dK = g^*(d\theta \wedge d\theta \wedge d\theta \wedge d\theta)$, where θ is the canonical 1-form on the cotangent bundle T^*M (Bishop–Goldberg [1], Section 6.3) and $g: TM \to T^*M$ is the bundle isomorphism that sends each vector to its physically equivalent 1-form. (c) Let L_L be the Lie derivative with respect to the Liouville operator L. In (b) show *Liouville's theorem*: $L_L \omega = 0$. Show $L_L[i(L)\omega] = 0$.

(d) Let $\mathscr{N} \subset \mathscr{L}^+$ be an imbedded 6-submanifold whose closure is an integration region intersected at most once by each integral curve of L, and let F be a photon gas on M. Then $|\int_{\mathscr{N}} Fi(L)\omega|$ is defined as the *total number of photons in* ("crossing") $\mathscr{N}$. Show this definition is consistent with the definition in Exercise 5.7.7b. Use various special choices of $\mathscr{N}$ to motivate the various interpretations given for F in Section 5.5. (e) Give a heuristic argument for our interpretations in Exercise 5.7.7a along the following lines. Under the flow of the Liouville operator no 6-volume gets lost (8.6.3c); therefore if no photons get lost no photons per unit 6-volume get lost (cf. Ehlers [1]).

8.6.4 GASES WITH NONZERO REST-MASS

Let $(M, \mathscr{M}, F)$ be a relativistic model. (a) Let $\kappa: \mathscr{E} \to M$ be a particle with rest-mass $m \neq 0$ and electric charge e. Suppose κ obeys the Lorentz world-force law with respect to F. Show $\kappa_*: \mathscr{E} \to TM$ is an integral curve of a vector field X on TM uniquely defined by $\Pi_* X = I$, $(\Pi^* D)_X I = (e/m)(\tilde{F} \circ \Pi)I$, where we use the notation of Section 3.0.3 and of Exercise 8.6.1. (b) In (a), show X is a second-order differential equation, is not in general a spray, (cf. Dombrowski [1]), and obeys $L_X[g^*(d\theta \wedge d\theta \wedge d\theta \wedge d\theta)] = 0$ (cf. Exercise 8.6.3). (c) Generalize our discussion of a photon gas to get a model that describes a large collection of particles all having the same rest-mass and electric charge.

9 Optional exercises: Newtonian analogues

9.0 Review and notation

Throughout this chapter the notation in Sections 0.1.4 to 0.1.10 will be used. Thus $T = \mathbb{R}$ is the Newtonian time axis, and Newtonian space is $\mathbb{R}^3$, supplied with all its structures. In our (oversimplified) version of Newtonian physics it becomes consistent to regard 19th century electromagnetic theory as part of Newtonian physics; for brevity we shall do so. The reader interested in Newtonian physics *per se* or in its history should consult the references given in Section 0.1.5; the emphasis here will be on the intuition it supplies.

9.0.1 Some physics notation

A C^∞ *time-dependent scalar field* $\phi\colon \mathbb{R}^3 \times T \to \mathbb{R}$ will be written $\phi(\vec{x}, t)$ or simply ϕ. The summation convention, for Greek indices $\alpha, \beta, \ldots \in (1, 2, 3)$, will be used throughout. $\vec{j}(\vec{x}, t) \equiv \vec{j} \equiv (j^1, j^2, j^3) \equiv j^\alpha \vec{e}_\alpha \equiv j^1 \vec{e}_x + j^2 \vec{e}_y + j^3 \vec{e}_z$, with $j^\alpha = j^\alpha(\vec{x}, t)$ $\forall \alpha \in (1, 2, 3)$, denotes a *time-dependent vector field*. Here $\vec{e}_1 = \vec{e}_x$ is the unit vector in the $u^1 = x$ direction of $\mathbb{R}^3$, and so on. Similarly, with ϕ as above, $(\partial^2 \phi / \partial x^\mu \partial x^\nu) \vec{e}_\mu \otimes \vec{e}_\nu$ is the *time-dependent tensor field* whose components in the standard basis are $\{\partial^2 \phi / \partial x^\mu \partial x^\nu\}$. When integrating: $d^3 x$ denotes an $\mathbb{R}^3$ volume element, $d\vec{A}$ an area element, $d\vec{l}$ a line element (cf. Alonso–Finn [1]).

9.0.2 Naive spacetimes

To facilitate comparisons of naive Newtonian physics and relativity, we shall always work with a spacetime $(\mathbb{R}^4, g)$, *naive* in the sense that all the distinguished structures of $\mathbb{R}^4$ are here implied. ∂_4 will always be timelike future pointing; often ∂_4 will be unit norm. The summation convention (Section 3.6.1) for Latin indices will be used throughout.

9.0.3 $\mathbb{R}^3$ and $\mathbb{R}^4$

Let ϕ and $\vec{j}$ be as in Section 9.0.1. ϕ can and will be regarded as a function on the naive spacetime (Section 9.0.2), with rule $\phi w = \phi(u^1 w, u^2 w, u^3 w, u^4 w)$ $\forall w \in \mathbb{R}^4$; similarly for each j^α. Thus $\boldsymbol{j} = j^\alpha \partial_\alpha$ can and will be regarded as a vector field on $\mathbb{R}^4$ with the special property $du^4(\boldsymbol{j}) = 0$. Conversely, suppose $\boldsymbol{J}$ is a vector field on $\mathbb{R}^4$. Then $\boldsymbol{J} = J^i \partial_i$, where $J^i \colon \mathbb{R}^4 \to \mathbb{R}$ $\forall i \in (1, \ldots, 4)$. J^4 can and will be regarded as a time-dependent scalar field $J^4(\vec{x}, t)$; similarly $\vec{j} = (J^1, J^2, J^3)$ is a time-dependent vector field. A stress-energy tensor $\hat{T} = T^{ij} \partial_i \otimes \partial_j$ thus becomes a time-dependent scalar field $T^{44}(\vec{x}, t)$, two equal time-dependent vector fields $T^{\alpha 4} \vec{e}_\alpha = T^{4\alpha} \vec{e}_\alpha$, and a time-dependent tensor field $T^{\alpha\beta} \vec{e}_\alpha \otimes \vec{e}_\beta$. And so on.

9.0.4 GALILEAN INVARIANCE

Show Newton's law $\vec{F} = m\ddot{\vec{r}}$ (Section 0.1.6) is invariant under the *Galilean velocity transformations* $\vec{x} \to \vec{x} + \vec{v}t$, $\vec{F} \to \vec{F}$, $t \to t$, $\vec{r} \to \vec{r} + \vec{v}t$, $m \to m$; $\vec{v} \in \mathbb{R}^3$. (These transformations do not carry $\mathbb{R}^3$ into itself pointwise; for example origin $\to \vec{v}t$. It is mainly by ignoring these transformations that we are here oversimplifying Newtonian physics.)

9.1 Maxwell's equations

9.1.0 Conventions

Throughout this section: the conventions in Section 9.0 hold; $(\mathbb{R}^4, g)$ is Minkowski space; $\vec{E}$ and $\vec{B}$ will be interpreted as the Newtonian electric and magnetic fields; $F = 2E^1 du^1 \wedge du^4 + \cdots 2B^3 du^1 \wedge du^2$ is a 2-form on $\mathbb{R}^4$ related to $\vec{E}$ and $\vec{B}$ in the usual way (Exercise 3.4.5); $\sigma(\vec{x}, t)$ is the Newtonian electric charge density (charge per unit volume); $\vec{j}(\vec{x}, t)$ is the Newtonian electric current density (charge per unit area per unit time; cf. Alonso–Finn [1]). Thus ∂_4 is a reference frame and $\boldsymbol{J} = J^i \partial_i = j^\alpha \partial_\alpha + \sigma \partial_4$ is a vector field on $(\mathbb{R}^4, g)$.

9.1.1 THE CONSERVATION CONDITION

Show that the divergence of $\boldsymbol{J}$ vanishes iff the Newtonian equation of continuity $\partial \sigma / \partial t + \vec{\nabla} \cdot \vec{j} = 0$ holds.

9.1.2 EQUATIONS FOR $\vec{B}$

Show $dF = 0$ iff the following two classical Maxwell equations hold: $\vec{\nabla} \cdot \vec{B} = 0$ and $\vec{\nabla} \times \vec{E} + \partial \vec{B} / \partial t = 0$.

9.1.3 EQUATIONS FOR $\vec{E}$

Show div $F = 4\pi \boldsymbol{J}$ iff the other two classical Maxwell equations hold—that is, $\vec{\nabla} \cdot \vec{E} = 4\pi\sigma$ and $\vec{\nabla} \times \vec{B} - \partial \vec{E} / \partial t = 4\pi \vec{j}$.

9 Optional exercises: Newtonian analogues

9.1.4 CHARGE CONSERVATION

Suppose the two classical Maxwell's equations (Section 9.1.3) hold. Give two proofs that $\partial\sigma/\partial t + \vec{\nabla}\cdot\vec{j} = 0$, one using only Newtonian physics, the other using Sections 9.1.3 and 9.1.1.

9.1.5 INTEGRATION REGIONS

Let $\mathcal{N}_0 = \mathscr{A} \cup_{i=1}^{3} \mathscr{B}_i \cup_{j=1}^{2} \mathscr{C}_j$ be the standard cylinder (Section 3.0.1), $\mathscr{D} \subset \mathbb{R}^4$ be a space-section, $\mathscr{C} \subset \mathscr{B}_3$ be topologically trivial. In this exercise you will not need to give proofs, just to draw some $\mathbb{R}^3$ and $\mathbb{R}^4$ pictures. "Show":

(a) $\mathscr{B}_1$ is the open unit ball $\subset \mathbb{R}^3$ at $t = -1$.
(b) $\mathscr{C}_1$ is the unit sphere $\subset \mathbb{R}^3$ at $t = -1$.
(c) $\mathscr{A}$ is the history of $\mathscr{B}_1$ during $(-1, 1)$.
(d) $\mathscr{B}_3$ is the history of $\mathscr{C}_1$ during $(-1, 1)$.
(e) If $\mathscr{D}$ is contained in a level hypersurface of u^4 then $\mathscr{D}$ is an oriented, topologically trivial, relatively compact, open region $\mathscr{D} \subset \mathbb{R}^3$ with smooth oriented boundary $\partial\mathscr{D}$ diffeomorphic to $\mathscr{S}^2$.
(f) If $\mathscr{C}$ is chosen appropriately it is the history during $(a, b) \subset T$ of an oriented, imbedded, relatively compact, 2-submanifold $A \subset \mathbb{R}^3$ with smooth oriented boundary ∂A diffeomorphic to $\mathscr{S}^1$.

9.1.6 MAXWELL'S EQUATIONS IN INTEGRAL FORM

Proposition 3.7.5 has many Newtonian analogues. We here state the simplest ones. In each case you are asked to give two proofs, one Newtonian and one using relativistic results, so that the various facets of Propositions 3.7.5a to c will become clear. Throughout the exercise: $\mathscr{C}_1$, $\mathscr{D}$ and A are as in Section 9.1.5b, e, and f; $t \in T$; and $(a, b) \subset T$. Thus $Q(\mathscr{D}) = \int_{\mathscr{D}} \sigma d^3x$ is the Newtonian charge in $\mathscr{D}$ regarded as a subset of Euclidean 3-space. Show:

(a) The continuity equation $\partial\sigma/\partial t + \vec{\nabla}\cdot\vec{j} = 0$ holds iff $(d/dt)\int_{\mathscr{D}} \sigma d^3x = \int_{\partial\mathscr{D}} \vec{j}\cdot d\vec{A}\,\forall t, \mathscr{D}$ iff $\int_{\mathscr{D}} \sigma d^3x\rvert_{t=-a}^{t=+b} = \int_a^b dt \int_{\partial\mathscr{D}} \vec{j}\cdot d\vec{A}\,\forall \mathscr{D}, (a, b)$. (Cf. Proposition 3.7.5c; intuitively: if charge appears or disappears in an $\mathbb{R}^3$ region it must cross the boundary.)

(b) $\vec{\nabla}\cdot\vec{E} = 4\pi\sigma$ iff $4\pi Q(\mathscr{D}) = \int_{\partial\mathscr{D}} \vec{E}\cdot d\vec{A}\,\forall \mathscr{D}, t$. (Cf. Proposition 3.7.5a; intuitively: Gauss's law that the total charge in an $\mathbb{R}^3$ region generates an electric field whose flux through the boundary provides a measure of the charge.)

(c) $\vec{\nabla}\cdot\vec{B} = 0$ iff $\int_{\partial\mathscr{D}} \vec{B}\cdot d\vec{A} = 0\,\forall t, \mathscr{D}$. (Cf. Proposition 3.7.5b; comparing this result and Section 9.1.2 with (b), one sees explicitly the sense in which $dF = 0$ corresponds to no magnetic charge.)

(d) $\vec{\nabla} \times \vec{E} + \partial\vec{B}/\partial t = 0$ iff $\oint_{\partial A} \vec{E}\cdot d\vec{l} = -(d/dt)\int_A \vec{B}\cdot d\vec{A}\,\forall A, t$ iff $\int_a^b dt \oint_{\partial A} \vec{E}\cdot d\vec{l} = -\int_A \vec{B}\cdot d\vec{A}\rvert_{t=a}^{t=b}\,\forall A, (a, b)$. (Cf. Proposition 3.7.5b; intuitively: Faraday's law of magnetic induction that a time-dependent magnetic flux through a surface generates an electromotive force in the bounding loop.)

(e) $\vec{\nabla} \times \vec{B} - \partial\vec{E}/\partial t = 4\pi\vec{j}$ iff $\oint_{\partial A} \vec{B}\cdot d\vec{l} = 4\pi\int_A \vec{j}\cdot d\vec{A} + (d/dt)\int_A \vec{E}\cdot d\vec{A}\,\forall A, t$. (Cf. Proposition 3.7.5a; intuitively: Maxwell's displacement current hypothesis that a time-dependent electric field in, say, a capacitor gap acts as an extra current.)

9.1.7 A GENERALIZATION

Find a causal box $\phi: \mathcal{N}_0 \to \mathbb{R}^4$ such that $\phi\mathcal{N}_0$ corresponds to part of the history of a solid ball in $\mathbb{R}^3$ which expands at half the speed of light. State, prove, and heuristically interpret a Newtonian analogue of Proposition 3.7.5c in this case.

9.1.8 PLANE WAVES

Show directly that the Newtonian analogues of the plane waves (Example 3.7.3)—namely, $\vec{E} = f(z - t)\vec{e}_x$ and $\vec{B} = f(z - t)\vec{e}_y$—obey the four classical Maxwell equations (Sections 9.1.2 to 9.1.3) with $\sigma = 0 = \vec{j}$.

9.1.9 STRESS-ENERGY

Let $\hat{E}$ be the stress-energy tensor of F.

Using the conventions in Section 9.0.3 show $\hat{E}$ becomes: the Newtonian electromagnetic *energy density* $U = (1/8\pi)(|\vec{E}|^2 + |\vec{B}|^2)$; the *Poynting vector* $\vec{S} = (1/4\pi)(\vec{E} \times \vec{B})$; and the *Maxwell stress tensor* $\vec{t}$, whose components are

$$t^{\alpha\beta} = -\frac{1}{4\pi}[E^\alpha E^\beta + B^\alpha B^\beta - \tfrac{1}{2}\delta^{\alpha\beta}(|\vec{E}|^2 + |\vec{B}|^2)].$$

9.1.10 CONSERVATION LAWS

In Section 9.1.9, suppose the classical Maxwell equations (Sections 9.1.2 and 9.1.3) hold with $\sigma = 0 = \vec{j}$. Use two different computations—one Newtonian and one involving the relativistic result div $\hat{E} = 0$—to show: (a) $\partial U/\partial t = -\vec{\nabla} \cdot \vec{S}$ (energy conservation); (b) $\partial \vec{S}/\partial t = -\vec{\nabla} \cdot \vec{t}$ (momentum conservation) (cf. Alonso-Finn [1]).

9.1.11 INITIAL VALUE THEOREM

Use the four classical Maxwell equations to prove the Newtonian version of Theorem 3.11.1. Give a second proof by specializing the theorem itself appropriately.

9.1.12 FIELDS OF A POINT CHARGE

(a) Show that the fields (Example 3.7.6) of a point charge are $\vec{B} = 0$ and $\vec{E} = -e\vec{\nabla}(1/|\vec{x}|)$.
(b) Show directly that these fields obey the classical Maxwell equations with $\sigma = 0 = \vec{j}$ for $|\vec{x}| \neq 0$.

9.2 Particles

9.2.0 Conventions

Throughout this section the conventions of Sections 0.1.4 to 0.1.10, 9.0, and 9.1.0 hold. (γ, m, e) will be a particle with rest-mass $m \neq 0$ and with $\gamma: \mathscr{E} \to \mathbb{R}^4$. Define $\mathscr{F} \subset T$, the Newtonian time interval for γ, by $\mathscr{F} = u^4\gamma\mathscr{E}$. Define $\vec{r}: \mathscr{F} \to \mathbb{R}^3$ by: $\vec{r}(t) = (u^1\gamma(u^4\gamma)^{-1}t, u^2\gamma(u^4\gamma)^{-1}t, u^3\gamma(u^4\gamma)^{-1}t)$, $\forall t \in \mathscr{F}$.

9 Optional exercises: Newtonian analogues

9.2.1 NEWTONIAN PARTICLES

Show $(\vec{r}, m)$ is a Newtonian point particle (Section 0.1.6) and $|\dot{\vec{r}}(t)| <$ speed of light $\forall t \in \mathscr{F}$.

9.2.2 CONSERVATION LAWS

Let $\mathscr{P}$ be a consistent particle set (cf. 3.8). Suppose each particle in $\mathscr{P}$ has non-zero restmass. Consider a collision event such that $\sum_{\text{in}} m = \sum_{\text{out}} m$ (Newtonian conservation of inertial mass). Express the consistency condition (Definition 3.8.4) in the form:

$$\tfrac{1}{2} \sum_{\text{in}} m |\dot{\vec{r}}|^2 = \tfrac{1}{2} \sum_{\text{out}} m |\dot{\vec{r}}|^2 + \text{RC} \quad \text{(Newtonian energy conservation)};$$

$$\sum_{\text{in}} m \dot{\vec{r}} = \sum_{\text{out}} m \dot{\vec{r}} + \text{RC} \quad \text{(Newtonian momentum conservation)};$$

where "RC" denotes "relativistic correction terms," small if $|\dot{\vec{r}}| \ll 1$ for each particle involved.

9.2.3 ENERGY AND 3-MOMENTUM

For $u \in \mathscr{E}$, let $\gamma_* u = e \partial_4 \gamma u + \text{p}$ be the usual decomposition into energy and 3-momentum (cf. Section 3.1.2). With RC as in 9.2.2, show:

$$e = \frac{m}{(1 - |\dot{\vec{r}}|^2)^{1/2}} = m + \tfrac{1}{2} m |\dot{\vec{r}}|^2 + \text{RC};$$

$$\text{p} \leftrightarrow \frac{m \dot{\vec{r}}}{(1 - |\dot{\vec{r}}|^2)^{1/2}} = m \dot{\vec{r}} + \text{RC}.$$

9.2.3 LORENTZ WORLD-FORCE LAW

(a) Show (γ, m, e) obeys the Lorentz world-force law with respect to F iff $(\vec{r}, m)$ obeys the classical Lorentz force law—namely, $m \ddot{\vec{r}} = e(\vec{E} + \dot{\vec{r}} \times \vec{B}) \circ \vec{r}$.
(b) Show that then $(d/dt)[\tfrac{1}{2} m |\dot{\vec{r}}|^2] = e \vec{E} \cdot \dot{\vec{r}}$. (Give two proofs, one Newtonian and one that uses relativity. Intuitively this equation says that the rate of change of the Newtonian kinetic energy is the power $e \vec{E} \cdot \dot{\vec{r}}$ the electric field imparts to the particle.)

9.2.5 ENERGY CONSERVATION

Suppose $\vec{B} = 0$, $\partial \vec{E}/\partial t = 0$, and Maxwell's equations hold. In both parts of this exercise you are asked to give two proofs—one Newtonian and one that uses relativity. Show:

(a) There is a function $\varphi: \mathbb{R}^3 \to \mathbb{R}$ such that $\vec{E} = -\vec{\nabla}\varphi$.
(b) If the classical Lorentz force law holds, then $\tfrac{1}{2} m |\dot{\vec{r}}|^2 + e\varphi \circ \vec{r}$ is a constant function on $\mathscr{F}$. (This is the energy conservation law for a particle in Newtonian electrostatics.)

9.2.6 SPIRALLING

Translate all of proposition 3.8.2 into Newtonian notation, starting from the classical Lorentz force law (Section 9.2.4). You may assume $|\dot{\vec{r}}| \ll 1 \, \forall t \in \mathscr{F}$ for brevity.

9.2.7 COLLISIONS

Suppose two Newtonian point particles $(\vec{r}, m)$ and $(\vec{r}', m')$ collide. Suppose two particles are outgoing, each having inertial mass $(m + m')/2$. Assume the RC terms in Section 9.2.2 are negligible. State what else needs to be known about the outgoing particles in order to determine both outgoing momenta uniquely from the collision conservation laws. [Hints: two appropriately chosen angles can be used; it is simplest to work in the center of mass frame (Exercise 3.8.8).]

9.3 Gravity

9.3.0 Conventions

Throughout this section the conventions in Section 9.0 hold. We will use a naive spacetime $(\mathbb{R}^4, g)$ (Section 9.0.2) with $g = -(1 + 2\phi)du^4 \otimes du^4 + (1 - 2\phi)\sum_{\alpha=1}^{3} du^\alpha \otimes du^\alpha$, where $\phi: \mathbb{R}^3 \to \mathbb{R}$ will be interpreted as a time-independent Newtonian gravitational potential (cf. Section 0.1.7 and the fine-print remarks in Section 1.3). We shall regard ϕ as so small compared to unity that all quadratic terms, such as ϕ^2, $\phi(\partial\phi/\partial x^\alpha)$, $\phi(\partial^2\phi/\partial x^\alpha \partial x^\beta)$, and so on, are negligible.

9.3.1 TIME INDEPENDENCE

(a) Show ∂_4 is a timelike Killing vector field and $(\mathbb{R}^4, g)$ is static.
(b) Show that, in the approximation indicated above, $Z = (1 - \phi)\partial_4$ is a reference frame. (The observers in Z regard themselves as "at rest with respect to the sources of the gravitational field"; cf. Section 7.2.)
(c) Show that in the approximation indicated a suitable diffeomorphism of a normal Schwarzschild spacetime gives a spacetime of the form of Section 9.3.0.

9.3.2 CURVATURE AND TIDES

Let $\hat{R}$ be the (0, 4)-tensor field physically equivalent to the curvature tensor (Section 1.0.2) of $(\mathbb{R}^4, g)$.

(a) Show, using Sections 9.0.3 and 9.3.0, that $\hat{R}$ corresponds to
$$-(\partial^2\phi/\partial x^\mu \partial x^\nu)\vec{e}_\mu \otimes \vec{e}_\nu,$$
the tidal force tensor of Newtonian physics (cf. Section 4.2).
(b) Show Ric $(\partial_4, \partial_4) = \nabla^2\phi$ (cf. Section 4.2).
(c) Show that, in the sense of the conventions in Section 9.0.3, the Einstein tensor G becomes $2\nabla^2\phi$.

9.3.3 CONSERVATION OF ENERGY

Let (γ, m, e) and $(\vec{r}, m)$ be as in Section 9.2.0. Suppose $|\dot{\vec{r}}(t)| \ll 1 \; \forall t \in \mathscr{E}$.

(a) Show that to first order of an appropriate formal approximation scheme, γ is a geodesic iff the Newtonian equation for motion in a gravitational field, $\ddot{\vec{r}} = -(\vec{\nabla}\phi) \circ \vec{r}$, holds.

9 Optional exercises: Newtonian analogues

(b) Show from Section 9.3.1 and independently from a Newtonian computation that then $\frac{1}{2}m|\dot{r}|^2 + m\phi \circ \vec{r} =$ constant.

(c) Generalize from (b) and Section 9.2.5b to show that under appropriate assumptions and approximations $\frac{1}{2}m|\dot{r}|^2 + e\varphi \circ \vec{r} + m\phi \circ \vec{r} =$ constant. The constant is interpreted as the total Newtonian energy of the particle.

(d) Let $\lambda: \mathscr{E} \to \mathbb{R}$ be a freely falling photon. Define a function e over λ by the following condition: $\forall u \in \mathscr{E}$, $eu =$ photon energy observed by the instantaneous observer $(\lambda u, Z\lambda u)$, Z as in Section 9.3.1. Show $e(1 + \phi \circ \vec{r}) =$ constant, the "Newtonian" equation for how the energy (and thus the frequency) of a photon changes as it climbs up or falls down in a gravitational field.

Glossary of symbols

$\in$	means "belonging to" or "is a member of"
$\forall$	means "for every" or (very rarely) "for all"
$\emptyset$	denotes "the empty set"
$\subset$	means "a subset of" or "is contained in"
$\Rightarrow$	means "implies"
$\Leftrightarrow$	means "is equivalent to" or "if and only if"
$\cong$	means "is approximately equal to"

For two real numbers a and b, $a \ll b$ means "a is much less than b."

For two points x and y in a spacetime, $x \ll y$ means "x chronologically precedes y" (see Section 5.0) and $x \leq y$ means "x causally precedes y" (also see Section 5.0).

As usual, "if and only if" is abbreviated to "iff."

Bibliography

The following are the basic references for this book:

Differential geometry
Bishop–Goldberg [1], Kobayashi–Nomizu [1].

General relativity
Hawking–Ellis [1], Misner–Thorne–Wheeler [1], Weinberg [1].

Cosmology
Longair [1], Peebles [1], Schatzmann [1].

M. Alonso and E. J. Finn [1], *Fundamental University Physics* I and II, Addison-Wesley, Reading, Mass., 1970.

W. Ambrose, R. Palais, and I. M. Singer [1], Sprays, *Ann. Acad. Brazil. Ci* 32 (1960), 163–178.

R. L. Bishop and R. J. Crittenden [1], *Geometry of Manifolds*. Academic Press, N.Y., 1964.

R. L. Bishop and S. I. Goldberg [1], *Tensor Analysis on Manifolds*, MacMillan, N.Y., 1968.

Y. Bruhat [1], The Cauchy problem, in *Gravitation*, L. Witten ed., Wiley, N.Y., 1965.

B. J. Carr [1], The primordial black hole mass spectrum, *Astrophys. J.* 201 (1975), 1–19.

B. S. DeWitt [1], Quantum field theory in curved spacetime, *Physics Reports* 19C (1974), 295–361.

C. DeWitt and B. S. DeWitt [1], eds., *Black Holes*, Gordon and Breach, N.Y., 1973.

R. Dicke [1], *The Theoretical Significance of Experimental Relativity*, Gordon and Breach, N.Y., 1964.

Bibliography

P. Dombrowski [1], On the geometry of the tangent bundle, *J. Reine Angew. Math.* 210 (1962), 73–88.

J. Ehlers [1], General relativity, in *Relativity, Astrophysics and Cosmology*, W. Israel, ed., D. Reidel, Dordrecht–Holland, 1973.

G. F. R. Ellis [1], Theoretical cosmology, in *General Relativity and Cosmology*, R. K. Sachs, ed., Academic Press, N.Y., 1971.

J. G. Emming [1], *Electromagnetic Radiation in Space*, D. Reidel, Dordrecht–Holland, 1967.

R. P. Feynman, R. Leighton, and M. Sands [1], *The Feynman Lectures in Physics*, I, II, III, Addison Wesley, Reading, Mass., 1966.

A. E. Fischer and J. Marsden [1], Linearization stability of nonlinear partial differential equations, in *Proceedings Symp. Pure Math.*, volume 27, Part II, Amer. Math. Soc. Publications, Providence, R.I., 1975.

K. O. Friedrichs [1], Symmetric hyperbolic linear differential equations, *Comm. Pure Appl. Math.* 7 (1954), 345–392.

R. P. Geroch [1], Spacetime structure from a global viewpoint, in *General Relativity and Cosmology*, R. K. Sachs, ed., Academic Press, N.Y., 1971.

J. E. Gunn and J. B. Oke [1], Spectrophotometry of faint cluster galaxies and the Hubble diagram: an approach to cosmology, *Astrophys. J.* 195 (1975), 255–268.

S. Hawking and G. F. R. Ellis [1], *The Large Scale Structure of Spacetime*, Cambridge University Press, Cambridge, England, 1973.

B. J. T. Jones [1], Cosmic turbulence and the origin of galaxies, *Astrophys. J.* 181 (1973), 269–294.

S. Kobayashi and K. Nomizu [1], *Foundations of Differential Geometry* I, Interscience, N.Y., 1963.

K. R. Lang, S. D. Lord, J. M. Johanson, and P. D. Savage [1], The composite Hubble diagram, *Astrophys. J.* 202 (1975), 583–590.

P. D. Lax [1], The initial value problem for nonlinear hyperbolic equations in two independent variables, in *Contributions to the Theory of Partial Differential Equations*, L. Bers, S. Bochner, and F. John, eds., Ann. Math. Studies Volume 33, Princeton University Press, N.J., 1954.

A. Lichnerowicz [1], *Théories Relativistes de la Gravitation et de l'Electromagnetisme*, Masson, Paris, 1955.

M. S. Longair [1], *Confrontation of Cosmological Theories with Observational Data*, I.A.U. Symposium 63, D. Reidel, Boston, 1974.

M. S. Longair and M. R. Rees [1], Observational cosmology, in Schatzmann [1], 1973.

M. A. H. MacCallum [1], Cosmological models from a geometric point of view, in Schatzmann [1], 1973.

M. A. H. MacCallum [2], Quantum cosmological models, in *Quantum Gravity*, C. J. Isham, R. Penrose, and D. W. Sciama, eds., Clarendon Press, Oxford, 1975.

P. Messiah [1], *Quantum Mechanics* (2 volumes), North Holland, Amsterdam, 1961, 1962.

J. Milnor [1], *Morse Theory*, Ann. Math. Studies Volume 51, Princeton University Press, N.J., 1963.

C. W. Misner [1], Quantum description of singularities leading to pair creation, in Longair [1], 1974.

C. W. Misner, K. S. Thorne, and J. A. Wheeler [1], *Gravitation*, Freeman, San Francisco, 1973.

V. Moncrief [1], Spacetime symmetries and linearization stability of the Einstein equation I, *J. Math. Phy.* 16 (1975), 493–498.

B. O'Neill [1], *Elementary Differential Geometry*, Academic Press, N.Y., 1966.

P. J. E. Peebles [1], *Physical Cosmology*, Princeton University Press, N.J., 1971.

R. Penrose [1], *Techniques of Differential Topology in Relativity*, SIAM Publications, Philadelphia, 1972.

R. Penrose and W. Rindler [1], *Spinors*, Cambridge University Press, Cambridge, England (to appear).

M. R. Rees [1], "Concluding Remarks" in Longair [1], 1974.

M. P. Ryan and L. C. Shepley [1], *Homogeneous Relativistic Cosmologies*, Princeton University Press, N.J., 1975.

R. K. Sachs and H. Wu [1], General relativity and cosmology, *Bulletin Amer. Math. Soc.* 83 (1977), (to appear).

A. Sandage and G. A. Tamman [1], Steps towards the Hubble constant VI, *Astrophys. J.* 197 (1975), 265–280.

E. Schatzmann [1], ed., *Cargese Lectures in Physics* Volume 6, Gordon and Breach, N.Y., 1973.

D. W. Sciama [1], *Modern Cosmology*, Cambridge University Press, Cambridge, England, 1971.

J. Silk [1], The spectrum of density perturbations in an expanding universe, in Longair [1], 1974.

H. Spinrad [1], article to appear.

R. A. Sunyaev [1], The thermal history of the universe and the spectrum of relic radiation, in Longair [1], 1974.

G. A. Tamman [1], The Hubble constant and the deceleration parameter, in Longair [1], 1974.

R. V. Wagoner [1], Cosmological synthesis of the elements, in Longair [1].

J. Weber [1], *General Relativity and Gravitational Waves*, Wiley-Interscience, N.Y., 1961.

S. Weinberg [1], *Gravitation and Cosmology*, Wiley, N.Y., 1972.

S. Weinberg [2], Unified theories of elementary particle interaction, *Scientific American* 231 (1974), July issue.

J. A. Wolf [1], *Spaces of Constant Curvature*, 2nd ed., Berkeley, 1972.

H. Wu [1], Decomposition of Riemannian manifolds, *Bulletin Amer. Math. Soc.* 71 (1964), 610–617.

Ya. B. Zeldovich [1], Creation of particles in cosmology, in Longair [1], 1974.

Index of basic notations

(A)–(L) below summarize some of the general conventions in our notational scheme. A list of the basic notations then follows; the Latin characters precede the Greek ones, and the numbers at the end of each item refer to the sections in which they first occur.

(A) Boldface characters, e.g., $\boldsymbol{G}, \boldsymbol{\Omega}, \boldsymbol{P}, \boldsymbol{\partial}$ are reserved for tensor fields on manifolds; in particular, boldface Greek characters, e.g. $\boldsymbol{\Omega}, \boldsymbol{\omega}, \boldsymbol{\Lambda}$, are reserved for differential forms.

(B) f and g are generic symbols for real-valued functions.

(C) x, y, and z are generic symbols for points in manifolds.

(D) γ and ζ are generic symbols for curves in manifolds; their tangent vector fields are denoted by γ_* and ζ_*.

(E) a, b, c, and d usually denote real numbers.

(F) $\boldsymbol{d}$ denotes exterior derivative, e.g., $\boldsymbol{df}, \boldsymbol{d\omega}$; $\wedge$ denotes exterior product, e.g., $\omega^1 \wedge \omega^2$.

(G) Characters with arrows above them, e.g. $\vec{x}, \vec{B}$, denote vectorial quantities in Newtonian physics.

(H) If S is an (r, s) tensor field in a spacetime, then $\hat{S}$ will generically denote either the purely $(r + s)$-fold contravariant tensor field or the purely $(r + s)$-fold covariant tensor field physically equivalent to S. If S is of type $(2, 0)$ or $(0, 2)$, then $\tilde{S}$ denotes the $(1, 1)$ tensor field physically equivalent to S.

(I) X, Y, Z are reserved for vectors in a tangent space of a spacetime; Y is usually reserved for lightlike vectors.

(J) (x, X) and (z, Z) are generic symbols for instantaneous observers in a spacetime; in Chapter 6, (z, Z) is used exclusively to denote the actual observer.

(K) $\mathscr{E}, \mathscr{F}$ denote intervals in the real line.

(L) If A is a subset of a manifold, A^- denotes it closure, ∂A its boundary, and A° its interior.

A intrinsic cross-sectional area of a galaxy (6.1.6)

A_Q $(1, 1)$ tensor field induced by a reference frame $\boldsymbol{Q}$ (2.3)

Index of basic notations

A_γ	world acceleration of the observer γ (2.1)
a_0	blackbody constant (5.5.4)
B	Schwarzschild black hole (1.4.2) (7.3)
D	Levi–Civita connection (0.0.8)
D	stress-energy tensor of a relativistic model (3.10)
$\mathscr{D}$	space-section (3.0.1)
d	distance within a level hypersurface of Einstein–de Sitter spacetime (6.3.3)
d	a special function (6.3.4)
∂_i	ith coordinate vector field in Euclidean space (0.0.9)
d_A	area-distance (6.1.6)
d_L	luminosity distance (6.1.6)
div	divergence (3.0.2) (3.6.2)
E	total energy of a particle in a normal Schwarzschild spacetime (7.3.8)
E	stress-energy tensor, usually of an electromagnetic field (3.3) (3.7)
e	electric charge (3.1.3)
e	energy of a particle relative to an observer (3.1.2)
$\exp_x$	exponential map at $x \in M$ (0.0.8)
F	Fermi–Walker connection on a curve (2.2)
F	photon gas (5.7)
F	electromagnetic field (3.4)
F_Z	photon distribution function for (z, Z) (5.5.1)
f	frequency relative to an observer (5.1.3)
G	Einstein tensor (1.0.2) (4.0)
$\mathscr{G}M$	isometry group of M (0.0.13)
g	metric tensor (0.0.7), usually of a spacetime (1.3)
h	Planck's constant (5.1.3); on rare occasions also a real-valued function
h	standard metric on the unit sphere (0.0.9)
$i(C)$	interior product with respect to the tensor field C (3.0.1) (3.6.2)
J	total angular momentum of a particle in a normal Schwarzschild spacetime (7.3.8)
J	charge-current density (3.5.1)
K	Kruskal spacetime (7.5)
K_x	generator of rotations on the unit sphere around the x-axis (7.0)
k	Boltzmann's constant (5.5.4); on rare occasions also a real-valued function
L	absolute luminosity (5.5.5)
$\mathscr{L}^+$	future lightcone in the tangent bundle of a spacetime (5.6)
$\mathscr{L}_x^+$	future lightcone in the tangent space M_x (1.2)
L_X	Lie derivative with respect to the vector field X (0.0.5)
ℓ	apparent luminosity (5.5.5)
M, (M, g), (M, g, D)	spacetime (1.3)
$\mathscr{M}$	matter model (3.5)
M_x	tangent space to M at x (0.0.4)

Index of basic notations

$(M, \mathcal{M}, F)$	relativistic model (3.5)
$(M, \mathcal{M}, z)$	cosmological model (6.2.5)
m	rest-mass of a particle (3.1)
$\bar{m}$	active mass of a Schwarzschild spacetime (1.4.2)
$(m, e, \boldsymbol{P}, \eta)$	particle flow (3.2.4)
N	normal Schwarzschild spacetime (1.4.2) (7.3)
$\mathcal{O}^3$	A group isomorphic to the rotation group of Euclidean 3-space (2.1.7)
P	Planck function (5.5.4)
$\boldsymbol{P}$	energy-momentum of a particle flow (3.2)
$\mathcal{P}$	a finite collection of particles (3.8)
P_z	direction-energy space of (z, Z) (5.5)
$(\boldsymbol{P}, \eta)$	particle flow (3.2)
p	pressure of a quasi-gas or perfect fluid (3.15)
p	3-momentum of a particle relative to an observer (3.1.2)
Q	reference frame (2.3)
R	coefficient of the metric tensor of a simple cosmological spacetime (1.4.3)
$\boldsymbol{R}$	an instantaneous observer's local rest space (2.1.4)
$\boldsymbol{R}$	curvature tensor (1.0.2)
$\mathbb{R}$	real number field (0.0.1)
$\mathbb{R}^n$	Euclidean n-space (0.0.9)
Ric	Ricci tensor (1.0.2)
r	function canonically defined on Schwarzschild spacetimes (1.4.2)
$\imath$	frequency ratio (5.4)
$\imath(x, z)$	cosmological frequency ratio (6.0.7)
S	energy spectrum of a photon distribution function (5.5.3)
$\mathcal{S}$	auxiliary 2-sphere (6.3.5)
S	scalar curvature (1.0.2)
$\mathcal{S}^n$	unit sphere in Euclidean $(n + 1)$-space (0.0.9)
$\mathcal{S}_z$	celestial sphere of (z, Z) (5.1.3)
$\boldsymbol{T}$	an instantaneous observer's local time axis (2.1.4)
$\mathcal{T}$	timelike vectors in the tangent bundle of a spacetime (1.2)
T	temperature of a Planck function (5.5.4)
T	stress-energy tensor, usually of a particle flow (3.3) (3.5)
$\mathcal{T}_x$	timelike vectors in the tangent space M_x (1.2)
TM	tangent bundle of the manifold M (0.0.4)
t	a special coordinate function on Schwarzschild spacetimes (1.4.2)
t	predicted Hubble time (6.3.3)
t_H	empirical Hubble time (6.1.7)
u	total energy density of a photon distribution function relative to an observer (5.5.3)
u^i	ith coordinate function in Euclidean space (0.0.9)
v	observed recession velocity (6.1.5)
W	neighbor of an observer in a reference frame (2.3)
Z	comoving reference frame (2.3.6) (6.2.3), but sometimes a general reference frame
$\mathbb{Z}$	ring of integers (0.0.1)
$\mathbb{Z}^+$	positive integers (5.6)

Index of basic notations

(z, Z)	instantaneous observer (2.1.3)
$\varkappa$	cosmological redshift (6.0.7) (6.1.5)
(γ, m, e)	particle of rest-mass m and electric charge e (3.1.3)
Δ	(7.4.2)
$\Delta \mathscr{S}$	integration region in the auxiliary 2-sphere determined by a galaxy (6.3.6)
δ	auxiliary (distance) function (6.0.12)
ζ	standard volume element of $\mathscr{S}^n$ (0.0.9)
η	world density of a particle flow (3.2)
Λ_x	volume element in $\mathscr{L}_x{}^+$ (5.6)
λ	photon, usually freely falling (5.1.2)
$[\lambda]$	light signal containing the freely falling photon λ (5.0.2) (5.2)
λ	wavelength relative to an observer (5.1.3)
μ	constant associated with Schwarzschild spacetimes (1.4.2) and Kruskal spacetimes (7.5)
Π	canonical projection of the tangent bundle (0.0.4)
π_Z	natural volume element on P_Z (5.5.1)
ρ	energy density of a quasi-gas or perfect fluid (3.15)
$\{\omega^i\}$	local basis of 1-forms (1.0.3)
$\{\omega_j{}^i\}$	connection forms (1.0.3)
Ω	solid angle (5.5.1); also, apparent size of a galaxy (6.1.6)
$\boldsymbol{\Omega}$	metric volume element of a spacetime (3.0.1)
$\{\boldsymbol{\Omega}_j{}^i\}$	curvature forms (1.0.3)

Index

Definitions are indicated by **boldface.**

"Aberration", 46, 132
Absolute luminosity, **146**
   of galaxy, **212,** 214, 215
Acceleration, **38,** 253
Active mass, **30,** 218, 237
Actual observer, 170, 173, **179**
Adjustable parameters, 182
Alexandrov
   basis, **251**
   topology, **257**
Amplitude
   of electromagnetic wave, **84**
   of gravitational plane wave, **243**
Apparent luminosity, **146,** 169, 214
Apparent luminosity in the visible, **147**
Apparent size, **147**
Approaching the big bang, **161**
Appropriate matter equations, 102–103, 112
Area-distance, **171,** 185
Area-radius, 224, 229
Associated orthogonal decomposition, of instantaneous observer, **45**
At rest, 139, 146, 170, 173, 180, 214
   observer, in stationary spacetime, **219**
"Atomic clock", 53
Average world velocity, **94**

Baryon number, 68
Basis, of a vector space, 2
   dual, 5
   ordered, 2
   orthonormal, **2**

Bending of light, *see* solar system
Bianchi identity, **79**
Big bang, 28, 163
Blackbody constant, **145,** 156
Blackbody radiation, 145
Black hole, 7, 218, 237–242
   Cygnus X1, as a candidate for, 241
   Schwarzschild, **30**
Boundary
   manifolds of standard cylinder, **63**
   of a set, 1
Brouwer's fixed point theorem, 105

Cambridge catalogue of radio sources, 171
Cauchy surface, *see* spacetime
Causal
   box, **63,** 269
   character, **20**–**24**
   curve, **22**
   1-form, **22**
   vector, **20,** 24
   vector field, **22**
Causality relation, **124,** 257
   conformal invariance of, 125, 202
   in Einstein–de Sitter spacetime, 166–167
Causally continuous spacetime, *see* spacetime
Causal past, **124**
   in Einstein–de Sitter spacetime, 165
Celestial sphere, **130**
Center of mass frame, **94**
Charge-current density, **74**
   electric charge of, **86**
   of matter model, **77**

281

Index

Chronological
  future, 210
  past, **124**
Chronologically follows, **125**
Chronologically precedes, **124**
Chronology condition, *see* spacetime
Chronology relation, **124,** 257
  conformal invariance of, **125,** 202
  in Einstein–de Sitter spacetime, 166–167
Closure of a set, 1
Collision event, **92**–95
  particle created in, **92**
  particle destroyed in, **92**
Collision-free superpositions, **104**
Comoving instantaneous observer, **43**
Comoving observer, **43,** 57
Comoving reference frame
  in cosmological models, 179–181
  in dust, **103**
  in Einstein–de Sitter spacetime, **33,** 57,160
  in perfect fluid, **108**
Complete
  matter vacuum, **104**
  Riemannian manifold, **5,** 28
  semi-Riemannian manifold, **5,** 28
  vacuum, **104**
Complete spacetime, *see* spacetime
Components
  of matter models, **104**
  of superpositions, **104**
  of tensors, **78**
Conformal
  diffeomorphism, **131**
  spacetimes, **87**
Conformally invariant, **125**
Connection
  compatability of, with a metric, **4**
  forms, **18**
  induced, **38**
  of a metric, **4**
  over a map, **37**
  symmetry of, **4**
Conservation
  of charge, 268
  of electric charge at collision, **92**
  of energy, 269, 270–271
  of energy in special relativity, 96–97
  of energy-momentum at collision, **92**
  of momentum, 269, 270
  of photons, 211
  of total energy, **97**
Conservation law for vector fields, 64–65
  differential, **64**
  integral, **64**

"Conservation" of energy-momentum, 96–98
  differential law of, **97**
Conserved
  photon gas, **156**
  world density of particle flow, **69**
Consistency condition of time measurement, **137**
Consistent particle set, **92**
  average world velocity of, at collision event, **94**
  center of mass frame of, at collision event, **94**
  conservation laws of, **92**
Constants of motion, 129, 225
Constraint equations, 98, **262**
Convexity, of causal regions, **26**
Coordinate function, **5**
Cosmic rays, 175
Cosmological, **164**
Cosmological constant, 177
Cosmological frequency ratio, **163,** 186, 194
Cosmological model, 176–184, **180**
  automorphism of, 185
  electromagnetic field in, 179, 208
  comoving reference frame in, **179**
  interpretation rule for, 179
  intuitive discussion of, 176–178
  matter model in, 159, 178, 197–201, 203
  quantum effects on, 210
Cosmological nucleosynthesis, 206
"Cosmological principle", 181
Cosmological redshift, **164**
  observed, **170**
Cosmological time, **161**
  in simple cosmological model, **200**
Cosmology, 110, 159–215
  data of, 168–175
  nonquantum general relativistic, 179
  stress-energy tensor in, 174
Coulomb force, 88
Crab nebula, 218
  pulsar in, 218
Curvature, 18
  form, **19**
  infinitesimal and negligible, 44–45
  operator, **18**
  Ricci, 18, 114ff, 254
  scalar, **18**
  tensor, **18**
Curve, **4**
  endpoint of causal, **92**
  future-pointing causal, **27**

inextendible, **4**
orientation-preserving reparametrization of, **4**
past-pointing causal, **27**
positive affine reparametrization of, **4**
reparametrization of, **4**
Cygnus X1, *see* black hole

Decay, 93
Deceleration parameter, **184**, 190
Density
  charge-current, **74**
  energy, **72, 108**
  energy-momentum, **74**
  world, **66**
de Rham decomposition theorem, 122
Derivation, **37**
de Sitter, W., 32
Deuterium, 206
Diffeomorphism, **3**
  conformal, **131**
Differential form, *see* form
Direction-energy space, **141**
Displacement current hypothesis of Maxwell, 268
Distance
  area-, **171**
  by apparent angular diameter, 171
  by apparent size, 171
  luminosity, **171**
  Newtonian, 172
  parallax, 171
  radar, 171
  spatial, 171–172, 215
Divergence
  of tensor field, **80**
  of vector field, **64**
Dominant energy condition, *see* stress-energy tensor
Doppler effect, 139
Dust, **103**, 110, 119–121, 181, 198–199, 200
  matter equations of, **103**

Early universe, 203ff
  primordial fireball in, 205–208
Eigenvalue, *see* linear transformation
Eigenvector
  of Einstein tensor, 33
  of normal stress-energy tensor, **105**
Einstein, A., 8, 9, 32, 42, 44, 131, 136, 179
Einstein addition law, **48**

Einstein–de Sitter gravitational field, **32,** 43
Einstein–de Sitter model, **182,** 184–195, 200
  energy density of, 185
  luminosity distance in, 210–215
  modified, 197–198, 200
  predicted Hubble time of, **186**
Einstein–de Sitter spacetime, **31**–34, 47, 49, 57, 133, 135, 160, 180–182, 220, 259
  characterization of, 180–181
  comoving observer in, **43,** 57
  comoving reference frame in, **33,** 57
  curvatures of, 32–33
  focusing effect in, 189–190
  isometries of, 161–162
  Killing vector fields in, 82
  light signals in, 162–163
  number counts of galaxies in, 192–195
  photons in, 133, 162–163, 164
Einstein field equation, 52, **111**–123, 179–180, 198, 200, 247, 259, 262–264
  linearized vacuum, **263**
Einstein tensor, **18,** 33, 35, 111, 180, 184
  eigenvector of, **33**
Electric charge
  in space-section, **86**
  of particle, **68**
  of particle flow, **86**
Electric field, in Minkowski space, **75**
Electromagnetic field, **74**–76
  electric vector of, **75**
  in cosmological models, 179, 208
  magnetic vector of, **75**
  source of, **83**
  stress-energy tensor of, **85**
Electromagnetic radiation, **131**
Electromagnetic stress-energy tensor, *see* stress-energy tensor
Electromagnetic wave, plane linearly polarized, **84**
Electromagnetism, 66–78, 83–106
Electron-lepton number, 68
Endless causal curve, **92**
Endpoint of causal curve
  future, **92**
  past, **92**
Energy
  density of a quasi-gas, **108**
  density of a stress-energy tensor, **72**
  of a particle, **67**

283

Index

Energy [*cont.*]
  spectrum of a photon distribution function, **143**
Energy-momentum
  density of a particle flow, **74**
  of a particle, **67**
Entropy, 27
Equation of continuity, 65, 267
Equilibrium spectrum, 145
Equivalence
  of tensors, **6**
  ϕ-, **6**
  physical, **17**
Equivalent models, 168
Exhaustive sequence, **149**
"Expansion of the universe", 58, 140
Exponential map, **5,** 126

Faraday's law, 268
Fermi–Walker
  connection, **51**–**52**
  parallelism, 131
Feynman, R. P., 7
Flat manifold, **18**
Focusing effect in Einstein–de Sitter spacetime, 189–190
Form
  closed, **3**
  connection, **18**
  curvature, **19**
  exact, **3**
Freely falling
  observer, **42**
  photon, 130
Freely falling particle, **67,** 115–117
Frenet–Serret theory of curves, 41
Frequency of photon, **131**
Frequency ratio, 137–140, **138**
  cosmological, **163**
Frequency, ratio of emitted to observed, **138**
Friedmann, A., 32
Function, over a map, **37**
Function, of rapid decay, **150**
  photon distribution, **142**
  Planck, **144**
  of rapid vertical decay, **151**
Future lightcone: in tangent bundle, **151**
  in tangent space, **26**
Future endpoint, **92**
Future-pointing
  causal curve, 27
  l-form, **26**
  vector, **13,** 26
  vector field, **26**

Galaxies, 168–169
  random 3-velocity of, 110
Galaxy
  absolute luminosity of, **212**
  elliptic, 169
  emitting, 210–214
  idealizations of, 188
  intrinsic cross-sectional area of, 187, 189
  model for the history of, **179**
  radio, 169, 195
  spiral, 169
Galilean
  invariance, 267
  velocity transformation, **267**
Gamma rays, **131**
Gauge transformation, **263**
Gauss' law of electric flux, 86, 268
Gauss Lemma, 40, **125**
General relativity, 7–8
  history and current status of, 9
  preview of, 12–16
Generator of rotations on 2-sphere, *see* rotations
Geodesic, **4, 5,** 125, 129, 254, 259
  inextendible, **5**
  maximal, **5**
  reference frame, **53,** 59, 114–117
  spray, **264**
  vector field, **5**
Geometric energy function, **126**–**127**
Global hyperbolicity, *see* spacetime
Global time function, **258**
Gravitational collapse, 117–123, 237, 259
Gravitational field, 27
  gradient, 18
  representative of, **27**
  trivial, **29**
Gravitational plane wave, 242–249, **243,** 261, 269
  amplitudes of, **243**
  curvatures of, 243
  linearly polarized, 243
  linearly polarized, spacetime polarization plane of, **249**
  monochromatic, 243
  monochromatic, frequency of, **243**
  spacetime propagation direction of, **244**
  spacetime wavefront of, **244**
Gravitational wave, 242, 245, 248
  detector, 247
Graviton, 68, 246, 247
Gyroscope, **50**
  axes, 50–52

Hamiltonian mechanics, 153
Helium creation epoch, 206
Here-now, **179**
Horizontal vector field, **264**
Hubble
  constant, **173**
  law, 168, **172**–173, 185
  time, empirical, **172**
  time, predicted, **186**, 208

Idealizations in cosmology, 169–170
Imbedding, **3**
Immersion, **3**
Improper integral, existence of, **150**
Incoming particle, **92**
Inextendible
  curve, **4**
  geodesic, **5**
Infrared radiation, **131**
Initial value theorem, 98, 102, 109, 110, 262, 269
Inner product, **2**
  index of, **2**
  Lorentzian, **2**
Instantaneous observer, **43**, 45–46
  associated orthogonal decomposition of, **45**
  comoving, **43**
  local rest space of, **45**
  local time axis of, **45**
  projection tensor of, **45**
  stationary, **43**
Integration on light cones, 148ff
Integration region, **141**, 268
  exhaustive sequence of, **149**
Interior (of a set), **141**, 149
Interior product operator, 62, **80**
Interpretation rule, for cosmological models, 179
Invariant subspace, *see* linear transformation
Isometry, **4**
  group, of spacetime, 259
  group, orbit of, **260**
  local, **4**

Killing vector field, **81**–83, 95, 125, 216, 223
Kelvins (degrees Kelvin), **144**
Kruskal spacetime, **237**, 242, 259
  characterization of, 261
  relation to Schwarzschild spacetimes, 239

Laplace, P. S., 11, 12
Levi–Civita connection, **4**
Lie parallel
  tensor field, 41
  vector field, **39**
Light, **131**
Lightcone, **22**, 148–149, 150–151
  future, in tangent bundle, **151**
  future, in tangent space, **26**
  in tangent bundle, **151**
  integration on, 148–152
  past, in tangent bundle, **151**
  past, in tangent space, **26**
  volume element on, 149
Lightlike
  l-form, **22**
  submanifold, **22**
  subspace, **20**
  vector, 20, 24
  vector field, **22**
Light signal, **133**–136, 162–163
  standard, in Einstein–de Sitter spacetime, **133**
  standard, in normal Schwarzschild spacetime, **226**
Linearized Einstein equation, *see* Einstein field equation
Linear transformation
  eigenvalue of, **251**
  invariant subspace of, 106, **251**
  self-adjoint, **20**
  skew-adjoint, **20**
Liouville
  operator, 264
  theorem, 265
Local
  isometry, **4**
  rest space of (z,Z), **45**
  time axis of (z,Z), **45**
"Logic vs. history", 46–47, 159
Longest timelike distance to the big bang, 166, 183
Lorentz, H. A., 9, 88
Lorentz
  contraction, **48**
  force law, 88, 270
  group, **260**
  world force law, **88**–95, 265, 270
Lorentzian
  inner product, **2**
  manifold, **4**
  vector space, **2**
Lorentzian metric, **4**
  wider, **257**

Luminosity
  absolute, **146**
  apparent, **146**
Luminosity distance, **171**, 185, 189, 193, 210–215

Mach's principle, 177
Magnetic field
  constant, **83**
  in Minkowski space, **75**
Magnetic induction, law of, *see* Faraday's law
Magnetic monopole, hypothesis of nonexistence of, 86
Manifold, 3
  complete, **5**
  Lorentzian, **4**
  Lorentzian, time orientability of, **25,** 255
  Lorentzian, time oriented, **26**
  orientable, **3**
  paracompactness of, 255
  piecewise C^∞, **62**
  Riemannian, **4**
  semi-Riemannian, **4**
  tangent bundle of, **3**
Mass
  active, **30,** 218, 237
  rest-, **67, 69**
"Mathematics vs. physics", 44–45, 159
Matter, 66–73, 76–78, 88–98, 103–110
Matter equations, 95–96, 102–103
  appropriate, 102–103, 112
  dust, **103**
  perfect fluid, **108**
  simple, **95,** 102
Matter model, **76,** 103–110
  charge-current density of, 77
  in cosmological model, 159, 178, 197–201, 203
  stress-energy tensor of, **77**
  superposition of, **104,** 178
Maximal
  geodesic, **5**
  spacetime, **29**
Maxwell's displacement current hypothesis, 268
Maxwell's equations, **83**–87, 98–102, 110, 112, 179, 267–269
  classical, 267
Maxwell stress tensor, 86, **269**
Mean relative-acceleration, **114**–117
  Newtonian, 115–117
Meson, pi-nought, 93

Metric
  induced, **24**
  isomorphism, **6**
  Lorentzian, **4**
  Riemannian, **4**
Metric tensor, **4**
  index of, **4**
Microwave photons, **173**
  in simple cosmological model, 197–198, 201
Microwave radiation, 152, 173–174
  isotropy of, 168
  Planck spectrum of, 174, 203, 207
Microwaves, **131**
Millimeter radiation, **131**
Minkowski, H., 9
Minkowski space, 29, **47,** 49, 75, 96–97, 133, 259
  characterization of, 260
  2-dimensional, **13,** 16, 134, 139
Morse theory, 122
Mu-lepton number, 68
Myers' theorem, 122

Neighbor
  of observer in reference frame, **54**
  3-acceleration of, **54**
  3-velocity of, **54**
Neutrino, 68, 93
Newtonian
  active mass, 11
  angle, relative to, (z,Z), **45**–46
  black hole theory, 11–12
  electromagnetic energy density, 86, **269**
  gravitational force, 11
  gravitational potential, 11
  length, relative to (z,Z), **45**
  mean relative-acceleration, 117
  physics, 10
  point particles, 10–11
  speed, relative to (z,Z), **45**
  velocity, relative to (z,Z), **45**
No escape, 127–128
Nonconservation of parity, 27
Normal
  at a point, for a basis of vector fields, **79**
  neighborhood, **126**
Normal Schwarzschild spacetime, **30,** 113
  directly incoming standard light signals in, **226**
  directly outgoing standard light signals in, **226**
  escape to infinity in, **223**

geometry of, 220–228
Killing vector fields in, 82, 223
spatial infinity of, 222
static absolute reference frames in, 224
Normal stress-energy tensor, 104–106, **105**
eigenfunction of, **105**
eigenvector field of, **105**
eigenvector of, **105**
Number density of photon gas, 157
Number of photons measured by (z, Z), **142**

"**O**bservable universe", 168
Observed cosmological redshift, **170**
Observed recession velocity, **170**
Observer, **41**–50, 52–59
actual, 170, 173, **179**
comoving, **43**
comoving instantaneous, **43**, 160
4-velocity of, **42**
freely falling, **42**
in a reference frame, 53
"infinitesimally nearby", 38, 54
in Schwarzchild spacetimes, *see* particle in Schwarzchild spacetime
instantaneous, 43
instantaneous comoving, **43**, 160
proper time of, 41
stationary, **43**, 219
world acceleration of, **42**
world line of, 41
world velocity of, **42**
See also instantaneous observer
Optical radiation, **131**
Orientation, of a manifold, 3
induced, **4**
time, **24**–27
Orthogonality, relativistic interpretation of, 49–50
Orthogonal vectors, 2
Orthonormal basis, **2**, 5
Outgoing particle, **92**

Parallel tensor field, 59
Particle, 66–69, **67**, 269–271
created at collision event, **92**
destroyed at collision event, **92**
electric charge of, **68**
energy momentum of, **67**
energy of, relative to (z, Z), **67**
freely falling, **67**
incoming, **92**
outgoing, **92**
potentially in particle flow, **69**
proper time of, **67**
random 3-velocity of, **94**
rest-mass of, **67**
total Newtonian energy of, 272
Particle flow, **69**–71, 178
charge-current density of, **74**
electric charge of, **71**
energy momentum density of, **74**
energy momentum of, **69**
particles potentially in, 69
rest-mass of, **69**
stress-energy tensor of, **72**
type of, **71**
world density of, **69**
Particle in Schwarzchild spacetimes
angular momentum in x direction, **225**–226
circling, **226**
going directly in, **226**
going directly out, **226**
going in, **226**
going out, **226**
hovering, **226**
maintains height, **226**
maintains direction, **226**
total angular momentum of, **225**
total energy of, **226**
Particle set, 92
consistent, **92**, 270–271
Past endpoint, **92**
Past lightcone
in tangent bundle, **151**
in tangent space, **26**
Past pointing, **26**
causal curve, **27**
Pauli exclusion principle, 218
Pauli, W., 93
Perfect fluid, 107–110, **108**, 118ff, 158, 178, 199, 200
comoving reference frame of, **108**
energy density of, **108**
matter equations, **108**
pressure of, **108**
rest-mass zero, **108**
Perihelion precession, **237**
Petrov type, of a spacetime, **254**
"Phase space", 143
Photon, **68**, 129–133, 162–163
absorbed, **130**
beam, **71**, 210, 213, 215
emitted, **130**
freely falling, 130
frequency of, **131**

287

Photon [*cont.*]
  received, **130**
  sent out, **130**
  spatial direction of, measured by (z,Z), **130**
  standard, **133,** 162–163
  wavelength of, **131**
Photon distribution function, **142**–147
  Planck, **144**–146, 152ff
Photon gas, **152**–158, 178, 264–265
  conserved, **156,** 157, 264–265
  number density of, **157**
  Planck, with temperature T, **153**
  spatially isotropic, **154**
  stress-energy tensor of, **154**
Photon occupation number, 147
Photons, specific intensity of, **143**
Photons, total energy density of, **143**
  in Einstein–de Sitter spacetime, 133, 162–163
  microwave, **173**
  total number of (in space-sections), **157,** 265
Physical equivalence,
  of tensor fields, **17**
  of tensor fields over a map, **37**
Physical theories, 8–9
Planck function, **144,** 156
Planck, M., 131, 145
Planck photon distribution function, **144**–146
  temperature of, **144**
Planck spectrum, 145
  of the microwave radiation, 174, 203, 207
Planck's constant, **131,** 143, 199
Plane electromagnetic wave, **84**
  amplitude of, **84**
  angular frequency of, **84**
Poincaré group, 260
Poincaré, H., 9
Poisson equation, 112
Positive affine
  function, **1**
  reparametrization of a curve, **4**
Poynting vector, 86, **269**
Predicted Hubble time, **186,** 189
  of a simple cosmological model, 208
  of Einstein–de Sitter model, **186**
"Present determines future", 70, 90, 98, 102, 157, 262
Pressure
  of perfect fluid, **108**
  of quasi-gas, **108**

Pressure gradient, 118, 119
Primordial fireball, 205–208
Projection tensor, relative to (z,Z), **45**
Proper time function:
  of proper time synchronizable reference frame, **53**
  on general spacetime, **259**
Proper time ratio, **138**
Pulsar, 7, 218
Pure radiation universe, **182,** 196, 209

Quantum effects, on cosmological models, 210
Quantum theory, 8–9
Quasar, 147, 169, 195, 207
Quasi-gas, **104**
  energy density of, **108**
  energy of component in, **110**
  pressure of, **108**
  random 3-velocity of a component in, **110**

Radar, 134–135, 230
Radiation, **131**
  blackbody, 174
  microwave, 173–174
  Planck, 174
  thermal equilibrium, 174
Radio waves, **131,** 152
Random 3-velocity, **94**
Rapid decay, function of, **150,** 152
Rapid vertical decay, function of, **151,** 152
Raychaudhuri equation, **122,** 254
"Receding", one observer from another, **139**
Redshift
  cosmological, **164**
  gravitational, 219, **220,** 229
  observed cosmological, **170**
Reference frame, **52**–59
  absolute, **219,** 224
  acceleration of, 253
  comoving, **33,** 57, **103, 108,** 160, 179–181
  expansion of, 253
  geodesic, **53,** 59, 114–117
  inertial, 59
  "infinitesimally nearby" observer in, 54
  irrotational, **56**
  local geodesic, **246**
  locally proper time synchronizable, **53**
  mean relative-acceleration of, **114**ff, 254
  observers in, **53**
  proper time synchronizable, **53**

rigid, **56**
rotation of, **253**
shear of, **253**
static, **219,** 224
stationary, **219**
synchronizable, **53,** 136–137
Reflections, 129
Relativistic model, **76**
Relativity, *see* general relativity; special relativity
Rest-mass, **67, 69**
Ricci:
  curvature, 18, 114ff, 254
  scalar, 18
  tensor, **18**
Ricci flat:
  manifold, **18**
  spacetime, *see* spacetime
Riemannian manifold, **4**
"Rigid meter stick", 59
Robertson–Walker model, 208
Robertson–Walker spacetimes, 32, **183**–184, 185, 208
  of negative spatial curvature, **183,** 190, 191, 208
  of positive spatial curvature, **183,** 191, 208
  of zero spatial curvature, 183
"Rotation", 50–52
Rotations
  generator of, on 2-sphere, **216,** 225–226
  spatial, in Einstein–de Sitter spacetime, **162,** 188

Schwarzschild black hole, **30**
  geometry of, 220–228
  Killing vector fields in, 82
Schwarzschild spacetimes, **30**
  characterization of, 261
  curvatures of, 222
  geometry of, 220–228
  see also normal Schwarzschild spacetime; particles in Schwarzschild spacetimes; Schwarzschild black hole
Second fundamental form, 122, **256**
Second order differential equation, 91, 265
Self-adjoint linear transformation, **20**
Simple cosmological model, 196–202, **200**
  microwave photons in, 197–198
Simple cosmological spacetime, **31,** 167–168, 180, 183, 200–201
  characterization of, 260
Simple matter equations, **95,** 102

Simply convex neighborhood, **126**
Skew-adjoint linear transformation, **20**
Solar system, 228–237
  bending of light in, 233–236
  idealizations of, 228
  orbit function of particle in, **231**
  particle orbits in, 231–233
  perihelion precession of planets in, **237**
  total bending angle of light signal in, 234, 236
Solid angle, **141**
"Solid angle" subtended by object, 147
Space form, 184
Spacelike:
  1-form, **22**
  submanifold, **22**
  subspace, **20**
  vector, **20,** 24
  vector field, **22**
Space-section, **64,** 157
  total number of photons in, **157**
Space slice, **161**
Spacetime, **27,** 256
  Cauchy surface in, **258**
  causally continuous, **258**
  chronology condition on, **257**
  complete, 28, 256
  conformal, **87**
  containing another, 28
  globally hyperbolic, **258**
  maximal, **29**
  Ricci flat, 31, 113–117, 262
  spherically symmetric, **261**
  stably causal, **258**
  static, **219,** 261
  stationary, 218–220, **219,** 261
  strongly causal, **257**
  see also black hole; Einstein-de Sitter spacetime; gravitational plane wave; Kruskal spacetime; Minkowski space; normal Schwarzschild spacetime; Schwarzschild black hole; Schwarzschild spacetimes; simple cosmological spacetime
Spacetime diagram, 34
Spatial direction (z,Z) measures for photon, **130**
Spatial distances, 171–172
"Spatial homogeneity", 169, 175, 192
Spatial infinity, 95, 222
Spatial isotropy, **47,** 175
"Spatial reflection", 129
Spatial rotation around *x*, **162,** 188

Spatially isotropic
  photon gas, **154,** 158
  spacetime, **47**
  stress-energy tensor, **107**–109
  tensor for (z,Z), **47**
Special relativity, 8, 29
Specific flux, **147**
Specific intensity of photons, **143**
Spectrum
  energy, **143**
  equilibrium, 145
  Planck, 145
  thermal equilibrium, 145
Spin, 68, 247
Spray, 264, 265
Stably causal, *see* spacetime
Standard cylinder, **63**
  boundary manifolds of, **63**
Standard light signal
  in Einstein–de Sitter spacetime, **133**
  in normal Schwarzschild spacetimes, **226**
Standard photon, in Einstein–de Sitter spacetime, **133,** 164
Star
  end-states of, 218
  history of, 217
  neutron, 218
Static
  reference frame, *see* reference frame
  spacetime, *see* spacetime
Stationary
  observer, *see* observer
  reference frame, *see* reference frame
  spacetime, *see* spacetime
Steady state cosmology, 195
Stokes' theorem, **62**
Stress-energy tensor, **71**–73, 104–106
  dominant energy condition on, **252**
  electromagnetic, **85,** 86, 253
  energy density of, **72**
  in cosmology, 174
  normal, **105**
  of matter model, **77**
  of particle flow, **72**
  of photon gas, **154**
  spatially isotropic, **107**–109
  strong energy condition on, **252**
  timelike convergence condition on, **123,** 158, 179, 183, 252
Strong energy condition, *see* stress-energy tensor
Strongly causal, *see* spacetime

Submanifold, 3
  extrinsic curvature of, **256**
  imbedded, 3
  totally geodesic, **228**
Subspace, 2
  invariant, **251**
Summation convention, 78–80
Superposition of matter models, **104**
  collision-free, **104**
  components of, **104**
  finite, **104**
Symmetric hyperbolic system, of partial differential equations, 101
Symmetry condition, of Levi–Civita connection, **4**
Synchotron frequency, 90
Synchronizable reference frame, *see* reference frame

Tangent bundle, of a manifold, 3
Tangent vector field, of a curve, 37
Temperature of a Planck function, **144**
Temperature of a system of particles, **146**
Tensor, 2
  algebra, **2**
  Einstein, **18,** 111, 180, 184
  metric, **4**
  Ricci, **18**
Tensor field, 3
  components of, **78**
  covariant constant, **59**
  derivation of, 37
  divergence of, **80**
  over a map, **36**–37
  parallel, **59**
  physical equivalence of, **17, 37**
  restriction of, **37**
"Test matter", 77, 174, 198
Thermal equilibrium, 145–146
Thermal spectrum, 145
Thermodynamics, second law of, 27
Thomas precession, **52**
3-momentum space, 147, 148
3-space, 177
"Tidal force", 112, 242, 271
Time-dependent vector field, **266**
Time development, in Ricci-flat spacetime, **262**
Time dilation, 48
Time function, 53
  proper, of proper time synchronizable reference frame, **53**
  proper, on general spacetime, **259**

Timelike
 1-form, **22**
 submanifold, **22**
 subspace, **20**
 vector, **20**, 24
 vector field, **22**
Timelike convergence condition, **123**, 158, 179, 183, 252, *see also* stress-energy tensor
Time orientability, 24–27, 255
Time orientable Lorentzian manifold, **25**
Time oriented, by vector field, **26**
Total energy density of photons, **143**
Total energy, in Minkowski space, 97
 conservation law of, **97**
Total number
 of particles, **69**
 of photons, **157**, 265
Trivial gravitational field, **29**
Twentieth century, model of the history of, 241
Twin "paradox", 42–43
Type, of particle flow, **71**

Ultraviolet radiation, **131**
Unit
 sphere, **5**
 vector, **2**
Units, physical, 9–10
Universe
 early, 203ff
 early, nonconservation of sources in, 195
 expansion of, 140
 inhomogeneity of, 195, 208
University of California, regents of, **27**

Vacuum, **113**, 115, 222, 237, 240, 243
 complete, **104**
Variational principle, 263
Vector
 causal, **20**
 norm of, **2**
 orthogonal, **2**
 unit, **2**
Vector fields
 conservation laws for, **64**
 derivation of, **37**
 divergence of, **64**
 Lie bracket of, 3
 Lie parallel, over a curve, **39**
 lightlike, **22**
 spacelike, **22**
 timelike, **22**
Visible light, **131**
Volume element, **3**
 metric, **61**
 on lightcone, 149

Wavelength of photon, **131**
Weak interactions, 27
Whitehead Lemma, **126**
World density, **66**
 conservation of, **69**
 of particle flow, **69**
World velocity, average, **94**
Wrong-way Schwarz inequality, **24**
Wrong-way triangle inequality, **26**, 259

X-rays, 131